Science for the Nation

Also by Peter J.T. Morris:

POLYMER PIONEERS: A Popular History of the Science and Technology of Large Molecules (1986)

With Colin A. Russell, ARCHIVES OF THE BRITISH CHEMICAL INDUSTRY 1750–1914 (1988)

THE AMERICAN SYNTHETIC RUBBER RESEARCH PROGRAM (1989)

Edited with H.L. Roberts and W.A. Campbell, MILESTONES IN 150 YEARS OF THE CHEMICAL INDUSTRY (1991)

Edited with Susan T.I.Mossman, THE DEVELOPMENT OF PLASTICS (1994)

Edited with Anthony S Travis, Harm G Schröter and Ernst Homburg, DETERMINANTS OF THE EVOLUTION OF THE EUROPEAN CHEMICAL INDUSTRY, 1900–1939: New Technologies, Political Frameworks, Markets and Companies (1998)

With Otto Theodor Benfey, ROBERT BURNS WOODWARD: Architect and Artist in the World of Molecules (2001)

Edited, FROM CLASSICAL TO MODERN CHEMISTRY: The Instrumental Revolution (2002)

Edited with Klaus Staubermann, ILLUMINATING INSTRUMENTS (2009)

Science for the Nation

Perspectives on the History of the Science Museum

Edited by

Peter J.T. Morris
Principal Curator and Head of Research, Science Museum, London, UK

First published 2010 by
PALGRAVE MACMILLAN

Palgrave Macmillan in the UK is an imprint of Macmillan Publishers Limited,
registered in England, company number 785998, of Houndmills, Basingstoke,
Hampshire RG21 6XS.

Palgrave Macmillan in the US is a division of St Martin's Press LLC,
175 Fifth Avenue, New York, NY 10010.

Palgrave Macmillan is the global academic imprint of the above companies
and has companies and representatives throughout the world.

Palgrave® and Macmillan® are registered trademarks in the United States,
the United Kingdom, Europe and other countries.

ISBN: 978–0–230–23009–5 hardback

This book is printed on paper suitable for recycling and made from fully
managed and sustained forest sources. Logging, pulping and manufacturing
processes are expected to conform to the environmental regulations of the
country of origin.

A catalogue record for this book is available from the British Library.

Library of Congress Cataloging-in-Publication Data
    Science for the nation : perspectives on the history of the Science
Museum / edited by Peter J.T. Morris.
        p. cm.
    Summary: "A engaging history of the British Science Museum, produced
to mark the centenary of its separation from the Victoria & Albert Museum
in 1909. Rather than simply a narrative history, it uses the key themes
of the Science Museum's evolution to explore major issues and tensions
common to the development of all science and technology museums" –
Provided by publisher.
    ISBN 978–0–230–23009–5 (hardback)
    1. Science Museum (Great Britain) – History. 2. Science and
state – Great Britain – London – History. I. Morris, Peter John Turnbull.
Q105.G7S36 2010
507.4'42134—dc22   2010002705

10 9 8 7 6 5 4 3 2 1
19 18 17 16 15 14 13 12 11 10

Printed and bound in Great Britain by
CPI Antony Rowe, Chippenham and Eastbourne

Dedicated to all Museum staff, past, present and future

Contents

Tables

Illustrations

Graphs

Plates

Acknowledgements

All the authors wish to thank Museum volunteer Eduard von Fischer for all his research efforts on their behalf in the Science Museum archives and The National Archives. Without his unstinting assistance, this volume would be far less valuable as a historical record. In a very real sense, it is as much his volume as ours. It has been a pleasure and a privilege to work with him. We are also very grateful for the efforts of Mark Bezodis and Deborah Bloxam of the NMSI Trading Company and Michael Strang and Ruth Ireland of Palgrave Macmillan in bringing the publication of *Science for the Nation* to fruition. The illustrations in this volume would have been impossible without the support of David Exton and John Herrick (photo studio) and Deborah Jones (Science and Society Picture Library). We wish to thank the National Potrait Gallery, London for permission to reproduce the photographic potrait of Sir Robert Morant (1902) and The National Archives at Kew for permission to reproduce the photograph of the 'Exhibition of Science' (1950). Peter Morris would also like to thank Paul Keeler, Golden Web Foundation, Cambridge, for permission to quote from his personal communications. We also wish thank Tony Simcock, archivist at the Museum of the History of Science in Oxford, for his assistance with matters relating to two former curators of that museum, Frank Sherwood Taylor and Francis Maddison. All the authors are grateful for the assistance of former members of staff including, in alphabetical order: Robert Anderson, Michael Ball, John Becklake, Sandra Bicknell, Sir Neil Cossons, Graham Farmelo, Frank Greenaway, Jane Raimes, Derek Robinson, John Robinson and Dame Margaret Weston. Robert Bud wishes to thank Ruth Barton for her help with his chapter on the origins of the Science Museum. Finally the editor would like to thank Chris Rapley, the Director of the Science Museum, and Tim Boon, the Chief Curator, for their encouragement and support, and John Liffen for sharing, once again, his unparalleled knowledge of the Science Museum's history and workings. The opinions and interpretation expressed here are those of the individual authors and do not represented the official view of the Science Museum. Similarly, any errors are solely the responsibility of the editor and individual authors.

Foreword

The Science Museum is a great national institution: its many roles and projects in the first century of its existence have been surprising and challenging, often as dramatic and as dynamic as those of the sciences and technologies it has put on show. For these reasons, amongst many others, the Museum's history extends far beyond the confines of its home in South Kensington. Indeed, *Science for the Nation* might at first glance seem a somewhat unexpected title for this history. Such a title could perhaps herald an account of how applied science has contributed to the country's economic and physical welfare, or an argument in support of scientific planning. These were, of course, both major themes of this book's celebrated near-namesake *Science and the Nation*, published by Penguin in 1947. Alternatively, the phrase might introduce a history of a major public centre for the generation of science, such as the National Physical Laboratory, or of a leading producer of scientists, such as the Museum's neighbour Imperial College. While the Science Museum has not often principally conducted scientific research nor directly trained scientists, it has certainly encouraged many young visitors to become scientists. In an even more fundamental sense, however, it has provided science for the nation. Both through presenting millions of visitors with images and realities of the sciences and by presenting technology as a form of applied science, the Science Museum has shaped many people's very concept of science. The Museum has also defined the icons of the Industrial Revolution and of the course of the sciences in Britain. The Arkwright spinning frame, the Rocket locomotive, the Wheatstone five-needle telegraph, the Bessemer convertor, the J J Thomson e/m apparatus, the Wright flyer and the Watson-Crick metal-plate model of DNA all owe their undoubted pre-eminence, at least in part, to the fact that they have been on show at the Science Museum.

As Robert Bud lucidly explains in the opening chapter, this crucial mediating role between the state and the public and this characterisation of the natures of science and of technology were no accidents. Nor were they simply a result of the legacy of the Great Exhibition of 1851. Rather, these aims were the deliberately intended result of thirty-five years of lobbying by groups of leading scientists and reformers. Having created the modern Science Museum at their behest, the British government ensured that the message presented by the Museum was congruent with its own aims. As Tom Scheinfeldt and Thad Parsons show in the following two chapters, the erstwhile 'peace museum' (itself an attempt to differentiate the Science Museum from the Imperial War Museum), became the showcase of Britain's military preparedness once the threat of global war became imminent. In his chapter on temporary exhibitions, Peter Morris shows how the Science Museum persuaded government ministries and state-owned industries to use temporary exhibitions to deliver the public potent messages, ranging from energy conservation to the promise of the microchip. This helped the Museum present itself as a moulder of public opinion for the state.

When in 2007 I became a Trustee of the National Museum of Science and Industry, of which the Science Museum is a major component, I swiftly became aware of how much more the Museum involves than what is put on show front of house. As in the vital processes of scientific inquiry and technological enterprise with which the Museum is so concerned, backstage labour and resource matter decisively to its everyday life and work: storage, conservation, education, outreach and audience research all play crucial roles in the activities of the Museum. The Museum's library and archives house so many of the treasures, materials and documents of central importance both to the national heritage and also to the Museum's successes. The Museum's projects extend in scope from those showcasing physical artefacts to the new worlds of digital media; from classical sciences to the threats and opportunities of the future world. Histories of many of these activities are covered in this volume and are clearly of concern well outside the institution itself. The authors have sought to relate the history of the Science Museum to the broader context of the changes in the world of museums, of technoscience and of transformations in policies of innovation, ideology and investment during the twentieth century. By reading about the history of the Science Museum, we learn both about the fate of public sciences and of museums in the modern world.

I am happy to commend this centenary history of the Science Museum, which tells how the embryonic institution of 1909, lacking either a permanent home or a clear future, became a dynamic and creative organisation in the splendid setting that can be visited today.

SIMON SCHAFFER,
Trustee of NMSI

Contributors

Scott Anthony is a Research Fellow at the University of Manchester. After writing a thesis on the Liberal public relations pioneer Sir Stephen Tallents at Oxford University, he contributed to numerous popular and scholarly projects on the British documentary film movement including the British Film Institute monograph *Night Mail*. He is also researching social, political and imaginative forms of technophilia in twentieth century Britain. Email address: scott.anthony@manchester.ac.uk.

Timothy Boon is Chief Curator at the Science Museum, London, where he is a member of the Museum's senior management team. He was curator of the Health Matters (1994), Making the Modern World (2000), Treat Yourself (2003) and Films of Fact (2008) exhibitions. His doctoral thesis at the University of London, 'Films and the Contestation of Public Health in Interwar Britain' was completed in 1999. He has published on the cinematic representation of public health issues, and science, and on curatorship. His continuing research focuses on several aspects of the cultural history of documentary films, museums and science engagement. His book, *Films of Fact: A History of Science in Documentary Films and Television*, was published by Wallflower Press in 2008. Email address: tim.boon@ScienceMuseum.org.uk.

Robert Bud is Principal Curator of Medicine at the Science Museum and Visiting Professorial Fellow in the Department of History at Queen Mary University of London. Earlier work on the history of educational developments in South Kensington conducted together with Gerrylynn K. Roberts was published in *Science Versus Practice: Chemistry in Victorian Britain* in 1984. Since then he has published several books on the history of instruments and on aspects of the history of biotechnology. He is currently working on a history of the dream of applied science from the French Revolution to contemporary times. Email address: robert.bud@ScienceMuseum.org.uk.

Anna Bunney is Curator of Public Programmes at the Manchester Museum, The University of Manchester. Anna worked at the Science Museum for thirteen years in a number of capacities including curatorial, exhibition and learning. She has also worked at the Museum of Science and Industry in Manchester and Salford Museum and Art Gallery. Her thesis for the Masters in the History of Science, Technology and Medicine from Imperial College forms the basis of this chapter. Email address: anna.bunney@manchester.ac.uk.

John Liffen is Curator of Communication Technologies at the Science Museum. He has been at the Museum for over forty years, working primarily in the communications and transport areas. From the beginning he has had a deep and abiding interest in the history of the Museum, in particular the development of its organisational structure and the growth of the collections. Email address: john.liffen@ScienceMuseum.org.uk.

Peter J.T. Morris is Principal Curator of Science at the Science Museum and Editor of *Ambix*, the journal of the Society for the History of Alchemy and Chemistry. In recent years he has published *Robert Burns Woodward: Artist and Architect in the World of Molecules* (with Otto Theodor Benfey) and *From Classical to Modern Chemistry*. Peter was given the Edelstein Award for lifetime achievement in the history of chemistry by the American Chemical Society in 2006. With interests ranging from chemistry in eighteenth century Oxford to the future of chemistry, he is currently writing a book on the development of synthetic rubber by IG Farben. Email address: peter.morris@ScienceMuseum.org.uk.

Andrew Nahum is Principal Curator of Technology and Engineering at the Science Museum, London. He recently led the curatorial team which created the recent special exhibition 'Dan Dare and the birth of hi-tech Britain', and previously directed the creation of the major new synoptic gallery at the Museum on the history of technology and science entitled 'Making the Modern World'. He has written extensively on the history of technology, aviation and transport for both scholarly and popular journals. His books include a study of Alec Issigonis the designer of the Mini and the Morris Minor cars in *Issigonis and the Mini* and Frank Whittle in *Frank Whittle: Invention of the Jet*. His interests include the historical, social and technical background to modern aircraft and vehicle design and their display within the museum context. Andrew is currently completing a technological and economic study of the British aircraft industry in the years following the Second World War. Email address: andrew.nahum@ScienceMuseum.org.uk.

Thad Parsons III was recently awarded an Oxford DPhil for his thesis under Jim Bennett at Oxford University on the collection and presentation of science and technology during the immediate post-war period. Besides the Science Museum, his research deals with the Festival of Britain, The London Planetarium, and several other National Collections. He is also interested in the presentation of early scientific instruments in modern museums. Email address: thad@rtp3.com.

Professor Chris Rapley CBE, Director of the Science Museum and Professor of Climate Science at University College London, was previously Director of the British Antarctic Survey (BAS), which he positioned firmly as the worldwide centre of excellence in its field. He is also well known as an expert in climate change science and the architect of the International Polar Year 2007–2008. Prior to his work at the BAS, he was Executive Director of the International Geosphere-Biosphere Programme at the Royal Swedish Academy of Sciences in Stockholm. This followed an extended period as Professor of Remote Sensing Science and Associate Director at University College London's Mullard Space Science Laboratory. He has been a principal investigator on both NASA and European Space Agency satellite missions and is a Senior Visiting Scientist at NASA's Jet Propulsion Laboratory in Pasadena, California. He is a Fellow of St Edmund's College, Cambridge, a visiting professor at Imperial College London and the University of East Anglia, and holds a D.Sc from the University of Bristol. In 2008, he was awarded the Edinburgh Science Medal for having made 'a significant contribution to the understanding and wellbeing of humanity'.

David Rooney is Curator of Transport at the Science Museum and former Curator of Timekeeping at the Royal Observatory, Greenwich. He has co-curated award-winning museum displays on the history of technology, and has published work on the cultural history of timekeeping systems and the place of network technologies in Victorian and later public life. He is the author of *Ruth Belville: The Greenwich Time Lady*. Email address: david.rooney@ScienceMuseum.org.uk.

Tom Scheinfeldt is Managing Director of the Center for History and New Media (http://chnm.gmu.edu) and Research Assistant Professor of History in the Department of History and Art History at George Mason University. Tom received his bachelor's degree from Harvard and his doctoral degree from Oxford, where his thesis examined inter-war interest in science and its history in diverse cultural contexts, including museums, universities, World's Fairs and the mass media. A research associate at the Smithsonian Institution Archives and a Fellow of the Science Museum, London, Tom has lectured and written extensively on the history of popular science, the history of museums, and history and new media. In addition to managing general operations at the Center for History and New Media, Tom directs several of its online history projects, including the September 11 Digital Archive (http://911digitalarchive.org) and Omeka (http://omeka.org). Tom blogs at Found History (http://foundhistory.org). Email address: tom@foundhistory.org.

Nicholas Wyatt is Library Manager at the Science Museum Library where he manages its services and collections at both its London and Wroughton sites. He has worked in the Library since 1990 in various capacities including cataloguer, collections services librarian and rare books librarian. He has a keen bibliographic interest in the history of science and technology and has contributed to a number of Museum publications including *Treasures of the Science Museum* and several editions of *Guide to the History of Technology in Europe*. Email address: nick.wyatt@ScienceMuseum.org.uk.

Sources and Conventions

The most important archival sources for this volume are the Science Museum's own records. The historical records and the actively used records are both held at the Science Museum Documentation Centre at South Kensington, referenced throughout this volume as SMD. This must not be confused with the Science Museum Archives which are located at the Science Museum Library Wroughton, near Swindon. Most (but not all) of the historical records are held in the so-called Z archive usually but not always prefixed by a Z, but this is administered by the staff of SMD and as far as any visiting researcher is concerned, they are effectively the same as the current records. To make an appointment to see any of the records cited as being in SMD in this volume, please email freedomofinformation@ sciencemuseum.org.uk. Another important archive for this volume is the Public Record Office at Kew, part of The National Archives, referenced throughout this volume as TNA: PRO in keeping with TNA policy. Please note that there are three separate runs of the Department of Education records, all with the PRO code ED. The majority are at the PRO (at Kew), but ED 79 has been returned to the Science Museum and can be found at SMD and ED 84 is held in the V&A Archives (at Blythe House next to Olympia).

Another important source for the history of the Science Museum are the Annual Reports. Originally the annual reports of the science collections in the South Kensington Museum were part of the Annual Report of the Department of Science and Art, published in the following year by HMSO. A run of these reports is held at the Science Museum Library South Kensington and may be consulted on request. The Science Museum's own annual reports began in 1913, when the Advisory Council was formed and were published in blue covers with the title 'Report of The Science Museum for the Year 19—' by HMSO in the following year. When the Advisory Council was reconstituted in 1930, the general format remained the same, but the reports were called 'Reports of the Advisory Council for the Year 19—'. After the Second World War, matters became more complicated. They were produced internally and consisted of two parts, 'The Year's Work in the Museum' and a short report from the Advisory Council. The degree to which they were produced together varied over time. The 'Year's Work' on its own was called 'Report of the Science Museum for the Year 19—' and the Advisory Council's report was called 'Report of the Advisory Council for the Year 19—' Sometimes they were issued together and sometimes (for example when the 'Year's Work' was distributed to junior staff) separately. They were bound separately from 1963 onwards. The Reports of the Science Museum were not produced after 1980 and the Advisory Council Reports ceased when the Council was replaced by the Board of Trustees at the end of 1983. The various formats are easily confused, even by experienced museum staff, and for this reason, no attempt has been made to distinguish them in this volume. All the annual reports are cited here as 'Science Museum Annual Reports for 19—' A run of these reports is held at the Science Museum Library South Kensington and may be consulted on request. There is also a set in SMD's Z Archive. For a longer note on this complex situation, please read Peter R. Mann, 'Working Exhibits and the Destruction of Evidence in the Science Museum', *The International Journal of Museum Management and Curatorship* 8 (1989), pp. 369–87, on pp. 384–5.

For clarity initial capitals have been used not only for Science Museum and Science (Museum) Library, but also for Museum, Library, Director and Keepers whenever they refer specifically to the Science Museum and Library, and the relevant staff. So Museum on its own with an initial capital always means the Science Museum. The Victoria and Albert Museum is called the V&A throughout this volume, except in quotations and certain historical contexts, and the modern title Natural History Museum is used even during the period when it was formally the British Museum (Natural History). Similarly the Patent Office Museum is given that title throughout its history, although it was formally the Patent Museum for a few years in the 1880s before it was absorbed by the South Kensington Museum. The Science Museum's objects have inventory numbers (of the type 1913-573), hereafter cited as Inv., e.g. Inv. 1913-573.

Introduction

Peter J.T. Morris

Yet another Anniversary

There have been three significant histories of the Museum, each of which was written by Science Museum employees or former employees to mark an anniversary. Frank Greenaway's account was published in 1951 to coincide with the centenary of the Great Exhibition at Hyde Park.[1] Only six years later, another volume was produced to celebrate the centenary of the opening of the Science Museum's ancestor, the South Kensington Museum.[2] More recently a former Director of the Science Museum, David Follett, produced a partial history of the Museum in 1978 in order 'to mark the fiftieth anniversary of the opening of the present lodgings of the Museum,' in other words, the formal opening of the East Block by King George V in 1928.[3]

So it appears that almost any year can serve as an anniversary year for the Science Museum. In addition to the three dates already mentioned, one could select 1876, when the exhibition of the Special Loan Collection of Scientific Apparatus began the process that led to the creation of a Science Museum separate from the Victoria and Albert Museum; or 1883, when the Patent Office Museum was amalgamated with the science collections of the South Kensington Museum; or 1885, when the Department of Science and Art first demarcated, rather prematurely, the South Kensington Museum's science collections as 'the Science Museum' and its art collections as 'the Art Museum'; or 1893, when Lieutenant General Edward Festing was appointed as the first Director of this 'Science Museum'; or, as in our own case, 1909, when an independent Science Museum finally came into being.

A different kind of museum

A museum can be defined as a collection of objects collected in the past (even if the collecting has continued up to the present), at least some of which are on display to the public. Clearly the Science Museum fits this criterion; it is a science museum, not a science centre. The relevance of a museum's collections can be either historical, as in the British Museum, or contemporary, as in natural

1

history museums or geological museums. Influenced by its counterpart in Paris, the Musée des Arts et Métiers, the Science Museum was initially contemporary in its approach, focussing on the display of contemporary techniques and methods in science and applied science. The Bell Committee in its report of 1911 set a new course for the new Museum, calling for the 'preservation of appliances which hold honoured place in the progress of science.'[4] The first galleries to have a historical orientation (rather than being completely historical in character) appeared in the new East Block in the mid-1920s. Thereafter the balance of contemporary and historical went in cycles, with Directors favouring one approach or the other. The historical approach became more favoured from the 1960s onwards as, increasingly, professional historians of science entered the Museum as curators and existing curators were encouraged to take historical degrees, thus creating a critical mass of historically minded curators. Even in this period, however, there were galleries and temporary exhibitions devoted purely to contemporary science. Thus, since the 1920s the Science Museum has never been wholly devoted to contemporary science or the purely historical, but a fruitful combination of the two.

Harking back to the Mouseion of Hellenistic Alexandria, museums are usually associated with research as well as collections. In the case of science-based museums, however, this can be either scientific research (as at the Natural History Museum) or historical research. One might add audience research, which examines how visitors approach and understand exhibitions, but this is of fairly recent origin, and the Science Museum has been one of the pioneers in this field. Initially the Science Museum did carry out scientific research, insofar as it was closely allied to the Royal College of Science. In the late nineteenth century, William de Wiveleslie Abney, a senior Civil Servant in the Department of Science and Art, and Festing carried out research on colour, photography and infra-red spectroscopy in a laboratory that was physically connected to the South Kensington Museum. This research nexus gradually disappeared as the Museum and College drifted apart in the early 1900s. Another Director, Herman Shaw, built up a reputation as a geophysicist, which partly enabled him to become Director, but it does not appear he carried out any scientific research while he was at the Museum. Nonetheless, nearly all the curators were professional scientists (or engineers) until the 1980s, and there was a chemical laboratory in the Museum until 1992, although it was not equipped for advanced scientific research and was rarely used.

In place of scientific research, the curators began to undertake historical research in the 1920s, and they were prominent early members of the Newcomen Society for the history of technology (founded in 1920) and the British Society for the History of Science (formed in 1947). It is perhaps not insignificant that the curators in the *Science* Museum were more prominent in the establishment of the technologically oriented Newcomen Society and that this society has remained closely associated with the Museum ever since. As the history of science and technology itself was not a fully fledged profession until the 1960s, these early efforts were essentially of an amateur nature. Curators were first encouraged to take professional qualifications in the history of science in the late 1960s, and the first professionally trained historians of science arrived in the late 1970s. Even so, the National Heritage Act of 1983 did not mention research as part of the Science Museum's mission, in contrast to the V&A. Since then, however, the Museum has developed a strong research programme in the history of science, of which this volume is one outcome.

Conventional museums such as the British Museum or the V&A are usually associated with adult visitors and experts. The Science Museum is now so firmly considered to be a children's museum – a few years ago the government minister in charge of the Museum referred to it as being the perfect place to take children on a rainy Sunday afternoon – that many visitors may be surprised to learn that it was originally conceived as a museum for teachers and experts. The importance of children in the Museum came about as a result of the sheer number of young visitors to the new East Block in the 1920s, the Museum basically bowing to visitor pressure. In the 1930s the impact of these young visitors was limited to the development of the Children's Gallery in the basement away from the main galleries. A more fundamental change came about with the recruitment of curators with some experience of schoolteaching after the Second World War and the development of the Education Service from the 1950s. In the mid-1980s, to meet the challenge of the new science centres, the Museum developed the first version of its interactive gallery, LaunchPad, and the importance of children to the Museum was firmly cemented.

It is sometimes said that people visit the Science Museum three times in their lives, once as a child, once as a parent and once as a grandparent. Because of this familiarity, be it once every quarter-century, the Science Museum and its collections have become part of popular culture. References to the Science Museum in the media or literature are immediately understood by the public. According to the author Michael Bond, for example, the Science Museum is Paddington Bear's favourite museum, although his paws are apparently too large to press the buttons on the interactives properly.[5]

So from the outset the Science Museum was different from other museums: it was not a 'serious' cultural museum like the British Museum or the V&A, or a 'serious' scientific institution like the Natural History Museum. Whereas the Keepers at the British Museum and the V&A have traditionally been public school-educated Oxbridge graduates, the Keepers at the Science Museum have come from a broader range of public and state schools and have attended a variety of universities, including Imperial College and Manchester, and technical colleges.[6] Furthermore, in the period between the First World War and the 1970s, they almost invariably worked in university laboratories or in industry for several years before joining the Science Museum, whereas their counterparts at the British Museum and the V&A up to the 1980s joined their museums soon after leaving university. There is also a striking difference between the directors of the British Museum and the V&A, who have been largely drawn from the ranks of their own Keepers, and their counterparts at the Science Museum, who have come from a variety of backgrounds, with only a minority being promoted from within. Yet only a small minority of these Directors have been professional scientists, and the most successful group have come from a military background. The atmosphere at the Museum until the mid-1980s was one of officers and other ranks. This hierarchical element and the lack of a common background have perhaps led to a greater distance between the Directors and their senior curatorial staff than would have been found at the more conventional museums. Certainly, in the past, the relationship between the Director and the Keepers was often fractious.[7]

Throughout its history, the Advisory Council (and its successor the Board of Trustees), the Directors and the staff of the Science Museum have continued to debate its role as a museum. Is it a science museum or a technical museum? Should it focus on using its objects to explain the principles of science or are they icons of science to be revered? Should the Science Museum concentrate on exhibiting contemporary science or is it a history of science museum? Should the major focus of exhibitions be the objects or hands-on interactives? Are they aimed at experts, adult visitors or family groups? There have been times when the debate on one or more of these issues seemed to have been finally settled, only for it to be revived, usually with the arrival of a new Director. This continual reopening of key issues has led to a better balance between the two extremes, for example, the recent development of the very successful 'Lates' events – late openings in the evening for adults – to balance the generally family-oriented offer from the Museum.

However, there has never been any doubt that the Science Museum's prime task was to present science – in one form or another and by various means – to the nation. This might mean presenting the scientific principles that lie behind iron and steelmaking, explaining the changeover to natural gas, displaying the history of telecommunications or providing a hands-on introduction to the principles of science in the Children's Gallery or LaunchPad. Whatever the route taken, the Museum has tried to engage the general public with science and thereby enable visitors to make up their own minds about the significance of science. This volume describes how the Museum has sought to carry out this role over the last century and more.

Where science, industry and the state meet

But the Science Museum is more than just a museum; it is a space in which the Museum's staff, other Civil Servants, industrialists, scientists and the public at large have interacted over the last century, during which science and technology have become a powerful cultural force and one of the state's major concerns. It is not insignificant that many of the leading actors in this engagement were also members of that most intellectual of gentlemen's clubs, the Athenaeum, and many of the events at the Museum, including the origins of several temporary exhibitions, probably stemmed from conversations on its premises near Piccadilly Circus, close to the current homes of the Royal Society and the British Academy.

Initially, the newly independent Science Museum emphasised the importance of science as an intellectual force for good in its own right, breaking away from the mid-Victorian emphasis on the role of science – alongside craft and design – as a handmaiden of industry. In particular, its early Directors were anxious that the Museum should not become a home for trade exhibitions. In time, however, they became concerned that the Museum was only attracting workers and middle managers and not the captains of industry. At the same time, the Museum came under external pressure to be more supportive of British industry. It was then recast at the end of the 1920s as the 'National Museum of Science and Industry', and temporary exhibitions were seen as the means of displaying the applications of science to industry. Yet, almost as soon as this new approach was introduced, it was found to be problematic. However much the Museum's staff might wish to rise above the level of the trade exhibition, it was the format most familiar and most acceptable to industry. Several of the temporary exhibitions in the 1930s were in effect nothing more than tastefully presented trade shows, for example, smoke abatement or television. Such exhibitions might have been acceptable to the Museum – not least in the hope that industry would then effectively create new permanent galleries – but for the problem of commercial neutrality. As part of the Civil Service the Science Museum traditionally had a strict policy of commercial neutrality, not favouring one company over another, nor promoting one company's products to the detriment of other firms. Companies taking part in the Museum's exhibitions sometimes tried to circumvent this rule and had to be brought back into line, not always successfully. Curiously, though, this rule does not appears to have been applied to children's toys, as the famous Hamley's toy shop was able to advertise in the catalogue of the Children's Gallery. Perhaps not surprisingly, the interest of industry in holding temporary exhibitions gradually waned. There was one section of industry, however, which had a less problematic relationship with the Museum and continued to play an important role in temporary exhibitions, namely nationalised industries, as they came under the Crown alongside the Science Museum. Government ministries, too, continued to support these exhibitions.

The role of the government and its servants in the Science Museum's exhibitions and galleries illustrates how the Museum was closely woven into the fabric of the British state apparatus. Not only is it still heavily funded by the state, but until 1984 it was part of the Civil Service, and a branch of the Board of Education and its successors until 1979, when it was transferred to the new Office of Arts and Libraries (which itself was brought back into the Department of Education and Science two years later). Although it was part of the education ministry, the Museum was also considered to be a branch of the Scientific Civil Service and, in 1934–1935, the Museum was nearly transferred to the Department of Scientific and Industrial Research (DSIR). And, at least during the 1950s, the upper echelons of the Scientific Civil Service were trawled for candidates whenever there was a vacancy for the Directorship.

This had implications for the staff themselves, who were employed as Civil Servants and for the Museum's exhibitions, which had to conform to the Civil Service's standards of political and religious neutrality and, as we have seen, commercial neutrality. But it also meant that the Science Museum – like any other part of the Civil Service – was an instrument of state policy. This can be seen

most clearly in the temporary exhibitions put on with the assistance of government ministries (especially the DSIR) and the armed forces, although the initial
impetus for these exhibitions usually came from the Museum. Perhaps even more
importantly, the state in the form of HM Treasury and HM Ministry of Public
Works controlled the physical development of the Museum. The conflicts and
delays caused by this external control can be seen most clearly in the construction of the Centre Block and the failure to set up a planetarium between the
1930s and the 1950s. The Science Museum's home in South Kensington in the
middle of the so-called Albertopolis was coveted by other state institutions, and
on occasions the Museum had to fight to stay in Exhibition Road or at least hold
on to all of its land. Hence the Museum has always been vulnerable to shifts in
government policy and public funding. In order to get the buildings it wanted,
and even to stay in South Kensington, the Science Museum had to fight its corner
within the Civil Service and mobilise its external supporters to argue its case in
Parliament and the Cabinet.

The Wright brothers first flew their heavier-than-air aircraft in December 1903
and Louis Blériot flew across the English Channel just one month after the Science
Museum was established. These famous events may not seem to have much to
do with the Science Museum, but the chapters in this volume illustrate how the
Museum has been closely tied throughout its history to the equally new field of
aeronautics, a key example of the intertwined relationship of science and technology, the state, the armed forces, industry and the Museum itself. The first temporary exhibition, in 1912, less than nine years after the Wright brothers' flight, was
on aeronautics. The Wright flyer came to the Science Museum in 1928 and stayed
here for twenty years. Even before the Blitz began in the summer of 1940, the
Museum mounted a temporary exhibition about 'Aircraft in Peace and War'. And,
after the Second World War ended, the Science Museum created an 'Exhibition of
German Aeronautical Developments'. No less than three chapters in this volume
refer to the *Sunday Times* article about the opening of the new Aeronautics Gallery
in 1963. By the 1960s, this relationship had been extended to space flight. The
crowds poured into the Museum in 1962 to see Colonel John Glenn's space capsule,
'Friendship 7'. The only Apollo spacecraft on display outside the United States, the
Apollo 10 command module, was lent to the Science Museum by the Smithsonian
in 1976. Even today, the Space Gallery and the Apollo 10 command module in
the Making the Modern World Gallery remain popular with visitors. It could be
argued (*pace* David Edgerton and Gerard De Groot[8]) that air and space flight have
attracted more attention from historians and the public than is strictly warranted
by their actual economic and social significance. The relationship between the
Science Museum and aeronautics has been both symbolic and symbiotic: aeronautics (and astronautics) was symbolic of the excitement (rather than the value) of
technology during the twentieth century, and the British public experienced this
excitement vicariously through the field's symbiotic relationship with the Science
Museum, which lent it significance and public exposure.

Thus the history of the Science Museum also sheds light on the ever-changing
relationship of science and technology with the state and society at large, from

Illustration 0.1   The Science Museum's first temporary exhibition, on aeronautics, in 1912

the very formal display of aeronautics in 1912 to the light-hearted yet informative 'Wallace and Gromit's world of cracking ideas' exhibition sponsored by the Intellectual Property Office (itself one of the forerunners of the Science Museum as the Patent Office) in 2009.

Overview of chapters

Although there was no attempt to create a formal division of chapters, this volume falls naturally into three sections: the chronological history of the Museum, the changing nature of its exhibitions and the development of its collections. This is not a narrative history of the Science Museum. Each author, even the writers of the 'chronological' chapters, was asked to take a distinctive theme which provided a fresh thought-provoking slant on the Museum's history. Readers of this history will not find a comprehensive account of the Museum's development in all its fine detail, but they will obtain a richer understanding of how one of Britain's most important curatorial institutions became the Museum we know today. For the first time, the reader can fully appreciate the complexity and drama of the Museum's gestation and comprehend the relationship between the ambitious plans for the Museum over the years and their concrete expression in the buildings and galleries we can see today.

The foundation of an independent Science Museum in 1909 followed a thirty-five-year campaign led by the astronomer Norman Lockyer, who can thus be considered the founder of the Science Museum. Robert Bud, in his chapter on the creation of the Science Museum, reveals how this campaign was brought to fruition at a time of scientific and technological enthusiasm on behalf of the government, which used the formation of the V&A as the excuse to found the Science Museum, but which was already committed to the ambition. The launch of the Science Museum as a free-standing institution in 1909 and the report in 1911 of the Bell Committee, which laid down a blueprint for the future of the Museum, were followed shortly afterwards by the outbreak of unprecedented violence in Europe. In his chapter on the early years of the independent Science Museum, Tom Scheinfeldt shows how the Great War and its aftermath had a profound effect on the moral and cultural imperatives of the Science Museum, which would set the stage for debates about the Science Museum's proper attitude toward pure and applied science, its relationship with industry, and its role in British government.

The experiences of the Second World War and the periods immediately surrounding it are essential to understand the modern Science Museum because of the vast changes that were caused both by the war and by the government immediately afterwards. In his chapter on the period around the Second World War, Thad Parsons illustrates how it is important not only to understand how the Museum survived the war but also to recognise how the Museum explained it. The period between 1950 and 1983 saw the Museum grow at unsurpassed speed. The construction of new buildings, the acquisition of new collections and the development of new museums in York and Bradford were the landmarks of an era of considerable institutional achievement. However, as Scott Anthony shows in his chapter on this period, the optimism of the 'white heat of technology' era was soon superseded by the reality of rapid deindustrialisation, the Museum increasingly found its appeal, purpose and structures of governance challenged. Thus, by the 1980s, some aspects of the Museum were judged to be in need of considerable reform. A museum that had championed change found itself struggling to resist the impact of the new political and social vision surging through Britain. Looking at this legacy of the 1970s, Tim Boon's chapter focuses on how the enduring dialectic between science education and history of science was played out in the last twenty years of the last century as the Museum adjusted to its more independent status after the National Heritage Act of 1983.

Throughout the Science Museum Library's history there have been tensions as it has had to serve the differing needs of Imperial College, the nation and the Science Museum. It has had to withstand shifting government and Museum priorities and even threats of dispersal or closure. Nick Wyatt, in his chapter on the Library, shows how it has renewed itself again and again and thus become a significant library for the history of science and technology. The tangible result of the plans and ambitions of Museum Directors and staff is the building itself. 'Bricks and mortar' hold great power in establishing, presenting and maintaining subject and status. Yet, in the case of the Science Museum, the results have rarely matched the original intentions, as the real world has collided with the theoretical intentions. By making four 'visits' to the Museum when change was in the air, David Rooney asks in his chapter on the Museum's buildings whether the visible structures shape our visit and whose vision is being articulated.

As the objects on display at the Science Museum were considered for many years to be different from those found in an art museum, little attention was given to the techniques of display beyond well laid-out cases and galleries. In his chapter on display methods at the Science Museum, Andrew Nahum examines the way narrative in time came to rule displays in a far more explicit way, and how the Museum learned to construct exhibitions as a particular, highly considered form of multimedia publication. Despite the large number of child visitors and the widespread perception outside the Museum that it is a 'children's playground', the Science Museum has rarely defined itself as an institution that exists for children. In her chapter on children and the Museum, Anna Bunney argues that the special provision for children is best understood in the context of the Museum reacting to children already visiting and in terms of segregating them from the rest of the Museum. Displays created for children were centred upon contextualisation in the real world, especially through the use of the history of science and technology. As a result, it was through children that realism and history were adopted in Science Museum displays. Temporary exhibitions have played an important role in the Science Museum's public offering since the opening of the East Block in the mid-1920s. In his chapter on temporary exhibitions and technological change, Peter Morris demonstrates how the curators actively sought important social and technological issues for these exhibitions to enable the Museum to present itself as a crucial channel of public communication.

In his chapter on the Museum's acquisitions, Robert Bud shows that the enduring issue in the development of the Science Museum's collections has been the process of documenting the story of progress from past to present to future. Using a self-conscious model of evolution as an organising principle, the process of acquisition was a constant renegotiation of relations between science, its application and industry. As a result of these acquisitions, the Science Museum now holds many more objects in its collections than can be displayed at South Kensington at any one time. In 'Behind the Scenes: Housing the Collections', John Liffen discusses why this came about and how it has entailed the establishment of progressively larger outlying stores. In its collections, its displays, its rhetoric and its relationships with similar institutions around the world, the Science Museum has fashioned itself as an international, as well as a national, institution. In his chapter on the international context, Tom Scheinfeldt indicates how the Science Museum's relationships with museums in Munich, Washington, Paris, and capitals around the globe have put it at the centre of the international science museum community.

A museum for the twenty-first century

The Science Museum today is very different from the museum that became independent of the V&A a century ago. The opening of the Wellcome Wing in 2000 with its focus on contemporary science and technology finally completed the development of the Science Museum according to the blueprint created by the Bell Committee of 1911. Both the staff and visitors are far more diverse than in 1909. The Museum will continue to evolve in order to remain relevant in an ever-changing world, so that it can continue to attract visitors and maintain its international reputation. Given the potential of new technology, the environmental challenges we face and the ever-increasing pace of social development, who can predict what the Museum will look like in 2109? What will never change, however, is the Science Museum's mission to explain the history and future of science and technology to the public at large and its commitment to excellence in all its activities. I hope you enjoy reading this book and finding out how the modern Science Museum was created.

Notes

1. Frank Greenaway, *A Short History of the Science Museum* (London: HMSO, 1951).
2. Anonymous, *The Science Museum: The First Hundred Years* (London: HMSO, 1957).
3. David Follett, *The Rise of the Science Museum under Sir Henry Lyons* (London: Science Museum, 1978).
4. *Report of the Departmental Committee on the Science Museum and the Geological Museum* (London: Board of Education, 1911), p. 5.
5. Paddington Bear (via Michael Bond), Evening Standard, 12 May 2009.
6. These conclusions were drawn from an analysis of the biographies of thirty-five Science Museum Keepers between 1890 and 1980, forty-one Keepers from the British Museum and fifty-one Keepers from the V&A Museum, almost entirely drawn from the online version of *Who's Who,* http://www.ukwhoswho.com/ (accessed 14 December 2009).
7. To give just one example here, Sir Henry Lyons and David Baxandall were said to fight like 'two Kilkenny cats' (Follett, *The Rise of the Science Museum*, p. 102).
8. David Edgerton, *The Shock of the Old: Technology and Global History since 1900* (London: Profile Books, 2006) and Gerard De Groot, *Dark Side of the Moon: The Magnificent Madness of the American Lunar Quest* (London: Jonathan Cape, 2007).

1

Infected by the Bacillus of Science: The Explosion of South Kensington

Robert Bud

Introduction

The Science Museum became an independent entity in the prosperous, if wet, summer of 1909[1] It was a step that had demanded forty years of subterfuge and political manoeuvre from a small group of men who could be regarded as 'scientific evangelists'. Without a large constituency behind them but with tenacity, daring and ingenuity, and a nose for publicity, these scientists and Civil Servants had been determined to build their country's scientific muscle. They fought not just for a museum, but also for a particular vision of the relationship between science and practice. Success had followed from adherence to a new creed of 'pure and applied science'. This explains their vision of the linkage of science and practice, which drew together the constituencies of industry and science, and provided an account of progress to integrate history and science. The very title 'Science Museum' is an indicator of their values, and the strength of the brand today a testament to their success.

The ultimate victory of such men as the astronomer Norman Lockyer, the chemist Henry Enfield Roscoe and the Civil Servant Robert Morant in building the newly independent institution, whose centenary we now celebrate, followed two waves of assaults a generation apart. The first, in the wake of the 'Devonshire Commission on Scientific Instruction' in the mid-1870s, won the name 'The Science Museum' and the concept of an institution which incorporated both pure and applied science, but no separate identity was then achieved, nor was a new building constructed. The second assault, early in the twentieth century, proved more successful. Independence was won and plans for a new building followed shortly. Even then, on account of the First World War, that building was not opened until 1928, fully forty-five years after the name had been chosen. If some of that delay was unavoidable, most had been occasioned by dissent based on vocal objection to the principles and values of the proponents, yet, through tenacity and public enthusiasm in a new age, even this was overcome.

Ambition for the recognition of science was fulfilled at the heart of a period which, in British history, has been associated with the 'Edwardian' age. Ironically for such a development, the phrase today connotes the decadent images of Hollywood's young men in striped blazers lazily punting up the Cam and the Cherwell, while their nation's strength diminished and the empire enjoyed an Indian summer.[2] Yet in American history this period has been called, more positively, 'The Progressive Era'. Big business was growing on the back of research, consumer spending and global demand. Experts, engineers and businessmen were asserting their importance and a lively press condemned corruption. Despite transatlantic differences and the nation's reputation for decadence, in Britain too the period was characterised by a 'national efficiency' movement.[3] The number of science graduates in the country tripled between 1900 and 1914.[4] Businessmen, experts and scientists were seen to hold the key to improved running of the country at a time of great uncertainty. The growth of unparalleled comfort for working people and the middle classes was accompanied by disruption in ways of living and thinking.

Rather more accurately than Hollywood's striped blazers, the mood was, in H.G. Wells's 1907 description, of 'petrol and progress'. Change was disorienting for Bert Smallways, his hero of *War in the Air*, a suburban greengrocer finding himself now selling foreign-grown apples imported even from the Americas. On a larger scale, British coal was being increasingly displaced as a source of power – even in the Royal Navy – by oil, from Mexico, the Dutch East Indies (now Indonesia), Russia, Romania and America. In this era of globalisation, the challenge of ruling the world's waves was becoming harder and the public was demanding the building of no fewer than eight super-battleships, the Dreadnought class. It was the decade of the founding of London's Imperial College, the rise of socialism and the Labour Party, the creation of old age pensions, union-organised mass dock strikes, the opening of Mr Selfridge's department store in Oxford Street, record players and telephones, popular cinemas, the introduction of the car and the aircraft – Blériot's famous plane was shown off in Selfridge's in July 1909 only four months after the shop had opened – and of Wells's fearsome fantasies of cataclysm. *War in the Air* begins with foreign apples and ends with global holocaust.[5]

In June 1909, three weeks before the founding of the Museum, London hosted the meeting of the International Commission of the Congresses on Applied Chemistry. Three thousand chemists from across the world converged on the Empire's capital. Winston Churchill as President of Board of Trade was elected an honorary Vice-President. The auditoria of the newly established Imperial College in South Kensington were used for the presentation and discussion of several hundred papers, while plenary sessions were held in the vast Royal Albert Hall nearby. At the opening of the conference the King – who had once studied chemistry at Edinburgh University – told delegates that 'the age of the rule of thumb' was over.[6]

The President of the Conference, the London-based chemist William Ramsay, had performed his ceremonial and professional duties to the acclaim of the press.[7] He was nonetheless relieved when the meeting was over because his principal concern that year was the future of radium, which he had learned about from his student Frederick Soddy.[8] In February the 'Radium Institute' had opened for the treatment of cancer, with himself on the board. In October he launched the Radium Corporation, a new company run by his son, to extract radium from waste from Cornish tin mines.[9] At a time when coal reserves seemed to be endangered,

Illustration 1.1   The era in which the Science Museum went from concept to reality was also one of fundamental economic change and widely experienced depression in which Marx's *Das Kapital* and trade union strikes evoked widespread anxiety among the middle classes. This cartoon of the 'gas monster' represents the anxiety about the gas stokers' strike which occurred in 1889–1890

he was promoting the serious consideration of the potential of atomic power.[10] Ramsay's plans were aborted when the mine was bankrupted after just five years and the emergence of petroleum as a global energy source delayed the immediate pressure for atomic power. Nonetheless, the pursuit of such concerns by the meeting's president, the text of the Royal opening and the King's invitation to visit

Windsor at its close demonstrate the commitment to the future role of science that was current at that time.[11]

Such special concerns of the early twentieth century have been underestimated in the history of the Science Museum as a consequence of seeing the Museum's history in terms of enduring debates over buildings and land. This orientation has emphasised the Museum's emergence as the result of the development of the federal mid-Victorian institution, the South Kensington Museum.[12] Such accounts do bestow antiquity stretching back to 1857 and the Royal blessing of that laudable Prince Consort, Prince Albert. Less flatteringly, but as a consequence of conflating the history of the South Kensington Museum with that of the V&A Museum, the Museum has been portrayed as the long-established but unfashionable remnant left behind after the transformation of the art and design collections of the South Kensington Museum into the V&A.[13] Both views emphasise royal creation and roots deep in the reassuring mists of mid-nineteenth-century England. Neither, however, does justice to the creativity of late Victorian and Edwardian culture, in which old institutions were radically refashioned, amalgamated and destroyed by governments and society driven by the challenges of responding to political, professional, scientific, industrial and technological change within, and to Germany without. Within a few years the First World War and the Bolshevik revolution in Russia would show the reality of those challenges.

Rather than a Victorian creation, the Science Museum might more accurately be described as the result of the failure of the vision behind the integrated South Kensington Museum. One can depict the institutional reorganisation of June 1909 in terms of a metaphorical demolition of the by-then obsolete Victorian construction, to be replaced by two quite independent institutions: the V&A Museum, devoted to the Arts, and the Science Museum. The rationale of both, as they were then created, had appeared long after the Great Exhibition had closed, and indeed Albert and even that great spider at the centre of the South Kensington web, Henry Cole, Secretary of the Department of Science and Art, had both passed away.

Certainly the history of the Museum was integrally connected with the overall development of the South Kensington site. In *Science versus Practice*, Gerrylynn Roberts and the present author saw that enterprise promoting new concepts of 'pure' and 'applied' science not just to interpret the affairs of the laboratory but also to describe the skills of the manufacturer.[14] As a museum that was expressly devoted to the display of 'pure and applied science', the Science Museum was established as an integral part of the polemical message of South Kensington, even more than of its teaching. We can watch the process of its promotion by seeing the emergence of the very title 'Science Museum'.

The South Kensington Museum: The building of a brand

In August 1851, Prince Albert sent his friend, Baron Stockmar, the outline of a proposal for the dispersal of the profits of the Great Exhibition, which had proved a huge success. The exhibition had been divided into raw materials, machinery, manufactures and the plastic arts. He proposed to purchase land in

South Kensington and devote it to the promotion of the 'industrial pursuits of all nations in these four divisions.' The systematic German prince also divided the way his people learned into four means:

> I find them to consist of four: (1) Personal study from books, (2) Oral communication of knowledge by those who possess it to those who wish to acquire it, (3) Acquisition of knowledge by ocular observation, comparison, and demonstration. (4) Exchange of ideas by personal discussion.[15]

In other words, the Prince intended South Kensington to be a vast machine that promoted understanding of technology of the day. He went on to establish appropriate institutions for each of these means of learning: libraries, colleges, museums and a great meeting room, respectively. Encouraged by his advisers Henry Cole and Lyon Playfair, who were at the same time conniving Civil Servants, courtiers and institutional inventors, he envisaged that these institutions would serve the four divisions of knowledge, whose subdivisions he even spelled out in detail. During the 1850s and 1860s, Albert's vision was given reality by the Department of Science and Art, headed by Cole and Playfair. Preparations were made for education in industrial design and naval engineering, a great meeting room – the Royal Albert Hall – was built and a museum constructed.

Albert had been articulate in envisioning South Kensington and the overall scope of the museums may have been specified, yet how the vision should be realised and of what the museums should be composed was left unclear. It was perhaps obvious to Cole what should go into the exhibition of plastic arts and industrial design, which was the natural development of an earlier exhibition displayed at Marlborough House. Remaining, however, were the sections of raw materials, manufactures and machinery, miscellaneously represented by collections relating to education in general, building, food, economic entomology, animal products, fish and models of naval machinery and ships. Although these mainly contemporary collections included some older items, none of them was consciously historical.

The main South Kensington Museum collections were, however, complemented by a collection of models of historic machinery assembled by the Clerk to the Commissioners of Patents, Bennet Woodcroft. In principle any inventor could deposit a model of his invention therein, and the collection was administered by the Commissioners of Patents quite independently from the Department of Science and Art, which owned the main museum. Nonetheless, Woodcroft was no mere petty bureaucrat. He actively sought out acquisitions that seemed to him desirable, whether or not they had been patented. Historian Christine Macleod has linked this enthusiasm for the museum's development to the need to preserve the patent system, which was then under attack. Woodcroft explained in an annual report that the collection should be both 'an archaeological collection referring to the lives of eminent mechanists' and also an illustration of 'the progress of mechanical inventions.'[16]

The collection was described generally as models of patented machines, but the collection was neither a comprehensive representation of patents, nor was it restricted to inventions issued with letters patent. It was also, occasionally, brilliant. Woodcroft was himself a pioneering historian of technology, even if the objectivity of his epochal history of steam ships is marred by its ending with his own invention of a variable pitch propeller.[17] As a museum director he had saved both the Rocket locomotive and the 1812 engine of the Comet, the first commercial European steamboat. His collection of portraits of inventors inspired the hugely successful book by Samuel Smiles, *The Lives of the Engineers.* His collection seemed so idiosyncratic that an 1886 committee suggested it might have been called 'The Woodcroft Museum'.[18] Thus even this collection had been accumulated in a way that seemed to many scientists and Civil Servants to be haphazard and unsystematic.

The iron buildings which housed the collections initially were both ugly and inadequate, even though they were based on designs by the Prince himself.[19] Popularly referred to as the 'Brompton Boilers', they were poorly insulated and poorly ventilated, below freezing in winter and stinking in summer.[20] Yet the public, starved of other cheap attractions both moral and entertaining, loved them. Between the opening in 1857 and 1881, more than five and a quarter million people had visited the Brompton Boilers.[21] When in 1874 a Society of Arts delegation to the Lord Chancellor, Lord Selborne, lobbied for better accommodation, Lord Selborne was sympathetic. Even if he had never visited the museum, he did think that a museum of inventions including patents would be a good idea.[22] However, there was no time to implement this enthusiasm: within a month the Liberal government was replaced by the Conservatives.

Many of the South Kensington Museum collections were progressively rehoused in a new building on the east side of Exhibition Road, opened in stages from the 1860s and still part of the V&A. Meanwhile the Patent Office Museum remained in its Brompton Boiler next door, though the visitors kept coming, exceeding a quarter of a million by the early 1880s.[23] By 1873, when Henry Cole retired, the turnstiles' report of a million visitors a year to the South Kensington site as a whole was already a matter of public note. The mathematical economist William Stanley Jevons satirically commented, 'At the South Kensington Art Museum they make a great point of setting up turnstiles to record the precise numbers of visitors, and they can tell you to a unit the exact amount of civilising effect produced in any day, week, month, or year.'[24] He might have believed that the numbers represented children on a day out, but those million or so visitors each year did represent a huge resource. The historian of the V&A, Anthony Burton, has shown how this captive market was sought by the fine art community, who took over the direction of the 'plastic arts', putting aside any atavistic longing for the promotion of contemporary industrial design.[25]

While the art collections were progressively developed to meet the needs of middle-class collectors, thinking about the science collections was evolving as well. Cole and Playfair saw them as the most visible part of their South Kensington complex, managed by the Department of Science and Art and growing hopefully on a French model. From the late 1850s the Department's objective was to develop South Kensington as the centre of a great and stable national network. France had shown the way. The grandes écoles of Paris were fed by provincial universities

and regional centres. Similarly Manchester, Dublin and Edinburgh could potentially feed their most advanced students to London, from where they would in
turn obtain their teachers. The central schools in South Kensington would be
supported by appropriate museums equipped with laboratories.[26] This plan had
the support of Disraeli's Conservative administration of the late 1860s. However,
the Tories fell from power in 1868, to be replaced by an equally ambitious but
quite differently motivated Liberal government. For Britain's new leadership the
educational priority had to be compulsory elementary education up to the age
of twelve. The higher training envisaged by the Department of Science and Art's
founding leaders, Henry Cole and Lyon Playfair, took second place.

The Devonshire Commission, the Loan Exhibition
and the evolving vision

The so-called Forster Education Act, which made elementary education compulsory, was passed in 1870. William Forster's realignment of priorities undermined
the vision of scientific and industrial education put forward by Playfair, seeming
to condemn it to never-ending postponement. Nevertheless, the existing plans
had already acquired momentum through a committee investigating technical
education under the chair of the steel magnate Bernhard Samuelson. As so often
in Britain, the response to the challenges articulated by the Samuelson Committee
was a further enquiry, the Royal Commission into Scientific Instruction. This was
headed by the progressive and wealthy industrialist William Cavendish, the 7th
Duke of Devonshire. A collateral descendant of Henry Cavendish, the discoverer
of hydrogen, the Duke had built up the great shipbuilding port of Barrow-on-
Furness on the west coast of Lancashire. He had inherited his title and his seat at
Chatsworth from his uncle, the childless sixth duke who had built what is still
one of the great private collections of Italian neoclassical sculpture in the country.[27] The secretary of his commission was Lockyer, astronomer and editor of the
then-newly launched science magazine, *Nature*.[28]

Altogether the Commission issued eight reports, each dealing with its own subject. The entire fourth report, published in 1874, dealt with museums. It recommended the formation of a museum of pure and applied science. 'We accordingly
recommend the formation of a Collection of Physical and Mechanical Instruments;
and we submit for consideration whether it may not be expedient that this
Collection, the Collection of the Patent Office Museum, and of the Science and
Education Department of the South Kensington Museum should be united and
placed under the authority of a Minister of State.'[29]

Those half-dozen lines in this report would precipitate in the short term the
world's first ever great exhibition of scientific instruments and a decade of debate
about what should happen next. Determining what 'should be united' might
mean and what the benefits of such a unification might be would take a generation to resolve.

Henry Cole, now retired from the South Kensington Museum and already frustrated by the direction towards fine art taken by its art collections, responded

Illustration 1.2   Sir Norman Lockyer at the time the Science Museum was founded, after forty years of campaigning

immediately. Between 1871 and 1873 he had organised a series of small international exhibitions on the west side of Exhibition Road. These exhibitions he now saw as superfluous and suggested they be abandoned. The buildings would therefore be free. His diary shows he discussed a higher education for industry with Playfair on 31 March.[30] Three weeks later, having seen the Devonshire Commission's recommendation, Cole submitted a memorandum to the 1851 Commissioners urging them to establish a museum presenting the application of science to industry, including apt examples of modern technologies.[31] Cole's vision was of an updated Patent Office Museum combined with modern works, shown moving as appropriate. There are elements here which are familiar even a century and a half later. However, the unifying idea for Cole was technology, an idea pioneered in Edinburgh by Playfair's friend George Wilson and closer to Woodcroft's original idea than to a 'science museum'.[32] Cole submitted his memorandum not directly to the 1851 Commissioners, but to a subcommittee of four people entrusted with recommending the development of the South Kensington site, which included his old accomplice Playfair.

Another direction was taken through the formation of the collection of scientific instruments.[33] This was agreed by the South Kensington Museum on the precedent of previous loan exhibitions on art.[34] It was located not in the South Kensington Museum itself, but in one of the arcades that had previously housed the temporary exhibitions. The 1851 Commission, its owner, made available the Western Arcade (of which but a wall now remains next to Imperial College Library).[35] The large steam engines, including those of Watt and Newcomen, borrowed from the Patent Office Museum, and ship models were displayed in a separate building known as the 'South Arcade', on the south side of the gardens, sited roughly where the centre of the Science Museum now stands (see Illustration 1.3).

The management of the exhibition was the responsibility of a series of subject committees, but there was also a need for a central curator. This post was given to Lockyer, who had formerly administered the Devonshire Commission. His appointment as a scientific specialist to perform this role qualifies him to be the first expert 'curator' of the Science Museum.[36] Within two years 20,000 scientific instruments had been assembled from sources across the world. In charmingly blunt language the President of the Royal Society, Joseph Hooker, described in his presidential address how the museums of Europe had been 'ransacked'.[37]

The collections displayed were frankly historical, including apparatus from Tycho Brahe and Galileo down to modern times, as well the Rocket and Puffing Billy locomotives and the Watt and Newcomen engines. A measure of the total number of exhibits is that it exceeded the number of scientific artefacts on display at the Science Museum at the beginning of the twenty-first century. The space occupied was about 40,000 sq. ft (4,000 square metres), still a very large space for a temporary exhibition.[38]

A first-rate lecture series accompanied the exhibition. The physicist James Clerk Maxwell gave a famous presentation on the nature of the scientific instrument.[39] The English catalogue amounted to 1,000 pages; it was also published in French and German editions. Nor was the exhibition of interest only to scholars. The

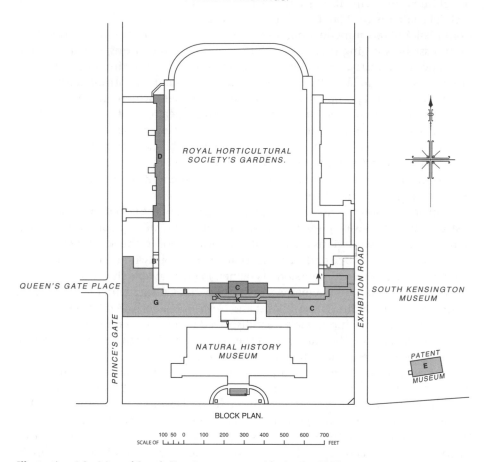

INTER–DEPARTMENTAL COMMITTEE ON THE
NATIONAL SCIENCE COLLECTIONS.
PRESENT BUILDINGS.

DRAWING NO 1.
TO ACCOMPANY REPORT OF
27TH JULY, 1885.

ROYAL HORTICULTURAL
SOCIETY'S GARDENS.

QUEEN'S GATE PLACE

EXHIBITION ROAD

SOUTH KENSINGTON
MUSEUM

PRINCE'S GATE

NATURAL HISTORY
MUSEUM

PATENT
E
MUSEUM

BLOCK PLAN.

SCALE OF   100 50 0   100   200   300   400   500   600   700   FEET

Illustration 1.3   Map of South Kensington, west side, in the 1880s

exhibition opened in May 1876 and by the end of September, less than six months later, almost a quarter of a million visitors had attended. The exhibition itself was therefore a major success in the short term.

From its opening day, the exhibition was also seen to mark a cultural watershed. *The Times* remarked that day:

> The world of art and of letters had advanced far long before science was considered to be anything more than a craze, more or less harmless. The natural consequence is that at the present day, governors and governed alike know much more about, and take much more interest in, art and literature and collections representing them, than about science.[40]

The enthusiasm engendered by the exhibition was immediately turned to advantage by a newly respectable young coterie of scientists who had risen through the Royal Society. Joseph Hooker, Thomas Huxley, Edward Frankland, and William Spottiswoode were all members of the so-called X-Club, which had been meeting monthly since the early 1860s.[41] This was not exclusively an X-Club move. Both Lockyer and the South Kensington director of science, Colonel of the Royal Engineers John F.D. Donnelly, were deeply involved in the discussions. Several of the X-Club members were also, in any case, professors in South Kensington. Certainly this was a group of people with numerous close affinities. Thus Hooker,

Illustration 1.4 From its earliest days the Science Museum interpreted machines in terms of science rather than for their significance to manufacturing or commerce. The 1876 Loan Exhibition of Scientific Apparatus held in South Kensington acted as a template for later developments This model of a pumping engine driven by steam made by James Watt was displayed at the 1876 'Loan Exhibition' in the section entitled 'Applied Mechanics'. The label accordingly specified its mode of action without reference to its industrial significance. The same description is still applied to the object today: 'Model of inverted cylinder direct action pumping engine with tappet valve motion'

President of the Royal Society, was also godfather to the son of Huxley, Secretary to the Royal Society. Their aim was to bend the museum's resource of collections, budget, buildings and visitors to ends they now perceived as critical.

Even at the very launch of the exhibition project there was talk at the highest levels about its potential as the basis for a museum. The possibility was, for instance, discussed at the board meeting of the South Kensington Museum that approved the exhibit.[42] This was even considered as 'semi-official assurances from the government' that the exhibition would be retained to form a British equivalent to Paris's Musée des Arts et Métiers.[43] Such oft-repeated hopes were greatly encouraged by the Commissioners of the 1851 Exhibition. Their development subcommittee, which had formerly received Cole's recommendation, now offered £100,000 for a building to house them for the benefit of 'the museum of science' without exactly specifying what that would be.[44] Within days, in the hope of strengthening the government's will, in June a distinguished list of 140 scientists, headed by Hooker as President of the Royal Society, signed a proposal to the government for a 'permanent museum of pure and applied science' including both the Patent Office Museum and the scientific instruments collections:

> The object of a Scientific Museum lies in the promotion of knowledge, and the establishment of the Scientific principles which must underlie all invention; and it would not only prove of great advantage to both Scientific investigators and the public if the two objects could be combined in one undertaking, but we believe that the objects of a Patent Museum would be better served by a Museum of the character here proposed than by a special collection, such as has hitherto subsisted.[45]

The letter suggested, therefore, an amalgamation of the existing Patent Office Museum collection with this new scientific instrument collection, requesting the permanent retention of as much of the collection as possible. Here we can see how the doctrine of 'pure and applied science' as a category embracing the artefacts of both fundamental research and manufacture acted as the legitimisation of the practical act by combining collections whose original inspiration had been rather separate.

So encouraged were the scientists that they went further than merely petitioning the government. Knowing that many of the objects borrowed would have to be returned if they were not retained by the government for a future science museum, and anticipating the slow pace of decision-making, they sought a £50,000 guarantee fund to underwrite the purchase of the collection. The Duke of Devonshire guaranteed £2,000, while the industrialists Warren De la Rue, Sir William Siemens and Joseph Whitworth subscribed £1,000 each. Even the Civil Servants Henry Cole and Norman Lockyer put themselves down for £200 each.[46] This move represents a decisive shift in the emphasis in South Kensington from a permanent equivalent to the temporary Great Exhibition, devoted to industry, to a scientist-led mission to underpin the communication of pure and applied science. The raising of money was slow, reputedly reaching less than £10,000 by the end of the year.[47] In any case, it came to be seen as premature for the committee to move to buy objects for the museum before its scope had been properly clarified.[48]

While Treasury was making up its mind, a letter was sent out by the Department of Science and Art. It urged lenders to leave their collections with Department 'pending the decision by Her Majesty's Government on the offer made by the Royal Commissioners for the Exhibition of 1851 of a building for the establishment of a permanent Science Museum.'[49] On the prompting of the Royal Commission, the fateful words 'Science Museum' had now been officially uttered. This usage was noticed and the letter drew an irritated reply by one 'FRS' in *The Times*. He complained prophetically, with typical Victorian vitriol: 'What might not a "Science Museum" be expected to contain? A huge and incongruous agglomeration of objects of the greatest diversity from the four quarters of the globe, which it would require a well-paid army of curators to arrange and keep in order.'[50]

The age of intellectual reconstruction

Five years passed, but a government now under financial pressures could blame the disunity of the scientists and no new museum resulted.[51] On the other hand, despite the failure of their purchase fund, the scientists continued to exert pressure. In 1881 the Lord President, Earl Spencer, appointed a formal committee comprising several of the leading signatories of the Royal Society memorandum (including the mathematician and publisher William Spottiswoode, Huxley, Lockyer, Donnelly, recently promoted assistant secretary of the Department of Science and Art, and his deputy Major-General Edward Festing). Their brief was to report whether the legacy of the Loan Collection constituted the core of a future museum. Unsurprisingly, they reaffirmed the plea of five years earlier, and cited as resources the offer of £100,000 by the 1851 Commission, on the one hand, and, on the other, unspent fees for patent rights and the Patent Office Museum. They also began to envisage a different kind of museum, one in which objects were not merely shown but also operated: 'one feature of such an exhibition might be to have specimens of recent inventions, either in themselves strictly scientific, or depending on and illustrating scientific principles, in actual operation during certain hours of the day, such, for instance, as Faure's storage battery, or some of the most recent forms of the electric light.'[52] Although some of the artefacts displayed in 1876 were purchased or copied by the Patent Office Museum, the government turned down the offer of cash from the 1851 Commission.

To explain the loss of momentum we must look not just to government parsimony or to the competing claims of the arts but, instead, to the geologists and a different model of science. A complex of institutions in Jermyn Street, off Piccadilly – the Geological Survey, the Geological Museum and the School of Mines – had frustrated the ambitions of the proponents of South Kensington for more than twenty years.[53] In the 1850s their leader Roderick Murchison had fought against the intrusion of an integrated science college in which the study of pure science would underpin all applications and specialisations such as mining. Murchison and his allies such as the Professor of Metallurgy, the chemist John Percy, had argued that a specialised technological practice such as mining should be recognised instead as a self-standing vocational discipline in its own right, which should be taught in an independent professional school supported by the appropriate applied science. Murchison had thus frustrated the visions of those, such as Cole and Playfair, for whom the Royal School of Mines would be the kernel of a major general science school. Then in 1871, just as the Devonshire Commission got underway, Murchison died. The balance of power shifted. Thomas Henry Huxley, professor of natural history at the Royal School of Mines and long-standing advocate of the school as a centre of the teaching of teachers, was a member of the Commission. The death of Murchison gave him and his allies the opening they needed.

Even in 1871 the geological faction within the Royal School of Mines was appalled by Huxley's treachery. A correspondent to *The Times* blamed a 'little "caucus" of philosophers, desirous of persuading the Chancellor of the Exchequer to assist in what is termed "the promotion of science".'[54] The Devonshire Commission's first report recommended that the School of Mines be moved to South Kensington. The facilities of Jermyn Street were lamentably constrained and mathematics teaching was totally missing. The geologists objected vociferously that the combination of museum, survey and teaching was essential to the vocational training they were giving. Mathematics was unnecessary, and so, for that matter, was natural history.[55]

On the issue of a college, Huxley nevertheless had his way. He and the professors of physics and chemistry, followed soon by geology, moved to South Kensington, taking over a building originally intended for another vocational college, the School of Naval Architecture which was transferred to Greenwich. At Huxley's behest, the Royal School of Mines was renamed the Normal School of Science in emulation of Paris's Ecole Normale to emphasise that now its remit would principally be the teaching of teachers. The emphasis shifted from the vocational to the scientific. For the moment, mining and metallurgy remained, with increasing bitterness, in Jermyn Street. In 1879 they too were ordered to move. John Percy, the metallurgy professor, made a stand, sending a comprehensive memorandum to the Chancellor of the Exchequer (helpfully including copies of all the previous *Times* correspondence).[56] When the move took place all the same, Percy resigned in high dudgeon. The controversy had demonstrated the divisions within the scientific community. Donnelly, head of the science branch of the Department of Science and Art, blamed the animosity of Percy for the lack of follow-through on a museum.[57]

Under the old, but now increasingly misleading, heading of the South Kensington Museum, a quite new institution was nonetheless slowly evolving, in which the 'Science' and 'Art' collections were becoming increasingly intellectually, institutionally and physically separate. Preparations were now being made for the expulsion of the Patent Office Museum from its site on the east side of Exhibition Road, as it was contaminating the now overwhelmingly arts-oriented provision destined for the site. It was also desired by a Patent Office reformed by the 1883 Patent Act. The Board of Trade suggested that it should be let go because it was principally not of commercial but of 'general historical or scientific interest.'[58] A last-minute amendment enabling the museum to be preserved as a 'Museum for Inventions' was foiled by a Parliamentary Committee chaired by Playfair, who was also now Secretary of the 1851 Commissioners.[59] Instead, the new plan, agreed with Treasury, was for it to be amalgamated with the South Kensington Museum to provide the core of the applied science collections on the west side, in short, to be combined to form a 'Science Museum'.

So when the Treasury turned to a committee of distinguished engineers to recommend what should be kept and what disposed of from the Patent Office Museum, the brief explicitly mentioned 'the Science Museum'. Published formally just before Christmas 1882, on 21 December, their remit read:

> The following gentlemen are invited to act as a Committee to advise my Lords as to what objects in the existing Patent Museum should be retained by the Science and Art Department for the various sections of the Science Museum.
>
> This Committee is also requested to favour my Lords with any suggestions that may occur to them as to the scope and development of the Mechanical Section of the Science Museum treated from the Scientific and educational point of view.[60]

The committee – which included Sir William Armstrong, the arms manufacturer, and Sir Joseph Bazalgette, the civil engineer famous for his sewers – surveyed the legacy of Bennet Woodcroft. It was not merely to be automatically adopted; rather, the collections were rigorously weeded according to new criteria. Again they expressed an evolutionary historical model, in forceful language: 'The principle of selection that we have followed has been to throw out such objects as have no historical interest, and are neither good examples of accepted practice or modern improvements, nor steps or links in invention.'[61]

The new-found emphasis on links or steps in invention represented the evolutionary perspective that was promoted by leading sponsors of the Museum such as Hooker and Huxley. At the same time, these were leaders in the new movement to explain biological diversity through the historical process of evolution by natural selection. It was, moreover, fashionable museologically. The Pitt Rivers ethnographic collection, from 1877 to 1884 a neighbour in South Kensington, was also arranged specifically on evolutionary lines, in its case to show how technologies had evolved in 'native' environments.[62]

These apparently repetitive discussions generated enormous irritation to proponents at the time, and frustration among those looking back, impatient for the foundation of the Science Museum. Nonetheless, they can be seen as a collective thinking through of what a science museum might be at a time when no such thing existed. Important issues were addressed: should the audience be students at the Normal School planning to be science teachers, or mechanics wishing to improve themselves? At its heart the question was that raised by Percy: should science or vocational training be the objective? On the margins of a minority report of the 1885 commission on the housing of the South Kensington Museum, the head of the office of works, Algernon

Bertram Mitford, pencilled his irritated response to departmental and scientist empire-building:

> I do not think the true measure is the production of great men but the convening of the artisan class with the scientific instruction appropriate for their servant trades = <u>Normal</u> School of Science.[63]

Mitford went on to suggest in the published version that the science collections could be loaned out across the country and the suitably modestly sized remainder could be easily accommodated on Exhibition Road's eastern side.[64]

These comments caused much anguish to the scientists. Donnelly declared the Science Museum 'to be hanging by a thread'.[65] In consequence the response was brutal and unprecedentedly public. Against Mitford's protests a denunciation of his minority report by Donnelly was also published.[66] Donnelly emphasised the popularity and number of visitors: he reported that the average annual number of visitors to what would constitute the Science Museum in recent years had been 270,000 to the Patent Office Museum and 153,000 to the science collections, giving a total of more than 420,000 a year. In a splendid piece of Victorian invective Mitford in turn questioned the figures and the attractiveness of uninterpreted scientific artefacts. His reference to bands and fountains would ring down the years:

> According to Colonel Donnelly's showing there must be an average of some 500 visitors a day, all of whom have the gift of invisibility. This is miraculous. But if it be possible that the figures given are absolutely and strictly correct, which is incredible, compare them with the numbers of daily visitors to the Inventions Exhibition, and the obvious inference is that exhibitions of scientific instruments and machinery are not attractive to the general public unless they be baited with exotic bands and illuminated fountains.[67]

Mitford's attack was the most threatening assault on the idea of the Science Museum at its most delicate moment. Other major issues discussed were the role of science and the role of the past. In 1889, Roscoe, Samuelson and Donnelly were again called upon to justify the contents of the now-consolidated science collection which an influential MP had considered, 'a quantity of old iron and worn-out models that ought to be consigned to the rubbish heap.'[68] The committee used the opportunity to confirm, on the contrary, that these actual machines presented 'a record of inestimable value for the history of English scientific invention.'

The Director of the Science Museum, Major-General Festing, testified in 1898 to the Select Committee on Museums of the Science and Art Department on the importance of showing scientific principles against current technological realisation:

> We do not profess that it is only the thing immediately up to date that is of interest for us to show; the thing immediately up to date is a thing to be shown at a commercial exhibition but we want to illustrate the progress and illustrate

scientific principles. A machine hundreds of years old may illustrate scientific principles just as much as one that was made yesterday. Therefore merely because a thing is antiquated you cannot say it is useless for the purpose of illustrating scientific principles.[69]

As Festing made clear here, and elsewhere in his testimony, no longer was the Museum to be seen as an encyclopaedic conspectus of the material culture of industry for which Prince Albert had once aimed. Indeed Festing denied that his was a 'technical museum' at all. Rather, it was the Science Museum.

At the same time, a major breakthrough of the 1898 hearing was to resolve the tension with the geologists of Jermyn Street, who also had a claim on 'science'. The new committee recommended the closure of the Jermyn Street buildings themselves, but, rather than forcing their amalgamation with their long-standing South Kensington antagonist, a separate Geological Museum in South Kensington was recommended. Thus, by the end of the nineteenth century, while there had been no construction, there had been much thinking and a new consensus had emerged.

A new series of developments can be seen stemming from 1893, when the science and art Collections, now cleanly divided by Exhibition Road, were granted separate directors. This was a period that began with yet another Select Committee enquiring into South Kensington, on the one hand, and, on the other, very different directions taken by the now increasingly separate art and science collections.

The art collections were under public pressure to professionalise their interpretation of art according to the canons of an increasingly commercially important art market. The *Burlington Magazine* was complaining about the standards of the South Kensington Museum. Concerns about its academic standards persisted even after the aesthete and former director of the Fitzwilliam Museum in Cambridge, John Henry Middleton, was appointed director in 1893. A former student of Platonic philosophy in the Great Mosque in Fez, Morocco, and friend of William Morris, he was dedicated to the serious study of antiquities and the aesthetics of the old. The museum's recent historian Anthony Burton describes the internal political pressures to which Middleton was subject, and his death by an overdose of morphine, to which he had resorted since childhood. His dead body was found slumped over his desk at the museum on 10 June 1896.[70] Middleton's successors ended their tenures less dramatically but showed an equal commitment to the arts and crafts movement, and an increasing aversion to modernism. The success of this move was celebrated by the spectacular design of the new V&A Museum, reminiscent of a cathedral and complete with cupolas, designed by Aston Webb and begun in 1899. Meanwhile the directors of the science collections would be technocrats and scientifically oriented Civil Servants for whom the past had to serve the present and the future.

The more the distinction between the ethos of the art and science collections was recognised, the better the chances of the scientists getting their own distinct museum. As Donnelly explained to Lockyer in 1892, the proposal for the new art building should be seen as bringing closer the day of the new science museum.[71] And yet for a decade after building of the new art museum had begun, and the new title Victoria and Albert Museum was entering popular discourse, that name was being applied in official documents and even on the noticeboard of the new museum, to encompass both the art and science collections.

The building of the Science Museum

The collapse of the federal identity, therefore, did not just happen. Rather, it was forced by a new pressure group of scientists led by that old champion of the Science Museum, Lockyer, who was a continuing lobbyist through *Nature* and the British Science Guild, and it was carefully engineered by the leading Civil Servant at the Board of Education, the dominant and domineering Morant. In the background was the establishment of Germany's Deutsches Museum in 1906.

The challenge of the Deutsches Museum highlighted the fruition of a shift in orientation over the previous generation. Formerly, the references in private correspondence and public papers had been to the model of Paris's Musée des Arts et Métiers for the new institution. In homage to French influence the nearby science college had been entitled the Normal School of Science. Increasingly, however, the reference point for developments in Britain was no longer to be France. In the wake of German victory in 1870, and in respect for Germany's new-found unity and power, German institutions, values and practices appeared to show the way. The general meaning was clear: research and the science that underpinned practice were to be valued. In the museum it was the scientific principles underlying the collections on display that would provide the criteria for selection and interpretation.

The meaning and significance of German competition and challenge are worthy of further consideration. They were constantly cited in debates on topics ranging from education and museums to old age pensions. Certainly appeals to the German menace could amplify a proposal's appeal and help win resources from a reluctant Treasury. Members of Britain's political and academic elite, many of them trained in Germany, were aware of the new standards being set across the North Sea. So Richard Burdon Haldane who had studied physiology in Göttingen before turning to the law and politics, warned his British contemporaries about the new standards they would have to meet. He drove through the reform of the army and the establishment of Imperial College as a public response to the German challenge. The chemists in particular used the semi-centenary of Perkin's discovery of mauve in 1906 to warn their countrymen of German chemical prowess.[72] Accordingly, historians have analysed the period in terms of 'cosmopolitan nationalism'.[73] Yet it is often unwise to see in German competition a complete explanation of the precise form of institutional innovation in Britain. The threat of the German bogey could be summoned up in support of projects with native roots.

To understand the local roots of the next stage of development, one must address the radical reform in the organisation of education in general that was ruthlessly instituted early in the twentieth century under the leadership of Morant. This

Illustration 1.5 Sir Robert Morant photographed in 1902. Sir Robert, Permanent Secretary in the Board of Education was the powerful Civil Servant who finally manoeuvred the formation of the Science Museum. © National Portrait Gallery, London

was a man who might have been the very model of the modern cosmopolitan nationalist, with detailed experience of education at Winchester College, in Siam and in Switzerland. As early as 1898 Morant concluded in his review of education in Switzerland that 'in the international struggle for existence' what he called 'concentrated brain power' would be crucial for survival.[74] His views would prove very influential when he became Permanent Secretary of the Board of Education in charge of both the museums and the education system in 1903.

Working with the Conservative administration of Arthur Balfour, Morant reformed the nation's elementary education system. A hodge-podge of local schemes, often reaching beyond the minimum leaving age of 12 were replaced by a national system of primary education up to 11, and secondary education between 12 and 18. Rather than a trade education in secondary school, which would exclude students from universities, Morant promoted grammar schools, whose scholarly excellence would enable even the poorest to reach university.

The nationalisation of parochial schools, and the introduction of a national inspectorate, would require progress through the political swamp of religious education, and his apparently cavalier exposure of poor standards in religious schools would later cost him his job. But Morant has been criticised in retrospect more for his apparent disregard of technical education.[75] Entrepreneurial elementary schools had introduced technical training at the higher age groups, which was swept away by Morant's systematisation of elementary education and focus on academic skills in secondary schools. It would be wrong, however, to see Morant as an enemy of an understanding of either science or technology.[76] His concern was that the vocational training of the working class should not be rigorously separated from the broader education of the middle classes. For him, therefore, the idea of a Science Museum which incorporated both the vocational and the general was of natural interest. In later years it would be recalled that he was a hero to Frank Pullinger, his Chief Inspector of Technical Schools, and Edmund Chambers, responsible for 'T' branch, the technology branch. John Dover Wilson, who had come to work for Pullinger and Chambers shortly after Morant had been transferred, later described the vision as 'education for industrial citizenship'. Dover Wilson was very clear on what Morant had meant by this:

> The industrial democracy that Morant dreamed of would welcome and understand the results of technical achievement, face them boldly, and declare that the works of man's hands, even in these grimy days, deserved the blessing which poetry, art and culture had previously conferred.[77]

Morant would move from the Board of Education in 1910 to establish the new National Insurance system. In his new role he would quickly exploit the enabling legislation to set up the Medical Research Council.

There were also local professional reasons for the increasing significance of science as the critical dimension both of instruments employed in its advance and in the machines which seemed to exemplify its principles. The thirty years from the Devonshire Commission to the final announcement of the museum saw sustained campaigns for better endowment of scientific research. They also saw a strengthening and growing professional scientific community anxious to win its proper place in society. Roy Macleod has shown that in the second half of the nineteenth century the number of scientific posts in British academia increased from 60 to over 400.[78] The growing claims for cultural space, and curriculum time, by the scientists provide a context for reading their warnings of the German threat. The claims for resources for a new museum were apparently driven on the back of respect for and, increasingly, fear of German competition, commercial, imperial and military. At the same time, there was also constantly in the background the search for a social and political status commensurate with a profession increasingly sure of its international role and status.

The pressures the scientists had exerted in the late 1870s and 1880s were matched and greatly exceeded by the British Science Guild early in the twentieth century.[79] This was founded as a lobbying organisation by Lockyer, frustrated by the dithering of the British Association. With its connotations of brotherhood and guild socialism, the British Science Guild was very much a child of its time. However, many of its leading members, including Lockyer and Roscoe, were veterans of the 1870s and 1880s. As Roscoe wrote wistfully to Lockyer in 1907 on reading his latest polemics on scientific education, they had been campaigning together for forty years.[80] Other leading members had known each other for almost as long. Hugh Bell, a leading steel magnate had been introduced to Lockyer in 1880 at the home of his father, the iron manufacturer and chemist Sir Lowthian Bell.[81] Of course, some of its members were younger, including Richard Glazebrook, a former student of James Clerk Maxwell, recently appointed Director of the new National Physical Laboratory.

In March 1909 the new Chancellor of the Exchequer, David Lloyd George, announced the 'People's Budget'.[82] He promised benefits for the working class and aid for agriculture. That aid, Lloyd George told the House of Commons in his budget speech, would be brought through chemists, perhaps the only time chemists have featured in the budget speech. The Guild felt it had much to celebrate. Recent years had been remarkably productive of new scientific institutions. The Guild's 1909 annual report reflected on progress in the previous two years, coinciding with Asquith's administration, which had seen the formation of the African entomological service, the Sleeping Sickness Bureau, the Advisory Committee for Aeronautics, contributions to Ernest Shackleton's Antarctic expedition and the Development Commission, and funding for conference attendees overseas and hospitality at home.[83] Regrettably still outstanding were two issues particularly of concern to Lockyer: the Science Museum and the solar physics laboratory.

However, the Board of Education was now determined to act on behalf of the science interest. A formal dance of documents was now choreographed by Morant. Lockyer launched a new energetic initiative by having printed a memorandum under his own name, calling on the government for the completion of the Science Museum project and on the 1851 Commission to renew its quarter-century-old offer of £100,000 for the project.[84] By mid-1908 Morant – negotiating with Lockyer and Roscoe on one hand and with the Treasury on the other – had won agreement that, if the offer were made, a Science Museum would be built. A parliamentary question by the science-supporting Oxford University MP William Anson and an appropriately positive answer on behalf of the Board of Education were arranged and announced in June 1908:

> I think it would be eminently desirable that there should be a science museum properly housed in immediate propinquity to the Imperial College of Science and Technology, and if the Commissioners of the 1851 Exhibition feel themselves in a position to co-operate I should be happy to bring the matter under the notice of my right hon. friend the Chancellor of the Exchequer; but it is obvious that any steps requiring the financial assistance of the Government could only be undertaken with due regard to the general calls upon the Exchequer.[85]

Within a year the problem of creating a distinctive science museum had been finally resolved. By June, Roscoe had agreed with Morant the text of a memorial addressed to the government and signed by all the leading scientists of the land, pleading for a new museum.[86] As Donnelly had predicted, the building of the new arts museum could now precipitate the final steps. A week before the King was to open the new museum, Morant struck. In a carefully worded memorandum which was effectively the birth certificate for the Science Museum, Morant reminded ministers that in approving the name 'Victoria and Albert Museum' the late Queen had intended the title only to apply to the Art Museum. For a decade the government had lived quite happily without that knowledge; now Morant used his skill to show there had to be a Science Museum by proving in great detail why the name 'Victoria and Albert Museum' could not possibly apply to anything but the art collections. As a result, he concluded that the collections housed on the west side should be simply called 'The Science Museum of the Board of Education'.[87]

The objective of a separate museum having been accomplished, the question of funding and housing came to the foreground. Roscoe's memorandum (which had already been negotiated with Morant) was formally submitted on 14 July with appropriate press notice. *The Times*, as Roscoe carefully noted, gave considerable coverage.[88] The Commissioners of the 1851 Exhibition (whose number included Morant, Lockyer and Roscoe) duly renewed Playfair's offer of half a century earlier of £100,000.

Now arose the trickier question of buildings and boundaries. The President of the Board of Education, Morant's political master, Walter Runciman MP, appointed a distinguished committee. The chair was Hugh Bell, the Middlesbrough steel magnate and long-time friend of Lockyer. The other eight members of the committee included five fellows of the Royal Society and two distinguished engineers. Richard Glazebrook, Director of the National Physical Laboratory, and the chemist William Ramsay were also members. Four of the members of the Bell Committee were also leaders of the British Science Guild. Bell was also remarkably connected

Illustration 1.6   View of the Science Museum from Exhibition Road around the time the Museum was founded, with the far more imposing Imperial Institute (with the Queen's Tower) in the background. The entrance to the Museum was at the canopy on the top right of the picture

to the ministers of the Board of Education. The President of the Board, Runciman, was a neighbour in the North East. His deputy, Charles Trevelyan, was Bell's own son-in-law and a scion of another great North East land-owning family. Just a year earlier, Bell had written to Runciman asking him to arrange political office for the young man, who had recently been elected as a MP.[89]

Bell happily discussed the museum issue 'at great length' with Morant, and consulted Runciman on 'the line he should take'.[90] The committee recommended little that was either new or surprising, but in its constitution and timing gave those conclusions a legitimacy that has endured for a century. Perhaps most enduring was its specification of the progress of building. There should be three stages, progressively filling the entire space between Exhibition Road and Queen's Gate.

One problem remained: the boundary with the Natural History Museum. The issue raised fierce debates between protagonists of the two institutions in the letter pages of *The Times* and amongst the Civil Servants. Initially there was support for the Science Museum from the powerful Office of Works, whose First Commissioner was Lewis Harcourt, himself grandson of the founder of the British Association for the Advancement of Science. Indeed, in October 1910, Harcourt put a paper to Cabinet. Pressing many buttons economic and nationalistic, the paper entitled 'Proposed new science museum' began:

> For many years the existing Science Museum at South Kensington has been a scandal. In consequence of the want of accommodation and the danger from

fire we have lost many valuable gifts and bequests, some of which have gone to Germany.[91]

However, at that moment Harcourt was succeeded by the much less supportive Lord Beauchamp, and it took all of Morant's guile to negotiate a reconciliation. When doubt was raised about what had actually been decided by a Cabinet subcommittee, Morant discovered that his minister was too busy to deal with the question![92] The problem was resolved by determining that the alcohol store next door at the Natural History Museum did not in fact constitute a fire hazard. Lockyer had lived long enough to be personally informed. Announcing a forthcoming visit to the now 75-year-old seer, Bell wrote in May 1911 of his eager anticipation of hearing Lockyer's views on a topic which he knew had been under consideration for a very long time and on which Lockyer had been a moving spirit.

One of Morant's last gifts to the Museum before he moved on to National Insurance was to appoint a new management to the Science Museum. As Director, he chose a Scottish physics professor and former Director of the Edinburgh Museum of Science and Art, Francis Grant Ogilvie, who had also been Secretary to the Bell Committee and previously an intermediary in the negotiations with Lockyer.[93] But he also appointed Henry Lyons of the Royal Engineers, as Secretary to the Advisory Council and effectively Deputy Director of the Museum, a position which led inevitably to his own Directorship. Lyons was not only an able man, but also an appointment which in a remarkable way affirmed the legacy of Lockyer and its endurance in the institution of the Science Museum.

For four years in the early 1890s, as a young surveyor in Egypt, Lyons had been seconded to work closely with Lockyer. The astronomer had developed a theory that the alignment of the temples of Egypt had astronomical significance and that therefore their religion was based on careful astronomical observation. In the 1920s, when the theory seemed to have been overthrown, Lyons would distance himself from it; however, in the 1890s he wrote frequently and warmly reporting on the progress of 'their' project.[94] Lockyer's 1894 volume, *The Dawn of Astronomy: A Study of the Temple-worship and Mythology of the ancient Egyptians*, was the outcome of their work. In 1906 Lockyer would get Lyons elected to fellowship of the Royal Society. In a grateful letter, Lyons, the future Director of the Science Museum, modestly expressed his own scientific limitations and graciously emphasised the personal debt he owed his mentor.[95]

The newly independent Museum did not have an easy infancy. The building was begun before the First World War began and the shell had been completed by the time war ended. However, in the deep post-war recession the government cast around for economies, and there was a serious risk that the Museum building, which had been so hard-won, would be commandeered for other uses. Fortunately for this history, such infanticide was prevented. The new President of the Board of Education in the coalition Liberal-Labour government of 1924 was the now Labour-affiliated Charles Trevelyan, who as a Liberal had been the Deputy President when the Bell Committee was appointed. Trevelyan insisted in Cabinet

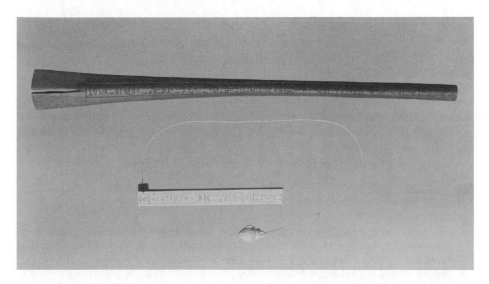

Illustration 1.7 Copy of a merkhet and bay, ancient Egyptian instruments used for astronomical timekeeping (the original is in the Royal Museum, Berlin). The merkhet is a bar with a plumb line used for measuring star transits; the bay is a forked stick used for alignment (Inv. 1913–573). This object which was acquired through Sir Henry Lyons is a relic of Sir Henry's interest in ancient Egyptian measurement fostered by his work with Sir Norman Lockyer. It is a relic of the enduring indirect influence of Sir Norman on the Museum

that the Science Museum project proceed.[96] Two years later a new Conservative government again considered reallocating the building. Nonetheless, in 1928, nearly fifty years after the 1876 Special Loan Exhibition, which had been administered by Lockyer, HM King George V, accompanied by the Director – Lockyer's protégé Henry Lyons – formally opened the East Block of the Science Museum.

Conclusion

The triumph of Lyons as he walked with George V was real. Attendance had grown fourfold during the 1920s. But this was not the realisation of the vision of George's grandfather Prince Albert. It was not a map of modern industries. The Museum through which they strode, introduced by the great steam engines of the past, documented the evolution of pure and applied science in the world. This, then, was truly a Science Museum, the hard-won symbol of victory in a struggle for cultural legitimacy that scientists and their protagonists had waged for half a century.

On the retirement of Lyons in 1932 the Museum could be seen to have been a great success. In a decade its attendance had increased fivefold, and on weekends its attendance was greater than any other in London. It was seen to be one of the most innovative in the entire Empire.[97] In private and in public, his Advisory Committee chairman Glazebrook proclaimed Lyons 'a genius'.[98] He was also a man who had begun his career, in his own words, spreading in Egypt 'the bacillus of science' that had been seeded by Lockyer.[99] As such he had brought to fruition one of the great achievements of 'Progressive Era' Britain and of the 'promotion of science'.

However, the interests and concerns that had been overcome in creating the Museum had not disappeared. The question of how much the Museum should deal with industrial practice and how much with science; how much with historical development and how much with the contemporary; how much for future men of science and how much with the mass public; were left as constantly renegotiated challenges in the future. Even if the names were forgotten, the ghosts of such defeated exponents as Cole, Percy and Mitford would look over the shoulders of Directors and curators for the next century, and longer.

Notes

1. The Museum was founded on 26 June 1909. For the weather and the business climate in the summer of 1909 see Wesley Clair Mitchell, *Business Cycles* (New York: Burt Franklin, 1970, first published in 1913) p. 79.
2. See, for instance, 'Finding Neverland', 2004; 'Dean Spanley', 2008; 'Angel', 2008. Where H.G. Wells's novels have been used as the inspiration for modern movies, however, such as *Island of Dr Moreau* or *The War of the Worlds*, their context has been updated to the late twentieth century.
3. G.R.Searle, *The Quest for National Efficiency. A Study in British Politics and Political Thought, 1899–1914* (Oxford: Oxford University Press, 1971).
4. Roy M. Macleod, 'The Support of Victorian Science: The Endowment of Research Movement in Great Britain, 1868–1900', *Minerva* 4 (1971), pp. 197–230.
5. H.G. Wells, *War in the Air* (London: George Bell and Sons, 1908).
6. 'The Congress of Applied Chemistry', *The Times*, 28 May 1909.
7. 'The Congress of Applied Chemistry. Closing Ceremony', *The Times*, 3 June 1909.
8. Ramsay correspondence, Ramsay papers, University College London Archives, vol. 14/1.
9. David I. Harvie, 'The Radium Century', *Endeavour* 23 (1999), pp. 100–5.
10. Ramsay chaired a subcommittee of the British Science Guild which investigated the potential of atomic energy. See 'Annual Report of the British Science Guild', 1909. Ramsay wrote to Emil Fischer on 19 July 1907 of his belief that progress depended on being able to concentrate and direct energy (William Ramsay Papers vol. 14/1 p. 156, University College London). This was of course independent of Einstein's work on relativity theory but in the empirical knowledge that radium created heat.
11. A month later the King laid the foundation stone of Imperial College's new buildings. 'Imperial College of Science and Technology', *The Times*, 8 July 1909.
12. 'Science Museum', in F.H.W. Shepherd, ed. *Survey of London: South Kensington Museums Area* (London: Athlone Press, 1975), vol. 38, pp. 248–56.
13. Anthony Burton, *Vision & Accident: The Story of the Victoria and Albert Museum* (London: V&A Publications, 1999).
14. Robert Bud and G.K. Roberts, *Science versus Practice: Chemistry in Victorian Britain* (Manchester: Manchester University Press, 1984). See also Bud and Roberts, 'Chemistry and the Concepts of Pure and Applied Science in Nineteenth-Century Britain', in E. Torracca and F. Calascibetta, eds *Storia e Fondamenti della Chimica* (Rome: Accademia Nazionale delle Scienza, 1988, pp. 19–33); Bud and Roberts, 'Thinking About Science and Practice in British Education: The Victorian Roots of a Modern Dichotomy', in P.W.G. Wright, ed. *Industry and Higher Education* (Milton Keynes: Open University Press, 1990, pp. 18–30). I should like to express my debt to Gerrylynn Roberts for the many conversations and discussions that underlie our shared publications.

15. 'Memorandum by the Prince Consort as to the Disposal of the Surplus from the Great Exhibition of 1851', published as an appendix to Sir Thomas Martin, *The Life of his Royal Highness the Prince Consort*, 2 vols (London: Smith Elder, 1876), vol. 2, pp. 391–2 and 569–73.
16. Quoted in Christine Macleod, *Heroes of Invention: Technology, Liberalism and British Identity, 1750–1914* (Cambridge: Cambridge University Press, 2007), p. 260.
17. Bennet Woodcroft, *A Sketch of the Origin and Progress of Steam Navigation* (London: Taylor, Walton and Maberley, 1848).
18. 'National Science Collections Copy of Report of the Interdepartmental Committee on the National Science Collections', *PP* 1886 (246), p. 9.
19. Shepherd, *Survey of London* (vol. 38), pp. 97–123.
20. Archives of the 1851 Commission, Imperial College, London, Deputation of Society of Arts to Lord Chancellor 17 January 1874, quoted in a letter from 'Bennet Woodcroft to Board of Management of the Commission appointed by Her Majesty for the Promotion of the Exhibition of the Works of All Nations (hereafter 1851 Commission)', Enclosure L, 'Minutes of the 105th Meeting of the Commissioners of the 1851 Commission', p. 13.
21. 'Report of the Commissioners of Patents for 1881', *PP* (1882) 374, p. 7.
22. Henry Cole Archives, National Art Library, V&A, Cole diary 1874, 17 January.
23. 'National Science Collections', p. 49.
24. William Stanley Jevons, 'The Use and Misuse of Museums', in his *Methods of Social Reform and other Papers* (London: Macmillan, 1883), pp. 53–81.
25. Burton, *Vision & Accident*; see also Timothy Stevens and Peter Trippi, 'An Encyclopaedia of Treasures', in Malcolm Baker and Brenda Richardson, eds *The Grand Design: The Art of the Victoria and Albert Museum.* (London: V&A Publications, 1997), pp. 149–160.
26. Bud and Roberts, *Science versus Practice*, pp. 133–5.
27. See James Lees-Milne, *The Bachelor Duke: William Spencer Cavendish, 6th Duke of Devonshire, 1790–1858* (Edinburgh: John Murray, 1991).
28. On Lockyer, see Jack Meadows, *Science and Controversy: A Biography of Sir Norman Lockyer* (London: Macmillan, 1972).
29. 'Fourth report of the Royal Commission on Scientific Instruction and the Advancement of Science', *PP* 1874 [C.884], para. 93, p. 14.
30. National Art Library, V&A, Cole diary 1874, 31 March.
31. Archives of the 1851 Commission, Imperial College, London, Cole Memorandum, Board of Management, 1851 Commissioners.
32. Indeed, Cole had already shown the south arcades to Bennet Woodcroft in February 1874. See Cole Diary 1874, 3 February. On George Wilson and technology, see R.G.W. Anderson, 'What Is Technology?: Education through Museums in the Mid-Nineteenth Century', *British Journal for the History of Science* 25 (1992), pp. 169–84; Geoffrey N. Swinney, 'Reconstructed Visions: The Philosophies That Shaped Part of the Scottish National Collections', *Museum Management and Curatorship* 21 (2006), pp. 128–42. I am grateful to Dr Swinney for the opportunity to discuss the nature of George Wilson's museum.
33. I should like to acknowledge the benefit of several conversations with Sook-Kyoung Cho while she was preparing her doctoral dissertation on 'The special loan collection of scientific apparatus, 1876. The beginning of the Science Museum of London and the popularization of physical science' [in Korean], unpublished PhD thesis, Seoul National University (2001). Regrettably I have been unable to read the thesis itself; the analysis here is my own and one for which I take responsibility.
34. Donnelly Minute, 27 January 1875, 'Board Minutes of the South Kensington Museum', TNA: PRO, ED 28/30.
35. See Ann Cooper, 'For the Public Good: Henry Cole, His Circle and the Development of the South Kensington Estate', unpublished PhD thesis, Open University, 1992, p. 284.
36. See Donnelly's pleas to Treasury over the recruitment of Lockyer, 'Board Minutes of the South Kensington Museum', pp. 146–59, TNA: PRO, ED 28/30.

37. Royal Botanic Gardens, Kew, Hooker archives 346/23, JDH/4/10/.
38. 'Loan Collection Of Scientific Apparatus', *The Times*, 13 May 1876.
39. Deborah Jean Warner, 'What Is a Scientific Instrument, When Did It Become One, and Why?', *British Journal for the History of Science* 23 (1990), pp. 83–93. See J.C. Maxwell, 'General Considerations Concerning Scientific Apparatus', in *Handbook to the Special Loan Collection of Scientific Apparatus, South Kensington Museum London*, 1876, pp. 1–21.
40. 'Loan Collection Of Scientific Apparatus', *The Times*, 13 May 1876.
41. Ruth Barton, 'X Club (act. 1864–1892)', *Oxford Dictionary of National Biography*, online edn, Oxford University Press, Oct 2006; [online edn, Oct 2008, http://www.oxforddnb.com/view/theme/92539 (accessed 14 February 2009)].
42. See notes by Sir Francis Sandon, Board Minutes of the South Kensington Museum. TNA: PRO, ED 28/30, p. 61.
43. Hooker draft of presidential address to the Royal Society, 15 December 1876, Hooker papers, p. 346/20, JDH/4/10/2.
44. General Scott, 1851 Commissioners to The Secretary of the Treasury, 21 June 1876, 'A Museum of Scientific Instruments and Objects', SMD, ED 79/23. See also 'Eighth Report of the Special Committee of Enquiry appointed at the 104th Meeting of the Royal Commission' 20 July 1877 Appendix A, Minutes of the 113th Meeting of the Commissioners of the 1851 Commission, p. 8, Archives of the Commissioners of the 1851 Commission, Imperial College London. Also see Hermione Hobhouse, *The Crystal Palace and the Great Exhibition. Art Science and Productive Industry. A History of the Royal Commission for the Exhibition of 1851* (London: Athlone Press, 2002, p. 204).
45. The scientists' memorial is reprinted in 'Correspondence between the Science and Art Department and the Treasury as to the Organization of the Normal School of Science and Royal School of Mines', *PP* 1881 [C.3085], Appendix D to Mr Donnelly's Memorandum, p. 28.
46. For correspondence about setting up the subscription, see Hooker papers, 346/9–11.
47. 'FRS', 'A Science Museum', *The Times*, 6 January 1877.
48. Mr. Spottiswoode to Mr. Macleod, 11 January 1877, SMD, ED 79/23.
49. 'The Scientific Apparatus Loan Collection', *The Times*, 3 January 1877.
50. 'FRS', 'A Science Museum'.
51. 'Science Museum', *Survey of London: volume 38: South Kensington Museums Area* (1975), pp. 248–56.
52. 'Report of the Committee on the Science Collections of the South Kensington Museum', 28 October 1881, TNA: PRO, ED 24/47.
53. See Bud and Roberts, *Science versus Practice*, pp. 90–1.
54. Y [a pseudonym], 'Royal Commission on Scientific Instruction and the Royal School of Mines', *The Times*, 24 August 1871.
55. See, for instance, Lewis Gordon, 'The Royal School of Mines', *The Times*, 4 September 1871.
56. John Percy, *Letter from Dr [John] Percy to the Chancellor of the Exchequer, and correspondence relating to the proposed removal of the Metallurgical Department of the Royal School of Mines from the Museum of Practical Geology, Jermyn Street, to South Kensington* (London: William Clowes, 1879).
57. Donnelly to Lockyer, 3 September 1891, Special Collections, University of Exeter Library, Lockyer papers, MSS 110; there is some evidence in support of Donnelly's complaint. Sir Francis Sandon had explained the government's unwillingness to support a new museum to Parliament in 1877 as being related to the 'divergence of opinion among scientific men on the subject'. See Supply—Civil Services and Revenue Department, Supplementary Estimates for 1876-7. HC Deb 26 February 1877, vol. 232, c1065.
58. Henry G. Calcraft, Board of Trade (railway department) to K.M. McKenzie, Principal Secretary to Lord Chancellor, 18 July 1882 in SMD, Z 262/1/2.

59. 'Report from the Standing Committee on Trade, Shipping, and Manufactures, on the Patents for Inventions Bill; With the Proceedings of the Committee', *PP* 1883 (247), p. 15.
60. 'Thirty-first report of the Science and Art Department of the Committee of Council on Education', with appendices, *PP* 1884 [C.4008] Appendix A, p. 5.
61. 'We however suggest the preservation of some few very old Machines of great historical interest, as for instance Watt's first Engine, Trevithick's Locomotive and Stationary Engines, Stephenson's "Rocket", and Hackworth's "Sans Pareil", and "Puffing Billy" Locomotives. We also propose to retain a good selection of Anchors, Screws, and Paddle Wheels of various forms and dates, including the "Rattler" Screw Propeller which was the first fitted by the Government to a Ship of War; a complete set of all the parts of a small arm Rifle in the various stages of manufacture; many interesting and instructive Models connected with Wool Combing and with the Cotton Manufacture; a series of Sewing Machines; and many other examples.' (ibid., p. 6).
62. On the history of the Pitt Rivers collection see Alison Petch, 'Chance and certitude. Pitt Rivers and his first collection', *Journal of the History of Collections* 18 (2006), pp. 256–66, and the papers in B.A.L Cranstone and S. Seidenberg, *The General's Gift – A celebration of the Pitt Rivers Museum Centenary 1884–1984*. JASO Occasional Paper No. 3 (Oxford: JASO/Pitt Rivers Museum, 1984)..
63. Mitford annotation on draft, annotation opposite page 33, TNA: PRO, Works 17/20/5, pp. 1–50.
64. 'Report by A.B. Mitford C.B.', in 'National Science Collections' (1886), pp. 35–40.
65. Donnelly to Lockyer, 23 September 1886, Special Collections, University of Exeter Library, Lockyer papers.
66. 'Memorandum by Col. Donnelly, R.E. on the separate report of Mr Mitford C.B.', 'National Science Collections' (1886), pp. 41–9; A.B.M. 'p.s.', 'National Science Collections' (1886), p. 40.
67. A.B. Mitford, 'Reply Colonel Donnelly's Memorandum', 'National Science Collections' (1886, pp. 51–4, 53–4).
68. Henry Roscoe, *The Life and Experiences of Henry Enfield Roscoe D.C.L., LL.D., F.R.S. Written by Himself* (London: Macmillan, 1906), pp. 297–8.
69. General Festing testimony, 'Second report from the Select Committee on museums of the Science and Art Department; together with the proceedings of the committee, minutes of evidence, and appendix', *PP* 1898 (327), para. 772, p. 55.
70. On Middleton see the treatment by Anthony Burton, in *Vision & Accident*.
71. Donnelly to Lockyer, 3 September 1891, Special Collections, University of Exeter Library, Lockyer papers.
72. A.G. Green, ed., *Jubilee of the discovery of mauve and of the foundation of the coal-tar colour industry by Sir W. H. Perkin* (London: Perkin Memorial Committee, 1906); A.S. Travis, 'Decadence, Decline and Celebration: Raphael Meldola and the Mauve Jubilee of 1906', *History and Technology* 22 (2006), pp. 131–52.
73. See Dominik Geperth and Robert Gerwarth, *Wilhelmine Germany and Edwardian Britain: Essays on Cultural Affinity* (Oxford: Oxford University Press, 2008). Also see Gerard Delanty, 'Nationalism and Cosmopolitanism: The Paradox of Modernity', in Gerard Delanty and Krishan Kumar, eds *The SAGE Handbook of Nations and Nationalism* (London: Sage, 2006), pp. 357–68.
74. R.L. Morant, 'The Complete Organisation of National Education of All Grades as Practised in Switzerland (1898)', *Special Reports on Educational Subjects*, Vol. 3, c. 8988, p. 24, cited in G.R. Searle, *The Quest for National Efficiency. A Study in British Politics and Political Thought, 1899–1914* (Oxford: Oxford University Press, 1971), chapter 7, p. 210.
75. See, for instance, the treatment of Morant by Michael Sanderson, *Education and Economic Decline in Britain, 1870 to the 1990s* (Cambridge: Cambridge University Press, 1999).

76. E.J.R. Eaglesham, 'The Centenary of Robert Morant', *British Journal of Educational Studies* 12 (1963), pp. 5–18.
77. J. Dover Wilson, *Humanism in the Continuation School* (London: HMSO, 1921), pp. 67–8.
78. Macleod, 'The Support of Victorian Science'.
79. On the British Science Guild, and its links to the early Science Museum committees, see Roy Macleod, 'Science for Imperial Efficiency and Social Change: Reflections on the British Science Guild, 1905–1936', *Public Understanding of Science* 3 (1994), pp. 155–93.
80. Roscoe to Lockyer 14 February 1907, Special Collections, University of Exeter Library, MSS 110, special collections, University of Exeter Library.
81. Hugh Bell to Lockyer, 27 November 1880, Lockyer papers, special collections, University of Exeter Library.
82. A transcript of Lloyd George's budget is available at:<http://www.number10.gov.uk/history-and-tour/prime-ministers-in-history/david-lloyd-george/1909-peoples-budget-transcript> and a recording is also available (accessed 14 February 2009). The setting of chemists and other experts in agricultural colleges was promised in the fourth paragraph.
83. British Science Guild, 4th annual report, 1910.
84. Norman Lockyer, 'Memorandum on the Proposed Science Museum', TNA: PRO, ED 23/523, on the perception of this as pressure on the 1851 commissioners see Ogilvie to Morant 8 July 1907, TNA: PRO, ED 23/523.
85. Morant to Murray 5 April 1908 TNA: PRO, ED 23/523. See also 'A Science Museum', *The Times*, 4 June 1908.
86. Roscoe to Morant, 17 June 1909, TNA: PRO, ED 23/523. For the memorial and its presentation see: 'Science Collections at South Kensington. Deputation to Mr Runciman', *The Times*, 14 July 1909.
87. Robert Morant, 'History of the Art Museum at South Kensington, especially as regards premises', Art Museum series no. 10, p. 4, TNA: PRO, ED 24/2. It is striking that there is also a copy of the same memorandum in SMD.
88. Roscoe to Morant , 17 June 1909, TNA: PRO, ED 23/523.
89. Hugh Bell to Runciman, 9 April 1908 and 26 April 1908, University of Newcastle Library, Runciman papers.
90. Bell to Runciman, 26 September 1910, Runciman papers.
91. Mr Harcourt to Cabinet October 1910 'Proposed New Science Museum', TNA: PRO, ED 23/523.
92. Schomberg McDonnell to Morant, 3 December 1910, TNA: PRO, ED 23/523. Despite his minister's busyness Morant wrote within a week to Runciman complaining about McDonnell's wilful misinterpretation of the meeting's decision.
93. 'Francis Grant Ogilvie', *The Times*, 15 December 1930.
94. Lyons to Dingle, 25 July 1925, Special Collections, University of Exeter Library, Lockyer papers.
95. Lyons to Lockyer 24 February 1906, Special Collections, University of Exeter Library, MSS 110.
96. David Follett, *The Rise of the Science Museum under Henry Lyons* (London: Science Museum, 1978).
97. S.F. Markham, 'Museums in the Empire', *The Times*, 22 September 1933.
98. Memorandum from Glazebrook to Royal Commission on the National Museums, 'The Organization of the Science Collections at South Kensington', Document 165, submitted 14 March 1929, TNA: PRO, AF 755/11.
99. Lyons to Lockyer, 25 November 1895, Special Collections, University of Exeter Library, MSS 110.

2

The First Years: The Science Museum at War and Peace

Tom Scheinfeldt

While the 1911 report of the Bell Committee laid down the administrative agenda for the Science Museum's early decades – an ambitious building programme, a major refocussing of the collections, a new emphasis on education – culturally the great fact of the decade following the Museum's independence in 1909 was the First World War. Although dwarfed today in the popular historical imagination by the Second World War, the 'War to End All Wars' was more deadly, at least as destructive to the national psychology, and at least as disruptive to British society and culture as its sequel. Approximately three-quarters of a million British men and boys fell in the First World War, nearly 7 per cent of the pre-war male population between the ages of 15 and 49. Their absence, as life attempted to return to normal after the Armistice in 1918, was conspicuous. 'Among the major combatants,' writes American historian Jay Winter, 'it is not an exaggeration to suggest that every family was in mourning.' Winter writes: 'Commemoration was a universal preoccupation after the 1914–1918 war. The need to bring the dead home, to put the dead to rest, symbolically or physically, was pervasive.'[1] When the Science Museum emerged with the rest of British society from the Great War, it was forced to pursue its mission, to argue its cause, to make its way in step with the new realities of this pervasive culture of mourning.

The War Museum and the peace museum

The battles of the Great War were fought overseas, but the Science Museum in London was very directly affected by the fighting. During the war years and those immediately following, the Science Museum maintained only limited operations and opening hours in the old Southern and Western Galleries. Originally built as temporary housing for the 1862 International Exhibition, the sequel to the Great Exhibition, the inadequacy of these lodgings was long and well recognised. The largely wooden structures presented a serious fire hazard to the Museum's collection, a situation which caused the Bell Report to recommend new accommodation at least double in size to be built to hold the collections.[2] New construction on Exhibition Road began in 1913, but the limited progress that was made before the outbreak of war would not immediately benefit the Science Museum. In 1918 wartime imperative allowed the War Office to commandeer this unfinished East Block along with parts of the old galleries. Only with the War Office's removal in 1922 did construction begin again in earnest, and, although the Science Museum gradually moved into the new space over the next several years, the East Block remained formally unopened until 1928. Despite much lobbying and the concerted efforts of the Science Museum's Directors and Advisory Council, the full extent of the building programme proposed by the Bell Report – including Centre and Western Blocks approximately equal in size to the East Block – effectively would not be completed until the opening of the new Wellcome Wing in 2000.

Even this meagre accommodation was threatened in the years immediately following the Armistice. Faced with an unprecedented loss of life and the ever-growing demands of mourning and memory, the War Cabinet founded the Imperial War Museum in 1917. Established by Act of Parliament in 1920 with the personal support of Prime Minister Lloyd George, the Imperial War Museum was opened by King George V on 9 June 1920. Like the Science Museum, the Imperial War Museum was inadequately housed. After unsuccessfully scouting several sites in

Illustration 2.1 The East Block under construction, with the Imperial Institute in the background, taken on 15th November 1916

central London, the Imperial War Museum finally settled for a short-term lease of the Crystal Palace in Sydenham. For the Imperial War Museum, the Crystal Palace lacked more than location. Its problems were both spatial and symbolic. The glass panes and painted girders of the Crystal Palace were completely unsuitable from a conservation standpoint. Additionally it seemed altogether too light and airy a setting for the Imperial War Museum's sombre subject matter: early photographs show artillery amidst potted palms and rose bushes.[3] With the Crystal Palace lease set to end in 1924, the Imperial War Museum sought new arrangements, and in 1922 the government provided the Imperial War Museum with the Science Museum's Western Galleries. Though the Science Museum began moving into the unfinished East Block in 1923, between 1924 and 1935 the Imperial War Museum and the Science Museum were uneasy bedfellows on what amounted to the same plot of land.

As one might expect, the new arrangement was not popular in the Science Museum. With the East Block still unfinished, the Science Museum was squeezed into the old Southern Galleries. The relocation of the Imperial War Museum to South Kensington reduced the Science Museum's effective gallery space to its nineteenth-century proportions and forced the removal of the entire Science Collections to storage.[4] The Science Museum could not remain silent, and the public controversy that ensued provides important insights into the perceived purposes of both museums. As they were forced to justify their worth, both revealed their aims and aspirations.

Unfortunately for the Science Museum, the Imperial War Museum had the powerful needs of mourning and commemoration on its side. Conceived of as an official location for post-war mourning, the Imperial War Museum stood as a state-sponsored recognition of the human loss of war. It also assuaged the common public anxiety that the dead might be forgotten. Unsurprisingly, the Imperial War Museum was met by tremendous public enthusiasm. The experience of war was universal in interwar Britain, and the War Museum received nearly universal support.[5]

It was unclear how the Science Museum could compete with this. To disparage the purposes of the War Museum was to dishonour the memory of the dead and needs of their bereaved, and the Science Museum could not attack the new museum directly. Unable to undo its rival, the Science Museum would somehow have to outdo the War Museum. This meant arguing its case in terms of war and peace, and construing its own work in terms of post-war necessities. This placed the Science Museum on an unfamiliar footing. During its accommodation dispute with the Imperial War Museum, the Science Museum was forced to compete by the War Museum's rules.

Advisory Council meetings were preoccupied with the problem. Formal appeals were made to the Board of Education, and informal appeals were published in *The Times* and *Museums Journal*.[6] Letters of support were obtained from prominent members of the scientific and industrial community, and sympathetic members of the museum community were organised in quiet opposition to the new museum. In each case, the Science Museum and its supporters argued their case in terms of the Science Museum's value to the 'national welfare' in post-war reconstruction. The strategy was to turn attention away from the destruction of wartime and towards the demands of the new peace. The Science Museum thus held itself out as a place to celebrate mankind's constructive achievement and to inspire peaceful progress. The Museums Association's 1924 Wembley British Empire Exhibition meeting heard one participant describe the Science Museum as 'the Cinderella of London museums':

> It was the most important of them all, yet it had been treated the worst. He was impressed with a sense of wonderment that it was possible at this time for a British Government to allow a museum of its character to have its floor space cut down to one-fourth of the minimum which had been laid down years ago, and to have its accommodation restricted in favour of a war museum. It was a curious anomaly that a museum which exhibited the means of destroying all the wonderful achievements of industry should be allowed to take the place that ought to be occupied by a museum which showed people how to better themselves and the whole world. He felt sure that if he asked the meeting for an emphatic opinion as to the propriety or otherwise of the War Museum's being at South Kensington it would object unanimously to its being there (Hear, hear).[7]

Such views were not confined to discussions of the Science Museum's accommodation problems. In fact, in the immediate post-war period, the concept of science museums being museums of global betterment became an important part of their public face and discourse. As early as November 1919 and well before the Science Museum's dispute with the Imperial War Museum, the *Museums Journal* published a piece entitled 'If a Museum of War, Why Not a Museum of Peace?'[8] Written by the British Association's General Secretary J.L. Myres, the article weighed the merits of museums of science and industry – such as the museums at South Kensington – with those of the new Imperial War Museum. It opened with a hint of irony. Not only was the glass and steel of the Crystal Palace inappropriate for the War Museum's collection, Myres argued, so too was the building's history incongruous with the War Museum's subject matter. More in the spirit of the Crystal Palace's history as 'a landmark in the history of invention', Myres suggested that the site be used to increase the nation's collections of science and industrial art. Indeed, Myres argued that the war's recent end provided a unique opportunity to preserve pre-war objects of art and industry. He wrote: '...the occasion of the national – indeed world-wide – stock-taking and spring-cleaning, after this "winter of our discontent" offers chances which will not recur of securing and assembling typical examples of 19th century craftsmanship.'[9]

Here Myres offered an alternative notion of post-war remembrance. Real commemoration and respect for the fallen, he maintained, required not only memories of war, but also memories of the 'material surroundings' of the peace that preceded it. The war must be remembered not only in terms of its losses, but also for the scientific knowledge that emerged from the crisis and the potentialities of 'post-war applications of military devices to civil life.' These applications must also be on display in any institution spawned by the war. Instead of a war museum, Myres argued, the nation should erect a peace museum to preserve and memorialise productive and progressive historical monuments.[10]

In its accommodation dispute with the Imperial War Museum and after, the Science Museum repeatedly marshalled these arguments. Forced to put its case

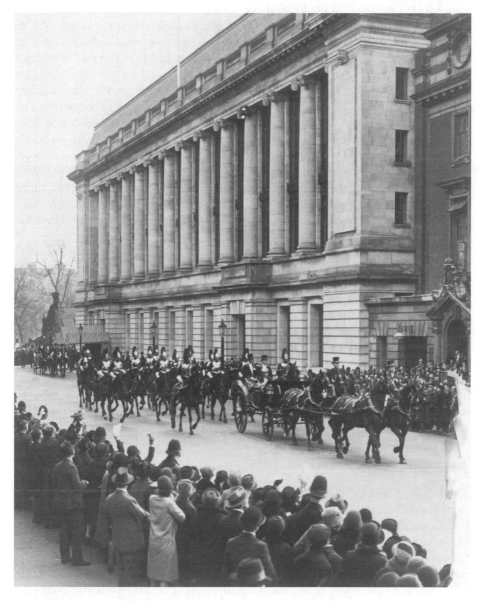

Illustration 2.2   HM King George V leaving the Science Museum after the formal opening of the East Block on 20 March 1928

in terms of war and peace – and hoping to distance itself and science more generally from the military technologies that had played such a destructive role during the war – the Science Museum took up the mantle of 'peace museum'. Thus, where the Imperial War Museum proposed a memorial based on the preservation of the experience of war, the Science Museum offered a memorial based on the history of scientific progress and subsequent hopes for reconstruction. Whereas the Imperial War Museum supported the nation's need to mourn, the Science Museum supported its hopes for peace. In the wake of the Great War, to support the Science Museum was to proclaim the promise of science's past, to promote the cause of peace and renewal, and to ensure that the fallen had not died in vain.

In the end, the accommodation dispute between the Imperial War Museum and the Science Museum was resolved in a way satisfactory to both parties. The Imperial War Museum soon secured a better site in Lambeth, and the Science Museum moved into the new galleries of the East Block. Moreover, in an unexpected windfall, the Imperial War Museum's initial move from the spacious quarters of the Crystal Palace to the cramped Western Galleries forced it to divest itself of its aeronautical collections, which the Science Museum happily received in exchange for its own sectioned firearms collections. Indeed, this happy trade is in itself powerful evidence of the two museums' preferred positions vis–à–vis issues of war and peace.[11]

Despite the happy outcome, the accommodation dispute with the Imperial War Museum is important on several counts. The threat of dislocation early in the period made the task of securing the additional space recommended in the Bell Report seem even more urgent, and, throughout the 1920s and 1930s, Advisory Council and Director's reports are dominated by the struggle for greater accommodation, especially the Centre Block. Just as importantly, however, this dispute draws attention to the powerful presence and operation of post-war memory in the interwar Science Museum. In subsequent years the Science Museum bolstered its claims to 'peace museum' status both administratively and intellectually, by constructing a notion of practitioner identity based around a calling to benevolent service and a history of science based on accounts of peaceful progress.

A museum officer's work

The interwar history of the Science Museum is dominated by two towering figures: the Chairman of the Advisory Council, Sir Hugh Bell – chairman of the Bell Committee, who thus gave his name to the Bell Report – and the Science Museum's Director, Sir Henry Lyons. Not only were Bell and Lyons central to the physical and conceptual development of the Science Museum, but in many ways their careers provided a template for the construction of identity among other Science Museum staff members during the period and even afterwards. In particular, each combined a career of scientific or industrial achievement with a potent strain of military or civil service. The combination of these characteristics as manifested in Bell and Lyons provided the Science Museum with a clear and tangible basis for practitioner identity during the 1920s and 1930s.

Hugh Bell was born in 1844 to a prominent Middlesbrough engineer and industrialist, Sir Lowthian Bell Bt. After studying chemistry in Edinburgh, Paris and Göttingen, the younger Bell returned home to work in his father's steel business. Hugh Bell found great success in the steel industry, growing his father's business substantially. In 1899 he helped to convert the private partnership of Bell Brothers into a public company and became managing director of Bell Brothers Ltd. In 1904 he succeeded his father to both the company's chair and the baronetcy. In 1923, when Bell Brothers merged with Dorman, Long and Co., Ltd, Bell became the enlarged entity's Vice-Chairman and eventually, just before his death, its Chairman.

Although he was first and foremost an industrialist, Bell was in the shadow of his father and, after his father's death, that of Arthur Dorman, only breaking free after Dorman's death in 1930.[12] Perhaps as a way of compensating for this, he held several posts in education and politics. His association with the Science Museum ran to more than twenty years, beginning with his influential work as chairman of the 1911 Board of Education Departmental ('Bell') Committee on the future of the Science Museum. He served on the Board of Governors of Imperial College and as chairman of the council of Armstrong College, Durham University's satellite college in Newcastle (now the University of Newcastle). He played a prominent role in party politics, first as a member of the Unionist Party and then as a Liberal, and sat three terms as Mayor of Middlesbrough over a period of nearly forty years. For decades he served as chairman of the Tees Conservancy Commission, using his position as an engineer, industrialist and public figure to implement much-needed navigational and floodplain improvements on the River Tees.

Bell's reputation was based as much on his commitment to public service as it was on his industrial achievement. While his *Nature* obituary applauded his successes in engineering and industry, it argued that Bell's greatest strength was his social commitment:

> Merely to recount the work of Sir Hugh Bell as a successful ironmaster would, however, be to give a very limited and one-sided view of his activities. He was, above all, a great public servant, and rendered services to the community of which it is difficult to speak too highly.[13]

More than just an engineer or industrialist, Bell was seen as one who put his engineering and industrial knowledge to good use, and it was this combination that made him an appropriate chair for the Advisory Council. At his death Bell's tenure as Advisory Council Chairman was applauded not only on the basis of his wise scientific counsel, but also on the basis of his ready service.[14] From the outset and until the end, Bell's qualifications were judged to rest on his being both competent scientist and able servant. His chairmanship was a powerful image statement, and Bell's model of scientific service was encouraged and affirmed in the Science Museum's other staff.

If Sir Hugh Bell provided an exemplar for the interwar Science Museum practitioner, then Sir Henry Lyons provided the paradigm. He is without doubt the central character in the story of the interwar Science Museum. From 1920 until 1933, Lyons served as the Museum's Director and from 1933 until 1943 as Chairman of its Advisory Council. During this time, Lyons oversaw the opening of new buildings, the growth of the collections and visitors, and a new public recognition for the Science Museum. As the President of the Board of Education Lord Irwin (better known as Lord Halifax) remarked upon Lyons's retirement:

> In 1920, when you were appointed, the museum was a small institution known only to a few ... You had the vision to realize the position the museum might take and the value it might be to science and industry. By your tact, energy, and ability you have made it what it is and is known to be – a treasure house of past achievements and an inspiring guide to future progress... We claim for the Museum a foremost place among institutions of its kind, and recognize that it is to you that this is due. You lay down that work with every good wish from all of us for your future happiness and prosperity.[15]

Lyons was born in London in 1864, the son of a career army officer. Following in his father's footsteps, Lyons attended the Royal Military Academy at Woolwich in 1882–1884. Commissioned a lieutenant in the Royal Engineers, Lyons entered a course of military engineering at Chatham in 1884–1886. From Chatham he was posted to Gibraltar, and finally to Cairo in 1890.

Lyons first became interested in science – specifically in geology – at a young age. He was elected to the Geological Society at only eighteen, and published his first paper in the Society's *Quarterly Journal* while still at Woolwich in 1887.[16] Later, working with the Royal Engineers at the Aswan Dam, Lyons published several works on the geology, geography and history of Egypt.[17] Finally, in 1896, Lyons was singled out to organise a Geological Survey of Egypt, and in 1906 he published the definitive treatment of Egyptian geomorphology.[18] Almost immediately Lyons was elected to fellowship of the Royal Society.

Returning from Egypt in 1909, Lyons accepted a post as a lecturer at the University of Glasgow. An undistinguished public speaker, Lyons remained at Glasgow for only two years, leaving in 1911 to become Assistant to the Director and Secretary of the Science Museum Advisory Council. His time in this position was cut short by the outbreak of war. In 1914 Lyons was recalled for service and in 1915 began work with the Meteorological Office in Gibraltar. By 1918 he had been given the acting rank of lieutenant colonel and was named Commandant of the Army Meteorological Services and Acting Director of the Meteorological Office. At the war's end Lyons left the Army with the rank of full colonel, and returned to the Science Museum in 1919. Within a year he was named Director, a position he held for nearly fourteen years.

Even during the war, and throughout his tenure with the Science Museum, Lyons was careful not to sever his ties to the scientific community. In 1915 he was elected president of the Royal Meteorological Society and president of Section E of

Illustration 2.3 Colonel Sir Henry George Lyons (1864–1944), Director of the Science Museum (1920–1933), and Chairman of the Advisory Council (1935–1940)

the British Association. In 1919 he became Secretary-General of the International Union of Geodesy and Geophysics. Later he served as treasurer of the Royal Society, Secretary-General of the International Council of Scientific Unions, President of the Institute of Physics, and founding editor of *Notes and Records of the Royal Society*. This remarkable combination of scientific achievement and long military and Civil Service was recognised and valued by his contemporaries. Upon Lyons's death, for example, Meteorological Office Director G.C. Simpson wrote: 'What a strange series of occupations: the army, geology, surveying, meteorology, and museum management: yet it contains the key to [Lyons's] great and undoubted success ...'[19] Royal Society President Henry Hallett Dale similarly stressed Lyons's combination of scientific and civic contributions, marvelling at 'so close a succession of duties, enterprises and achievements.'[20] As a measure of Lyons's own regard for this combination, Dale pointed to Lyons's abiding interest in the administrative history of the Royal Society, the end result of which was his posthumous publication of *The Royal Society, 1660–1940: A History of Its Administration under Its Charters*.[21] According to Dale, the work was as much an exercise in community-building as it was in scholarship.[22] It was also the kind of scientific service that made Lyons a model for practitioner identity in the Science Museum.

In fact, the interwar Science Museum was led mostly by men like Lyons who had applied scientific training and skills to public, and, more specifically, to military positions and projects. Take, for instance, the period's other directors. Although Lyons's predecessor Francis G. Ogilvie began his career teaching applied physics in his home town of Aberdeen and became Principal of Heriot-Watt College in Edinburgh, he also was a lifelong volunteer in the Royal Engineers and served as Assistant Controller of the Trench Warfare Research Department and then the Chemical Warfare Department during the First World War. Lyons's successor, Colonel Ernest E.B. Mackintosh, was initially a career officer with the Royal Engineers and later a governor of Imperial College and Vice-President of the Royal Institution. Members of the Advisory Council possessed similar backgrounds, among them Colonel Sir S.G. Burrard, Surveyor-General of India, Sir Maurice Fitzmaurice, chief resident engineer of the Aswan Dam, and Sir Philip Watts, Director of Naval Construction in the Admiralty.[23] Even members of the ordinary staff boasted strong public and military service credentials. For example, beginning with Lyons, the Museum maintained a policy of hiring only former warrant and non-commissioned officers as gallery attendants – a policy designed to reinforce an ethic of scientific service.[24] Finally, the many new members of staff who were hired in the interwar period without such credentials were instructed in the fundamentals of the accompanying ethic. Attempting to codify the standards of a 'Museum Officer's Work', Mackintosh recommended they cultivate a combination of 'recognized consultative authority' on matters scientific with service on 'committees', 'promotion boards' and the 'Civil Service Benevolent Fund'.[25]

Thus, by both example and decree, the Science Museum officer was intended to occupy a position of scientific service as an Army officer would occupy a position of military service. However, unlike officers in the armed forces, who were charged with devising and deploying destructive military technologies, Science

Museum officers were charged with describing a more progressive vision of science and technology. In this way, the ideal of scientific service embodied by Lyons and described by Mackintosh in 'A Museum Officer's Work' resonated with the post-war rhetoric of the 'peace museum' as a place of public betterment and future progress.

Constructing the history of science in the interwar Museum

Today the Science Museum maintains a vibrant programme of scholarship in history of science, technology and medicine. Yet, as Xerxes Mazda has shown, history has not always been part of the Science Museum's mandate. In fact, it was only in 1911 that the Bell Report 'suggested for the first time that one role of the Museum should be the preservation of historic objects' and it was only during the interwar period that history really took hold in the galleries, mostly to the detriment of displays of current practice.[26]

For instance, in the early post-war years, the Science Museum continued with a publication programme started before the war. This *Catalogues* series presented more or less straightforward inventories of the collections, including only the most basic accession and cataloguing information. By the late 1920s, however, this series was replaced by a new *Handbooks* series. In each *Handbook*, the first part presented the history and significance of the collections while a much-abbreviated second part offered the kind of curatorial information found in the earlier *Catalogues*. The replacement of the *Catalogues* by the *Handbooks* represents a shift in emphasis for the Science Museum, away from the technical role it originally intended and towards a historical role in marking mankind's technical development. Whereas the *Catalogues* set out to illustrate the Museum's collections and their technical functions, each *Handbook* 'intended to serve as an introduction to the study of the history and development' of a particular branch of science or technology, making only 'special reference to that collection in the Science Museum' which would illuminate this historical perspective.[27]

Yet, as both Mazda and Robert Bud have pointed out, during this time history in the Science Museum was valued not so much for its own sake, but rather 'as a powerful means of providing insight into culture.'[28] The subordination of pure history to the purposes of culture was vital to the Science Museum's continued cultural relevance and was part and parcel of the kind of history on offer. The interwar Science Museum presented the history of science as the cornerstone of the history of civilisation, as a history of mankind's largely uninterrupted progress. Importantly, in the wake of the Great War, it was also a history free from strife and upheaval, focusing as it did on the advances of human knowledge and industry rather than the degradations of war. In 1933, Lyons described this brand of history in an unpublished paper entitled 'Technical Museums: Their Scope and Aim':

The technical museum differs wholly from its forerunners, the museums of art and archaeology, in the principles which underlie its policy. Its key-note is development and it aims at illustrating man's striving after a greater efficiency from the earliest stages of his civilization. Now he is taking advantage of a better material, at another time a new source of power becomes available, then some increase in knowledge may enable him to break new ground and to improve upon the tools or the methods which he had inherited. His progress, therefore, is to be recorded from the earliest stages of his civilized development up to the present time; the earliest tools and processes which he employed are shown in relation to the later and more advanced types which he produced as his knowledge and skill increased... [29]

The Children's Gallery, which was instituted personally by Lyons in 1930–1931, offers a good example of how this history was told inside the Science Museum. The gallery's transport display illustrated the 'development of transport by land, sea and air from prehistoric times to the present day.' Its communications display provided a history 'from the primitive beacon to the modern automatic telephone exchange.'[30] Throughout the Children's Gallery the story told was one of peaceful scientific progress pointing to a brighter future: 'The history of science is closely bound up with the history of mankind like one of the strands of a heavy cable, and a knowledge that leads to a better understanding of the present, and often points the way to the future.'[31]

Designed as much as an introduction for adults as it was for children, the Children's Gallery was representative of the brand of history of science on display at the Science Museum.[32] Its story of progressive scientific and technological advancement was very much part of the public rhetoric surrounding the Imperial War Museum accommodation dispute. Working together, these different channels of communication told a story of progress that the Science Museum hoped would provide it with the relevance and immediacy of a 'peace museum' in the cultural marketplace of interwar Britain.

In the immediate wake of Armistice and amidst a pervasive culture of remembrance, therefore, it made good sense for the Science Museum to appeal to the more peaceful aspects of science and technology. But the Science Museum was not slow to change these emphases as broader societal concerns and cultural imperatives around it shifted. In the later part of the interwar period, as attention swung from the last war to the next, new pressures from industry and government began to undermine the priorities and values of the 'peace museum', and the Science Museum was obliged to reposition itself as an instrument of industrial power, national security and military strength.

What the Museum does for industry

Beginning in the late 1920s and early 1930s, the Science Museum and its staff came under increased pressure to make the Museum 'of more practical use to those actually engaged in industry as distinct from students.'[33] Increasingly, industrial interests were asking, 'What does the Science Museum do for Industry?'[34] The Science Museum did not have a good answer: more than once Director Henry

Lyons responded that he had 'no clear idea of what would be of value to them.' This became an increasing source of tension and embarrassment to the Museum as the decade wore on.[35]

A case in point is the investigation of the Royal Commission on National Museums and Galleries, which was established by the government in 1927 to investigate the operation and provision of the various national museums and to make recommendations on how their work could be improved and harmonised according to a national strategy. The Royal Commission was specifically interested in the Science Museum's provisions for industrial meetings. Although suggested by the 1911 Departmental Report, funding for conference facilities was not included in the building of the East Block, and at the time of the Royal Commission's hearings there was only a half-finished 200-seat lecture hall. Without funding, its completion was uncertain, and this emerged as a great concern among the Royal Commission's interviewees. The testimony of Sir Frank Edward Smith – soon-to-be Secretary of the Royal Society and long-time Advisory Council member Sir Richard Glazebrook's National Physical Laboratory colleague – is representative. When asked if he considered 'a Lecture Hall a requisite adjunct' to the Museum's facilities, Smith responded: 'That is so. If the Museum is going to perform its proper functions – and I do think that assistance to industry is one of its functions.' Similarly, the Royal Commission was worried about the constitution of the Advisory Council and its lack of industrial representation. According to Smith, the Science Museum Advisory Council should be constituted by men with industrial connections. He said:

> I had in mind ... that the Council of the Science Museum might have on it certain representatives of the larger technical Societies of London, say the Civil Engineers who might nominate a representative, and the Mechanical Engineers. If those people took their duties seriously, it would to a considerable extent ensure that the Museum was fairly up-to-date with regard to engineering and civil engineering and other branches.[36]

Director Lyons's testimony was also substantially occupied with questions about the constitution of the Advisory Council. Lyons repeated that the Science Museum's usefulness to industry was unclear. 'The problem is one of perception,' said Lyons. 'I rather feel that a great many people in industry still look upon the Museum as one that is primarily an educational one only and which stops short of being a Museum of Industry.' According to Lyons, the solution to this problem was to better co-ordinate the Science Museum's contacts with industry, and as a practical matter to reconstitute the Advisory Council:

> There is an Advisory Council and they give us advice about anything we like to ask them, but it seems to me that if there were some body which represented groups of industry or groups of technical institutions more directly then probably we might be more effectively kept up to the mark than we are now.[37]

In its final report, the Royal Commission took these suggestions to heart, rec-ommending that the Advisory Council be reconstituted and enlarged to make it 'more fully representative of scientific and technical institutions and industrial groups' and its powers extended 'so as to assign to the Advisory Council a more active part in the management and development of the Museum.'[38] These recom-mendations were made in 1930, and by 1931 the Advisory Council could already report that their number had been enlarged from twelve to thirty-two.[39] Nearly all the new members represented industrial concerns. Tellingly, this new Advisory Council came more and more to favour in its reports the Museum's longer title, the 'National Museum of Science and Industry'. More visibly, in 1933, the Science Museum launched a new programme of 'special exhibitions', intended first and foremost to provide industry with competent labour.

Lyons retired in the same year and, fortunately for the special exhibitions pro-gramme, Mackintosh took a personal interest in the new initiative, personally crafting a handbook for his staff.[40] This was not the only convenient person-nel change for the special exhibitions programme. Just two years earlier, upon Bell's death, Glazebrook was appointed Chairman of the Advisory Council. Like Mackintosh, Glazebrook was positively disposed to the special exhibitions pro-gramme. In its report for 1932, the first full year of Glazebrook's tenure, the Advisory Council took up for consideration 'the future policy of the Museum especially in relation to its utility for industry.' Noting first the success research organisations such as his own National Physical Laboratory had enjoyed host-ing small-scale temporary displays at the Science Museum, and the fact that the Science Museum was 'much less-well known to those occupying leading positions in the various branches of [industry],' the council resolved to 'afford facilities to an industry, or group of industry, for showing informatively its products'.[41] The first of these special exhibitions was the 'Plastics Materials Industry' exhibition of 1933.[42]

Organised almost independently by industrial concerns and without substantive input from Science Museum staff, the plastics exhibition was roundly considered a tremendous success. The Advisory Council was impressed by the exhibition's strong attendance figures and by the sales of its companion handbook and bibli-ography, reporting that 'the benefit accruing was undoubtedly realised by mem-bers of other industries who visited the Museum, with the expected result that enquiries as to facilities for similar exhibitions have been received from other industrialists.'[43]

A sequel was opened the following year, this time by the rubber industry. Apparently it met with even greater success, hosting over 300,000 visitors in its short run. On display was a recreation of an Asian rubber plantation, complete with diorama, films of the Empire's happy labourers planting and harvesting, and a set of live rubber trees sent specially from Malaya. Alongside this display was a display of the manufacture and commercial uses of rubber in the home, in automobiles and in hospital. The Museum was surrendering more and more control to industry. This time the Rubber Growers' Association, the Research Association of British Rubber Manufacturers, the Rubber Institute of Malaya and

the London Advisory Committee for Rubber Research provided their own sala-
ried guide lecturers and wrote their own handbook, which was published and
sold by the Museum.[44] In addition, 'facilities were provided for a series of visits by associations and institutions, at which the Rubber Growers and Rubber
Research Associations were "at home" to their guests.' When it finally closed six
months after opening, the complete exhibition was taken by the Rubber Growers'
Association and transferred to Manchester and Edinburgh.[45]

The increasing importance of special exhibitions put the Museum in an awkward position. On the one hand, exhibiting industries provided the Science
Museum with increased attendance, media attention, new exhibits, and even
soundproofing for its lecture hall (from the Anti-Noise League for its 'Noise
Abatement' exhibition).[46] On the other hand, as a public institution, the Science
Museum had to be wary of the appearance of impropriety. As industry played an
ever greater role in the galleries – in everything from financing to design to staffing – concerns mounted that the Science Museum might take on 'the semblance
of a Trade Show' or even become 'an ordinary trade exhibition'.[47] Increasingly the
Science Museum was coming to depend on industry for its stall in the changing
cultural marketplace of the late 1930s.

'Science in the Army'

If the special exhibitions programme highlights the new importance of industry
in the 1930s Science Museum, then it also demonstrates the extent to which this
industrial turn was tied to the changing political climate of the time. By promoting
national industrial strength and imperial power abroad, special exhibitions helped
to shore up the Science Museum's position as an important national institution in
the quickly changing geopolitical climate of the late 1930s. Gone from these instal-
lations were the transcendent historical themes of the early 1920s, replaced by the
promotion of national industrial strength and the display of imperial hegemony.
Having contrasted itself with the Imperial War Museum in the early 1920s, the
Science Museum by the mid-1930s had taken up the imperial banner.

Memoranda regarding the future of the Science Museum's ship model collection seem an unlikely place to look for evidence of this change. Yet the ship
model collection was a source of particular concern for the Science Museum, and
the settling of its future signals a turning point for the Museum. Two decades
after its dispute with the Imperial War Museum, the Science Museum would find
itself in another dispute with yet another national museum, this time with the
newly formed National Maritime Museum. A proposal to transfer its ship model
collection to Greenwich required the Science Museum to fine-tune its new rhetorical appeals to industry, empire, patriotism and war.

The National Maritime Museum in Greenwich was established by Parliament in
1934, although renovations to Queen's House delayed the installation of collections for some years. Yet, from the very founding of the new museum, the Science
Museum's attention to its nautical collection markedly increased. Special exhibitions highlighting 'Native Boats', 'The R.M.S. Queen Mary', 'British Fishing Boats',

'Chinese Junks' and 'The Centenary of Transatlantic Steam Navigation' all took place in the years between 1934 and 1939. Additions to the collection received new attention in the 'Acquisitions' section of Advisory Council reports. Finally, as the Maritime Museum began to demand the transfer of the collection, the Advisory Council became increasingly preoccupied with the ship models situation.[48]

Coincident with the death of shipping curator Geoffrey Laird Clowes, a member of the wealthy and influential Laird shipping family, the issue finally came to a head in late 1937. In early November a request was sent by representatives of the Maritime Museum to Lord Stanhope, Chairman of the Standing Commission on Museums and Galleries, asking for help in dealing with the controversy.[49] In the weeks following, the Standing Commission asked the Science Museum to submit a memorandum laying out its position on the matter. Dated December 1937, this paper sets out the Science Museum's long-held, historical view of the ship model collection. It reads:

> The aim of the Collections is to illustrate the history, development and purpose of the ship as a structure and as a machine ... The ship model collections of the Science Museum are therefore more concerned with the technical than with the purely romantic or historical aspect.

The paper goes on to discount the notion of any 'serious overlapping or competition' with the Maritime Museum, reminding the Standing Commission that 'it is normally assumed that the newcomer makes itself conversant with the scope of the established museums.' It regrets that the Maritime Museum went directly to the Standing Commission without consulting its fellows at the Science Museum first. And it rejects outright the notion that the admission of private objects into the Science Museum's galleries has somehow compromised the Museum's standing as a national institution. The paper concludes with a reiteration of the value of the Science Museum and its motives in keeping the ship model collection:

> It will be seen, therefore that the activities of the Science Museum and the Maritime Museum will be progressively divergent as time goes by; the one basing its collections on technical achievements and touching closely on the life of the people by its encouragement of vocational study; the other recording 'the history of our national maritime adventure'.

In fact this is probably the last, and ultimately unsuccessful, statement of the Science Museum's earlier hopeful vision. A few months later the Standing Commission judged in favour of the Maritime Museum, sending most of the ship model collection there. The Commission's decision appeared in their report for 1938, and began with a clarification of the museums' respective spheres of concern. With regard to the Science Museum, the Commission acknowledged the universalism of the Science Museum's mission. It writes:

> The Science Museum seeks by its general collections, which cover an immensely wide field of scientific endeavour, to aid in the study of technical development

and to illustrate the applications of science to industry. In the sphere of ship models it displays an historical conspectus, including a representation of craft from the earliest time to the present day. This display is not restricted to developments in this country only but is international in character.[50]

With regard to the Maritime Museum, the Commission recognised the essentially national character of its mandate and that, unlike the Science Museum, its ship model collection should illustrate the work of 'each type of ship in relation to its work both in peace and in war, in fostering commerce and in trade defence.'[51]

In the end, it seems it really was a choice, as the Science Museum had reckoned in its statement, between a history of progressive and peaceful technical achievement on the one hand, and a 'romantic' history of national commercial and military superiority on the other. In the immediate run-up to the Second World War, the Standing Commission chose the latter – as, ultimately, did the Science Museum. In fact, by the time of the Commission's decision, the Science Museum had already started to turn away from the ideals of the 'peace museum'.

This turning away was most apparent in the staging of a special exhibition in 1938–1939, entitled 'Science in the Army'. Organised in co-operation with the War Office and the Army, 'Science in the Army' set out to illustrate 'the working partnership of the British Army with Science and the technical training and research now essential in a modern army.'[52] The opening ceremony was chaired by the leading British industrialist Viscount Nuffield, and the ceremony was performed by the Secretary of State for War, Leslie Hore-Belisha. The exhibition followed the same administrative lines as earlier industrial special exhibitions, with the War Office supplying most of the exhibits and producing the accompanying printed material. After a brief historical overview of military technology – from spear to Spitfire – the exhibition was more or less a recruitment exercise, following the career of a new recruit as touched upon by technology: 'how a raw recruit is turned into the up-to-date soldier.' Absent is any notion of human unity through scientific progress. The exhibition stresses the unique superiority of British military science. Historical references are included to communicate the essential difference of modern warfare, the weakness of the old and the strength of the new. Absent as well is the earlier notion that the history of science is the history of peace. The exhibition's handbook concludes with King Solomon's proverb – 'A man of knowledge increaseth strength.'[53]

Conclusion

As late as 1930 the ideals of the peace museum endured – universal, progressive and hopeful. But these ideals would fade with the increasing influence of industry and the new attention to national and imperial politics in the late 1930s. Finally, as the climate turned away from post-war hope and towards pre-war preparation, from the productive, unifying collection of the Science Museum to the nationalist and military collection of the Maritime Museum, the Science Museum would take up the cause of 'Science in the Army'.

The Science Museum's progress from peace museum to 'Science in the Army' in its first three decades as an independent entity is interesting not only because it provides a unique window upon the priorities and pressures of the early institution, but also because it illustrates ongoing tensions within the Museum. The questions asked of the Science Museum's work in 1919 and 1939 are still asked of curators and critics of today's exhibitions. What is the proper balance between displays of pure versus displays of applied science ('LaunchPad', 'Making the Modern World', for instance)? What is the place of patriotism in the galleries ('Exploring Space')? How much influence should industry be given over the content of the galleries ('Food for Thought' and 'Challenge of Materials')? In these and countless other ways, ideas first articulated in the early twentieth century – of the Science Museum as transcendent and peaceable on the one hand and useful and potent on the other – continue to inflect perceptions and discussions of the institution today.

Notes

1. Jay Winter, *Sites of Memory, Sites of Mourning: The Great War in European Cultural History* (Cambridge: Cambridge University Press, 1998), pp. 2, 28.
2. *Report of the Departmental Committee on the Science Museum and the Geological Museum* (London: Board of Education, 1911), p. 5.
3. Gaynor Kavanagh, *Museums and the First World War: A Social History* (Leicester: Leicester University Press, 1994), pp. 144–50.
4. Science Museum Annual Reports for 1921–1922.
5. Gaynor Kavanagh, 'Museum as Memorial: The Origins of the Imperial War Museum', *Journal of Contemporary History* 23(1) (January 1988), pp. 77–97. See also Winter, *Sites of Memory, Sites of Mourning*, p. 80.
6. Published appeals on behalf of the Science Museum in this matter include Sir Hugh Bell, 'The Science Museum', *The Times*, 18 December 1923; Sir Hugh Bell, 'The Science Museum', *Museums Journal* 23 (1924), pp. 200–1; J Bailey, 'The Science Museum', *The Times*, 18 December 1923; J Bailey, 'The Science Museum', *Museums Journal* 23 (1924), pp. 201–2.
7. 'Discussion', *Museums Journal* 24 (1924), p. 118.
8. J.L. Myres, 'If a War Museum, Why Not a Peace Museum?', *Museums Journal* 19 (1919), pp. 73–6.
9. Ibid., 73.
10. Ibid., 75.
11. Science Museum Annual Report for 1924. See also the testimony of Imperial War Museum Director, Charles ffoulkes, before the Royal Commission on National Museums and Galleries, *Oral Evidence, Memoranda and Appendices to the Final Report* (London: Royal Commission on National Museums and Galleries, 1929), p. 96.
12. J.K. Almond, 'Sir Thomas Hugh Bell', in David J. Jeremy, ed., *Dictionary of Business Biography*, vol. 1 (London: Butterworths, 1984).
13. Henry Louis, 'Sir Hugh Bell, Bart., C.B.', *Nature* 128 (1931), pp. 96–8.
14. Louis, 'Sir Hugh Bell, Bart., C.B.', 98.
15. *The Times*, 12 October 1933.
16. Henry Lyons, 'On the London Clay and the Bagshot Beds of Aldershot', *Quarterly Journal of the Geological Society of London* 43 (1887), pp. 431–41.
17. See, for instance, Henry Lyons, *The History of Surveying and Land-Measurement in Egypt* (Cairo: National Printing Department, 1907).
18. Henry Lyons, *The Physiography of the River Nile and Its Basin* (Cairo: National Printing Department, 1906).

19. G.C. Simpson, 'Obituary, Sir Henry Lyons, F.R.S.', *Nature* 154 (1944), p. 328.
20. H.H. Dale, 'Henry George Lyons', *Obituary Notices of Fellows of the Royal Society* 4 (1944), pp. 795–809.
21. H.G. Lyons, *The Royal Society, 1660–1940: A History of Its Administration* (Cambridge: University Press, 1944).
22. Dale, 'Henry George Lyons'.
23. For instance, in 1919 the Council included only two academic scientists, Sir James J. Dobbie and Sir Richard Glazebrook. The remaining members included three civil engineers, three military engineers and three industrialists, along with Bell and Lyons. Throughout the 1920s, occasional vacancies were replaced with men of expertise similar to that of their predecessors. When the Advisory Council was reconstituted and enlarged upon the recommendation of the National Museums and Galleries Commission in 1931, the proportion of pure scientists was further reduced as the Council's representation by industry was vastly expanded.
24. Henry Lyons, 'Director's Report', June 1930, SMD, Z 193/1. It is interesting that not only did this policy ensure the attendants' qualifications as scientists and public servants, but it also helped ensure that the Science Museum would remain an entirely male domain in the interwar years. A comparison can be made with other Great War-era museums, including the Imperial War Museum. In these institutions, women were active not only in the galleries but also in building collections, staging exhibitions and raising funds. See Kavanagh, *Museums and the First World War*.
25. E.E.B. Mackintosh, 'A Museum Officer's Work', 1 March 1939, SMD, Z 184.
26. Xerxes Mazda, *The Changing Role of History in the Policy and Collections of the Science Museum, 1857–1973* (Science Museum Papers in the History of Technology No. 3 (London: Science Museum, 1996). The quotation is taken from p.3.
27. This statement is repeated in most of the Handbooks. See, for instance, Alexander Barclay, Handbook of the Collections Illustrating Industrial Chemistry (London: Science Museum, 1929). For a typical Catalogue, see *Catalogue of the Collections in the Science Museum, South Kensington: With Descriptive and Historical Notes and Illustrations, Mathematics* (London: Science Museum, 1920).
28. Robert Bud, 'History of Science and the Science Museum', *The British Journal for the History of Science* 30(1) (March 1997), p. 47.
29. Henry Lyons, 'Technical Museums, Their Scope and Aim', 1933, SMD, Z 183/1.
30. Lewis W. Phillips, *Outline Guide to the Exhibits* (London: Science Museum, 1937), p. 5.
31. Ibid.
32. In fact, Lyons himself considered the gallery's definition as a 'Children's Gallery' to be too narrow. In one report he wrote that the 'so-called Children's Gallery' would be better understood as 'the first of a graduated series of introductory collections designed specifically for the general public of all ages....' See Henry Lyons, 'A Memorandum on the Development of the Science Museum From 1920 to 1933', August 1933, p. 6, SMD, Z 183/1.
33. *Final Report, Part II* (London: Royal Commission on National Museums and Galleries, 1930), p. 47. The suggestion was submitted by the Federation of British Industries and was taken up by the Royal Commission on National Museums and Galleries and by the Science Museum Advisory Council. The Federation of British Industries would play a significant role in increasing the influence of industry in the Science Museum, receiving a seat on the Advisory Council and submitting nominations for several others. See, for instance, *Sub-Committee on the Representation of Industrial Concerns on the Advisory Council, Report* (London: Board of Education, 1931).
34. See, for example, the letter from J.S.M. Ward of the Federation of British Industry in a letter to Lyons dated November 1928 (SMD, ED 79/30), in which he said 'I am particularly interested in the remark that you receive enquiries such as "What does the Science Museum do for Industry?".'

35. Henry Lyons, 'Letter to Sir Richard Glazebrook', 2 November 1928, SMD, ED 79/30. See also Lyons's letter to the head of the British Federation of Industries: Henry Lyons, 'Letter to J. S. M. Ward', 19 November 1928, SMD, ED 79/30.
36. *Oral Evidence, Memoranda and Appendices to the Interim Report* (London: Royal Commission on National Museums and Galleries, 1928), pp. 157–61.
37. Ibid., pp. 150–7.
38. *Final Report, Part II*, p. 45.
39. Science Museum Annual Report for 1931. See also 'Science Museum Advisory Council Terms of Reference, Draft', 1931, SMD, Z 200/1.
40. E.E.B. Mackintosh, 'Special Exhibitions', March 1939, SMD, Z 108/4.
41. Science Museum Annual Report for 1932.
42. Science Museum Annual Report for 1933.
43. Ibid.
44. Science Museum Annual Report for 1934. See also *Rubber Exhibition (November 1934–April 1935): A Brief Account of the History of Rubber from Its Source to the Finished Product and a Descriptive Catalogue of the Exhibits, Compiled by the Rubber Growers' Association* (London: Science Museum, 1934).
45. Science Museum Annual Report for 1935.
46. Science Museum Annual Report for 1935. The Advisory Council points out that 'as an integral part of the exhibition and after a technical test of the acoustics of the lecture theatre, four firms kindly undertook to treat certain panels on the ceiling and back wall with sound-absorbing material.' Also see *Noise Abatement Exhibition* (London: Anti-Noise League, 1935). This exhibition is also covered in Chapter 10 on temporary exhibitions.
47. Mackintosh, 'Special Exhibitions': Henry Lyons, 'Museum Policy', 11 November 1932, SMD, Z 186.
48. See account of events described in 'Science Museum Advisory Council, Notes for New Members', 1943, SMD, Z 200/2.
49. Described in Lord Stanhope, 'Letter to Sir Evan Charteris', 11 November 1937, SMD, Z 193.
50. 'Memorandum on the Water Transport Collection at the Science Museum, with Particular Reference to Ship Models', December 1937, SMD, Z 193.
51. *Second Report* (London: Standing Commission of Museums and Galleries, 1938), pp. 30–1.
52. Science Museum Annual Report for 1938.
53. Anthony Armstrong, *Science in the Army: A Brief Account of the Scientific Training and Technical Work of the Soldier To-Day, as Illustrated by a Special War Office Exhibition Held at the Science Museum-November 1938–February 1939* (London: HMSO, 1938).

3
The Science Museum and the Second World War

Thad Parsons III

While the Second World War has been extensively studied, the role of the National Collections, particularly the Science Museum, has, comparatively, been neglected. During the war, the policies regarding the protection of the collections evolved from simply hiding items in seemingly secure locations to a system of protection and conservation. Once items were out of the London buildings, the museums, particularly the Science Museum, were used for a variety of purposes. After the war, all the National Collections were faced with severe constraints, but the Science Museum, which had escaped relatively unscathed, was able to reopen quickly in 1946. Besides management of their previous collections, the war also provided a collecting opportunity, in which several National Collections, including the Science Museum, became involved. Whether it was the threat of bombers or the subject of exhibitions, aviation was a central theme for many of the developments during the period.

Before the development of powered flight, Britain was well protected by the Royal Navy, but flight changed the perceived balance of power. On 17 December 1903, Orville Wright achieved the first successful controlled flight in a powered heavier-than-air machine. By the time this was achieved, the Wright brothers were the focus of international attention. In 1901, reports of the Wrights' achievements reached Europe and in 1902, Ferdinand Ferber, a Frenchman, had constructed a Wright glider based on the vague details he received. Then, in early 1903, both Ferber and Octave Chanute popularised their inaccurate versions of the Wright flyers and called for greater aeronautical research in France.[1] Between 1904 and 1906, contracts for planes were negotiated with Britain, America, France, Germany, Italy, Russia, Austria, and Japan. Then in 1907, Charles R. Flint, a well-known international arms dealer, became the Wright brothers' agent.[2] The public realisation of the power of the airplane quickly followed. In 1908, H.G. Wells published *The War in the Air* – the first fictional description of a future air war.[3] During the interwar period, fictional depictions of the next Great War were popular – Noel Pemberton Billing, pioneer aviator and co-founder of Pemberton Billing Ltd (later known by its telegraph address, Supermarine), wrote a play called 'High Treason' in 1928, which was turned into a movie in 1929; H.G. Wells published his second version of the future war, *The*

Shape of Things to Come, in 1933; and titles such as *The Gas War of 1940, War Upon Women: A Topical Drama, Four Days War*, and *Menace: A Novel of the Near Future* were published throughout the 1930s. While the individual plots varied greatly, they all shared one common feature – a central, devastating surprise attack on a major city (or cities):

> The familiar outlines of the Houses of Parliament were no longer visible, and where Big Ben had reared its imposing height was a mass of smouldering ruins. It was incredible that he would never see it again, never see what had been to him and to thousands of others, one of the monuments of English life. ... With a shudder John hurried on. As he reached the Chelsea Embankment he met a wild, bedraggled figure moaning.
> 'Are you hurt?' John asked. 'Can I help you?'
> 'Nobody can help me. To think that those flying inhuman monsters have demolished that wonderful building and all it contained, all it meant to England.'
> 'Yes,' agreed John. 'Parliament will never sit there again.'
> 'Parliament!' shouted the man. 'Who in hell cares about Parliament? That is what I'm talking about,' and with a wave of his hand pointed to the ruins of the Tate Gallery.[4]

Preventing this scene, where the landmarks of daily life and the country's cultural history were destroyed, was a top priority in the years immediately before the Second World War. For the civilian population, aviation was going to define the coming conflict, and for many government officials, including museum professionals, preparing for the air war was a priority. At the Science Museum, home to the Wright Flyer, the focus on aviation was a boon; the interest in aeronautics played to a strength of the collection, and showed the usefulness of the Museum during a time of crisis.

War on the horizon

Adolf Hitler became Chancellor of Germany on 30 January 1933. Within months, preparations began for the second war in Europe of the twentieth century. These preparations included the national museums and galleries of the United Kingdom, hereafter referred to as the National Collections. In 1933, at a meeting of the Standing Commission on Museums and Galleries, it was agreed to begin air raid precaution (ARP) planning to safeguard the nation's cultural treasures.[5] As part of these early ARP plans, each museum was asked to prepare a list of objects to be protected. By 1934, the first lists had been created; based on monetary value, most items included were worth over £5,000.[6] The first step to protect objects was the agreement to move the 'cream of the collections' into isolated country houses, where bombing was improbable.[7] Over the next few years, plans slowly formed as both the government and museums sought to safeguard their treasures. In spring 1938, as ARP plans neared completion, a small committee of the directors of the

Table 3.1   ARP Categories

Category	Description of items	Storage type/ priority	General examples	Science Museum examples
A	Culturally significant, easily handled, and/or the most valuable	In-house bomb pro of storage or immediate pre-war evacuation	Items made of precious metals or containing precious gems	Wright Flyer
B	Valuable but difficult to handle or of lower value but easier to handle	Tube system, secure area of a museum, or country house outbuildings at announcement of war	Sculptures and pottery	Automobiles, bicycles, and carriages
C	Considerable size but fragile or perishable	Evacuation to country houses as soon as possible	Large carpets and tapestries	Model ships
D	Large and/or immovable	Protected in-situ	Raphael Cartoons at the V&A	Large engines in East Block

major museums was established. They were charged by the Standing Commission to protect the National Collections from the ravages of war.

To fulfil that duty, the directors prepared detailed guidelines for the inevitable war (see Table 3.1 for details). The guidelines divided items into four categories, outlined their appropriate storage, and specified their evacuation priority. Each museum applied the guidelines to its own collections, and by the autumn of 1938 lists had been organised. The lists were only one part of the evacuation plan; packing and transport posed logistical and organisational difficulties. The most pressing concern was the provision of safe storage, and, despite the Office of Works being the 'central co-ordinating authority' for storage, many museums made independent arrangements because they thought the Office of Works could not provide sufficient storage.[8] Through the Office of Works, the Science Museum was assigned only two country houses for their Category A and B items: Basing Park and Herriard Park, both in rural Hampshire.[9] Despite the apparently inadequate provision of safe storage, all major arrangements had been finalised by mid-1938. In addition to accommodation, museums and the Office of Works made arrangements for packing and transport. Requisitioning packing material was difficult for two main reasons: no national institution had attempted to pack its entire collection and, with war approaching, supplies of some materials were limited. To simplify the task, the Science Museum, like some other museums, decided to break the tradition of custom-built cases and, instead, stockpiled standard-sized boxes for either one large item or up to 50 small items.[10] The Science Museum, like most other museums, relied on the Office of Works to arrange for van companies to transport the material, while the British Museum and the National

Gallery made separate arrangements for rail transport. One consistent feature of the transport plans was that they were overly optimistic – a feature recognised at the time and confirmed during the plans' first test. Despite their optimism, the railroads would be busy and virtually useless for the non-essential purposes of the museums. On top of that, the Science Museum Director, Col. Ernest Mackintosh, realised that his road transport was likely to be a few lorries for several days and not several lorries for a few days, as promised.[11]

At the same time during the 1930s, the Science Museum was involved with completing the building plan that was outlined in the Bell Report of 1911. The first schemes for the Centre Block were not created until 1925, but failed because of the inactivity of the Advisory Council, the Board of Education and others.[12] In 1927, the Royal Commission on National Museums and Galleries was created with the purpose, among others, of examining the conditions of the buildings containing the National Collections. One would think that the Commission would have immediately benefited the Science Museum, but the production of its first report postponed any action on the Centre Block until it was published in 1930.[13] In part two of the final report, the expansion of the Museum was 'an urgent national need.'[14] In October 1933, the Treasury agreed, in principle, to the construction of the Centre Block, but it was doubtful whether construction could start until after 1934–1935.[15] So, after a wait of over twenty years, it appeared that the Centre Block was at last going to be built.

Besides the Treasury's budget schedule, several issues cropped up during the mid-1930s that delayed construction. The most important unresolved issue was the question of the planetarium. Mackintosh had expanded Lyons's original plan for a 30-foot dome to a 70-foot dome on the third floor, with the dome supported by the roof of the Centre Block.[16] In July 1935, the positioning of the proposed planetarium became an issue because its inclusion and positioning had never been officially approved, despite Lyons's decision to include it in 1931. By the end of 1935, official approval had been granted by the Board of Education, with the planetarium positioned in Mackintosh's preferred location.[17] The approval was a positive step forward but put the Museum in a Catch-22 situation. The Treasury would not make provisions for the planetarium in the budget without assurances that it would be housed, and the Office of Works would not include the dome in their plans without Mackintosh being able to secure the equipment and furnishings for it. This resulted in the 'eleventh hour' demands by the First Commissioner of Works, William Ormsby-Gore, in January 1936 that funding for the planetarium projector and fixtures, such as seats and carpets, be found by the end of February.[18] This, and manoeuvring by Treasury Ministers, rescheduled the Centre Block construction until, semi-officially, 1938–1939.[19] In 1937 a plan for the redevelopment of South Kensington was created, quite to the surprise of Mackintosh, because he had not been informed of the plan until after it had been drafted. The plan was a large and complicated scheme involving seven institutions: Royal College of Art, the V&A Museum, the Science Museum, Imperial Institute, Imperial College, Natural History Museum, and the Royal College of Music. Within the plans, the Science Museum was provided with approximately the space allocated by the Bell Report,

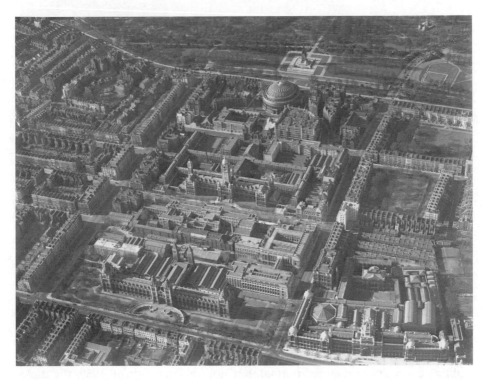

Illustration 3.1   Aerial view of the Science Museum, the neighbouring museums and the Royal Albert Hall, 1939.

© English Heritage. National Monument Record

but the Museum was dissatisfied for two reasons: the plan changed the layout proposed by the Bell Report and allowed for no future expansion.[20] This scheme, along with most government construction, ground to a halt when rearmament began.

The first real test of the museums' preparation came in the autumn of 1938. After the invasion of Austria in March 1938, Hitler turned to the Sudeten question. By mid-September, the crisis was deepening and it appeared that war was imminent.[21] Under these circumstances, the museums' readiness was put to the test. The largest problems occurred with transport, because items needed to be transported safely, securely and in a timely manner. On 15 September, the V&A discovered, much to its dismay, that the vans assigned were too small. The larger tapestries were too long for the vans and so could not leave London.[22] Even when items were able to leave museums, there were concerns about security and protecting the nation's cultural treasures while in transit. This crisis ended quickly when, on 30 September, Chamberlain returned to London and declared that there would be 'peace for our time'.[23] Luckily for the National Collections, Chamberlain's declaration allowed them to improve their plans.

Measures were immediately taken to correct the inefficiencies and inadequacies discovered with the transport, and improved plans ensured that adequate manpower would be available to pack and move the collections. In addition, John Forsdyke, Director of the British Museum, raised the issue of security. He was worried about the security of cultural treasures in transit and while in storage in remote locations. Lacking other options, he proposed to arm warders and staff assigned to the houses. Legally, museum staff could carry firearms, but the Subcommittee of Directors felt it unnecessary, because the presence of staff would be a sufficient deterrent.[24] The Science Museum did not have this deterrent because it did not staff its houses. Instead, it relied on the fire and security arrangements of the occupants, despite the obvious problems. To offset the danger, staff made regular inspections and maintained friendly contact with all owners.[25] Maintaining friendly contact was not difficult, since many of the owners had volunteered their homes to protect the national treasures, but some owners were more interested in protecting their homes from other dangers, such as military use or as a refuge for evacuated children, than in protecting the National Collections.[26] Security was only one problem with the country houses. The quality of storage was also an issue. Concerns about humidity, temperature and pests were expressed at the Subcommittee of Directors throughout late 1938 and 1939.[27] These discussions foreshadowed the dangers that country houses posed to the collections they sheltered.

The phoney war

Since the crisis passed quickly, no large-scale evacuation occurred and the visiting public's experience remained normal, but not for long. During 1939, the diplomatic situation worsened. On 24 August, the secret German and Soviet agreement was revealed. On 1 September, Poland was invaded, and the UK was drawn into war with Germany because of its guarantee to the Polish government. On 3 September, Chamberlain declared war.[28] When the German–Soviet Pact was announced, the Science Museum was ordered to enact its plans as war was imminent. Between 24 August and 1 September, trucks made twenty-six journeys to the houses. One large case, twenty-six oil paintings, and the books were stored at Herriard Park, while 147 cases were stored at Basing Park. This included all category A items, some B and C items, plus about 20,000 old or rare library books.[29] The evacuation was successful but very limited.

The vast majority of the Science Museum's collection remained in South Kensington and, until the public declaration of war, the visitor experience remained relatively normal. If raids had immediately followed the declaration of war, bombs would have inflicted disastrous damage on the National Collections. The Museum closed on 3 September to safeguard the remaining collections. The old Southern and Western Galleries were emptied, because the South Kensington Fire Brigade had deemed them a fire hazard for years. In addition, staff emptied the top two floors of the new East Block. They transferred all of the objects, most with their showcases, into the lower three floors of the East Block, because the concrete construction provided the best available protection. All available floor space was used and spare space in the cases was filled with objects from fixed cases left in the vulnerable portions of the Museum. Curators took steps to prevent collateral damage from flying debris. Small and/or fragile objects were wrapped and carefully placed into the display cases. Staff pasted windows and cases with paper to minimise flying glass, and protected objects left in situ with three-inch wooden covers. By November 1939, most of this work was completed and the Museum staff began regular inspections of the objects.[30]

By the time the President of the Board of Education, Lord Stanhope, ordered the Museum's closure, it only had a skeleton staff to maintain the collections and staff the Library, which was open as normal. On 2 September, Mackintosh left the Museum to become Commandant of the School of Military Engineering, leaving Herman Shaw as Acting Director until Mackintosh's return in July 1940. As well as the Director, forty-three staff left for positions in the Services or other government departments before the Museum closed. After closure, more staff left and the remaining staff were generally well past the normal age of retirement. They were, in the words of Mackintosh, a 'somewhat senile staff'. By the war's end, there were 115 on staff – nearly half the pre-war high.[31] Reduced staff levels had an effect on the Museum's collections, especially on the maintenance of the collection. Specifically, a maintenance backlog developed that would not be rectified until the mid-1950s.

The expectation was that Britain's entry into the war would be greeted by intense and immediate German bombing. Fortunately, the waves of bombers did not appear, and for eight months there was little visible activity. Inactivity created its own problems. Apathy, boredom and a false sense of security became common among all classes of people, including museum staff.[32] During this period of inactivity, many museums around London reopened. Most museums refurbished and reopened portions of old exhibitions, but three of the National Collections did something different. The National Gallery reopened without its collection to house lunchtime classical concerts. Concerts started on 10 October 1939, and during the war, a total of 1,698 daily concerts were given. In addition, a canteen serving light meals opened in May 1940 – the origin of the gallery's permanent café.[33] In February 1940, the Science Museum and the British Museum opened new exhibitions. All of the openings had one thing in common: to alleviate boredom. The cultural blackout and lack of military action created a sense of national ennui and many, such as pianist Myra Hess, who organised the lunchtime concerts, thought that the best solution was reopening cultural institutions. The reopenings were popular, but it was only a safe solution while air raids were light and infrequent.

At the British Museum, the Treasury put restrictions on the reopening. All exhibition proposals had to have Treasury approval, because items had to be 'not of such value that damage to or destruction of the exhibited specimens would seriously affect the strength of the main collection,' and 'no material increase' in expenditure for conservation, maintenance, or Air Raid Precautions could be incurred.[34] On 22 February 1940, less than six months after closure, a small exhibition of printed books, manuscripts, prints and drawings opened.[35] The much larger War Exhibit, created by the Department of Prehistory and Europe, followed shortly. It included archaeological material, from Sutton Hoo and elsewhere; a variety of porcelain, pottery and ceramics; metalwork from jewellery to lamps; and a collection of scientific material – six clocks, forty watches, seventeen weighing apparatuses, thirty-eight scientific instruments, a few calendars, and several counting apparatuses.[36] The exhibition summarised the British Museum's entire collections. The danger faced by the War Exhibit caused it to be known as the 'Suicide Exhibition'.

On 14 February 1940, the Science Museum opened an exhibition entitled 'Aircraft in Peace and War'. The exhibition, as denoted by the title, focused on the uses of aircraft throughout the twentieth century. The non-technical story presented was different from the Museum's normal story, and the staff realised they were presenting a social history. To present a social history of aircraft, the subject was divided into four periods: the Historic (1903–1914), the Great War (1914–1918), Civil and Military Aviation (1919–1939), and Aeroplanes of the RAF (1940). The majority of the items selected for the exhibition were models from the permanent aeronautical collections. Over 100 models were displayed. They were supplemented by a few full-scale items. These were pieces of modern equipment used by the RAF and loaned from the Air Ministry or the aeronautical industry. To give a sense of action and to complement the objects, photographs were displayed, many provided by the popular aviation magazine, *The Aeroplane*. Partially because of the demands of war, it was not a static exhibition – items frequently moved into and out of the display. Whether by purpose or accident, the changing exhibition helped fulfil the purpose. It provided a topical display, covering issues related directly to people's daily life, and educated visitors about the uses of aircraft during the short history of powered flight.[37]

One of the few staff left at the Museum during the war was Bernard Davy, who quickly went to work organising the exhibition. Davy joined the Museum in 1920 and spent his career working with the Aeronautical Collection.[38] In preparing the exhibition, Davy had, because of his careful pre-war work, models in the permanent collection of the majority of aircraft used by the RAF at the outbreak of war. As a result of his experience and contacts, he was able to request accessions and loans for 'Aircraft in Peace and War', even during this difficult period. From late 1939, Davy pressed his governmental and industrial contacts to contribute.[39] The correspondence between Davy and his contacts illustrates the difficulty that had to be overcome to present contemporary war materials.

To expand the knowledge about visitors' daily experience of equipment used for the war effort, Davy wanted to display a model barrage balloon with its complete support equipment. For this, he contacted the Balloon Development Establishment (BDE) at the Air Ministry. The BDE was unable to provide the model because of war work. Instead, A. Eldridge, an experienced Civil Servant at the BDE, constructed the model using official plans during his free time. The Museum paid £15 for the model, an amount that, in the opinion of Davy, barely covered the cost of the materials. It was installed in the exhibition on 27 March 1940 and provided the desired explanation of a common London wartime scene. At the beginning of planning, it was envisaged that a full-scale aircraft would be a central feature of the exhibit. To find it, Davy first turned to Vickers-Armstrong Ltd, with

Illustration 3.2   The 'Aircraft in Peace and War' exhibition, 1940

the hope of borrowing a Spitfire that was displayed at the Brussels Exhibition in July 1939. It was mounted in a manner appropriate for the Museum, but it was not available. Vickers recommended that Davy approach the Air Ministry.[40] Unfortunately, it was not possible to locate an aircraft in time for the opening of the exhibition because of the demand for planes for active service, but the Air Ministry later loaned a Hurricane fighter that was released from active service for the exhibition.[41]

Davy also used his industrial contacts for the exhibition. Among the models added to the permanent collection was the Whitley V (Inv. 1940-11) from Armstrong Whitworth Aircraft Ltd. On the other hand, Davy was denied a model of the new Boulton & Paul Defiant bomber in June 1940 because the details shown on the model violated the Official Secrets Act.[42] Davy also sought loans from industry. He contacted the makers of the important (and eventually iconic) aircraft engines. On 12 February, Napier delivered exhibition models of the Dagger and Rapier engines, and on 16 February a Rolls Royce delivered a Merlin engine to the Museum. Later in March, Napier replaced the exhibition engines with production engines, a Dagger Series III owned by the Air Ministry and a Rapier Series VI owned by Napier, and gave the Museum notice that the engines could be withdrawn for active service.[43] These industrial loans represented the majority of the full-scale exhibits on display.

About 80,000 visitors viewed the exhibition over the first four months, and the decision was made to extend it indefinitely.[44] Its rapid and successful production highlights the fact that relevant collections were up to date. Quite unlike the traditional view of the Museum as a repository for old and out-of-date technology, Davy managed a cutting-edge aeronautical collection, which, despite the lack of

Illustration 3.3   Rolls Royce Merlin engine mark III, c.1940

exhibition space and funds for accessions, was able to represent modern aviation. The Hurricane fighter and the aircraft engines gave visitors a direct connection to the airmen on active service. The importance of this connection is highlighted in Davy's correspondence, because he mentions the fighter's service history at every opportunity. The immediacy and sense of action embodied by these items were important because they helped to connect an essentially historical exhibit to the conflict that engulfed Europe. The Museum's ability to integrate contemporary material into exhibitions allowed the Museum to showcase its educational potential to the government.

The Blitz

During June 1940, southern England came under bombardment, and in August London and its suburbs were bombed. On the night of 7 September, a large raid hit London's docklands, marking the beginning of over two months of continuous bombing. The increasing intensity of bombing caused many museums to rethink their displays. At most museums, this involved closure and further evacuation. 'Aircraft in Peace and War' closed in September 1940 when serious air raids in South Kensington damaged the windows and glass roofs of the East Block.[45] This was the end of exhibits at the Science Museum for the duration of the war. At the British Museum, the small exhibition of printed materials was withdrawn in September 1940 because of the danger, but the War Exhibit remained open.[46]

Around the same time that 'Aircraft in Peace and War' closed, the Natural History Museum was hit by two incendiary bombs and an oil bomb. The resulting fire damaged significant portions of the building and damaged the furniture, fittings and collection of the General Herbarium, with 15 per cent of the specimens being destroyed or damaged by fire or water. Further damage occurred in July 1944 when a flying bomb hit the Western Galleries. This bomb destroyed 155 of 162 display groups of the famous Nesting Series of British Birds and required the removal of 130 tons of broken glass from the museum. On 10 May 1941, the British Museum experienced its most violent attack of the war, when incendiary bombs caused several fires. This attack broke a Clazomenian sarcophagus lid, burned 125,000 library volumes, and destroyed the majority of the War Exhibit.[47] Some of the scientific instruments had been removed, but the vast majority of those on display were destroyed. This included twenty-three scientific instruments, all six clocks and all forty watches.[48] By May 1941, every room of the Imperial War Museum had been damaged, and in December 1944 a rocket laid waste to most of the building. The Tate, long recognised to be in a dangerous position, was first damaged on 16 September 1940, and by January 1941 most of the roof was gone. By the Second World War's end, all thirty-four exhibition rooms were damaged.[49] Outside the Science Museum, on the Exhibition Road frontage of the V&A, evidence can still be seen of the bombs that, like many others, brought down significant amounts of glass and caused minor structural damage to both museums.

Compared with other institutions, the Science Museum escaped major damage. To paraphrase Mackintosh, the Museum licked its sores and thanked its luck that matters were no worse. In January 1941, Mackintosh broke his shoulder and was away for four months. During this absence, he seriously considered the protection of the Museum's collection and decided that more comprehensive measures were needed. On his return, Mackintosh set a goal of evacuating 50 per cent of all collections from South Kensington.[50] The Science Museum was one of the last National Collections to consider an upgrade to the protection of the treasures entrusted to them.

Improving storage

The story of Britain's cultural treasures during the Second World War is typical of the transformation that many aspects of life went through during the war. It went from an amateurish attempt to hide treasures in remote locations to an efficient system of protection and also conservation of the treasures. This change not only ensured the survival of the country's cultural treasures, but also led to many changes in the country's galleries and museums in the years after the war.

At the beginning of planning in the 1930s, the main worry on the Directors' minds was aerial bombing. Defending against bombing was the main purpose of all storage. As time passed, various members of the Subcommittee of Directors realised that their storage provided inadequate protection. Even the country houses, once seen as the salvation of the National Collections, were deemed unsafe. As air raids increased and the RAF defence strengthened, many realised that some houses were under German flight paths and were at risk of accidental bombing.[51] Like the museums, which favoured remote locations, many other tenants used either the houses or their grounds. They were perfect for refugees, training camps and military bases. The houses themselves caused issues surrounding their lack of environmental controls. The Science Museum realised this early, when quick removal of the rare books stored in the old kitchen at Herriard Park to the Cambridge University Library was necessary because of damp.[52] Museums also had to deal with the inhabitants. At Montacute House, the V&A had to a battle a severe moth problem.[53] The National Gallery was worried about Lord Penrhyn's St Bernard damaging their paintings at his north Wales castle.[54] Owners were also a problem. From the beginning of wartime preparations, owners looked out for themselves and their properties more than for the national good. In short, the idea to use country houses turned out to be a poor decision for many museums.

During the summer of 1940, several directors decided that the best storage option was to convert disused quarries into special underground facilities. Between 12 August and 18 September 1941, the entire collection of the National Gallery was transported to the Manod slate quarry in northwest Wales, the first specially built underground facility.[55] By January 1942, the Westwood Bath Stone Quarry in Wiltshire was ready for occupation by the British Museum and the V&A. A total of thirty-one other institutions also placed items in Westwood, including the Bodleian Library, the Fitzwilliam Museum, the Free French government, a number of cathedrals, and other National Collections.[56] Unfortunately, the Science Museum was unable to find space in a quarry because Mackintosh's decision to move the majority of the collection out of South Kensington came after the majority of the quarry space had been claimed. Thus, the Museum was forced into country houses, but Mackintosh had one advantage; he was able to make arrangements based on the experiences of the other museums.

Eventually, over 60 per cent of the Museum's total collections were accommodated in over thirty locations, including industrial warehouses, military depot quarries, and houses formerly occupied by other institutions.[57] After the Second World War, the Science Museum was able to find underground storage in a shelter near Stockwell and in a disused underground factory at Warren Row, near Reading.[58] This required the development of a complex system of storage and conservation, and the development of an extensive record-keeping programme to support that system. An examination of a couple of aspects of the complex system highlights the changes that occurred during the war.

Finding locations to store locomotives, cars and coaches was difficult. Barns did not protect the historic items sufficiently and coach houses had become scarce. Nine, mainly 'Class C', houses were selected for these large items. They provided storage in stables/outhouses, but conditions were unfavourable and the objects were vulnerable and exposed. Since vehicles were generally not protected by packaging, staff inspected 'Class C' houses more frequently than other houses, and the regular condition reports created during the inspections allowed the Museum to track the conditions of the items and conserve them, if necessary. Despite difficulties with removal and transport, 'Puffing Billy', Stephenson's 'Rocket' and the Rocket replica were moved to Brocket Hall (in Hertfordshire) and Ramster (in Surrey). Another example of the accommodation was Henley Hall, where five vehicles and nine cycles, including the 1888 Benz motor-car and the 1885 'Rover' safety bicycle, were stored in the house's outbuildings until 31 May 1951.[59] The transport collection shows the scale of the evacuation and the introduction of systematic condition reports, two items without precedence at the Science Museum.

The Wright Flyer, which Mackintosh called the Museum's most precious object, underwent a series of complicated manoeuvres during the Second World War. It was dismantled at the outbreak of war and packed into the original cases, which were placed in the safest part of the Museum's basement. Shortly afterwards, it was redisplayed until concerns for its safety were raised, some coming from the USA. This meant that safer accommodation was needed, and the houses were not considered safe. So Mackintosh requested space at Westwood, but this was denied because the Flyer's crates would not fit through the quarry's doors.[60] The War Office offered space at a depot quarry. Within months, mould appeared on the cases, and, because the Office of Works refused to provide the money to construct an air-conditioned room for protection, it had to be moved again. Luckily for the Museum, space was found in the Naval Depot Quarry, Corsham. The Wright Flyer stayed at Corsham until the end of the war, when it returned to London.[61] During the course of the war, it made four major moves, underwent several minor adjustments, and even faced the possibility of being shipped back to the United States. Because of the Flyer's importance, it received more attention than a standard object, but it highlights the pattern of movement experienced by much of the collection.

The process of providing better protection for collections marked the change from a series of amateurish attempts to professional procedures to protect, conserve and monitor. Surprising as it may seem to a modern museum professional, the Science Museum did not keep duplicate copies of any of the Museum's essential records, including the Main Inventory and Registers. If a calamity had happened, such as the fires experienced by other museums, and records were lost, it would have been an irreparable loss for the Science Museum. As Mackintosh realised it was imperative for copies to survive, the curators kept meticulous records during the second evacuation, made duplicates, and stored copies off-site.[62] Without the lists, Mackintosh realised that collecting the Museum's possessions after the Second World War would be impossible.[63] This change in attitude affected more than collection management during the war. The experience with the quarries was so positive that it led to changes in museums after the war. The experiences of climate control and the conservation of items possible under ideal conditions created a push for climate control in London. While many National Collections developed specialised conservation departments shortly after the war, the Science Museum left conservation to the workshops throughout the vast majority of its history. The Second World War helped to create the professional atmosphere that is expected at a museum, including such basics as regular condition and conservation reports and duplicated inventories.

Contributing to the war effort

Each of the National Collections contributed to the war effort. The National Gallery held its concerts. The Imperial War Museum's staff helped the civil defence and military authorities, Ministry of Home Security, War Office, Air Ministry, Ministry of Information, Foreign Office, and Allied governments. They provided photographs, posters, and other information for didactic and propaganda uses. Additionally, some of the Imperial War Museum's collection was used during the summer of 1940, when eighteen army and naval guns were pressed into service.[64] Parts of the V&A, from 1941 to 1944, became a school for evacuated children from Gibraltar, while the restaurant was used as a canteen for RAF cadets being trained in South Kensington.[65] The Tate was put to many uses, but, most famously, the gardens were converted into Victory Gardens.[66] While the Kew Gardens contributed greatly to the 'Dig for Victory' campaign, the laboratory at Kew was central to the search for vegetable replacements for scarce raw materials, and its pharmaceutical expertise saved numerous lives.[67] As with other museums, both the evacuated buildings and the resources of the Science Museum were used.

The buildings of the Science Museum were put to various uses. For a short period, the old Southern Galleries were used as a victuals dump. In June 1941, a radio repair corps occupied the Museum's mess hut. Then, in January 1942, the No. 7 Radio School, teaching 1,000 students, took up residence in the Museum.

Officially, the occupation was until July 1944, but in reality they were there 'somewhat longer'. The lecture theatre was important. It was in constant demand throughout the war, because of its capacity and projection equipment. It was used by the radio school, other RAF schools in the area, the local ARP organisation, professional groups, and to present public programmes. It was in such demand that, for the first time, a schedule was kept to prevent conflicts.[68] The buildings were not the only contributions made by the Museum.

The Library, arguably, provided the Museum's largest contribution. Despite volumes being evacuated for safety, the Library was open for the entire Second World War to provide access to its scientific and technical material. The Library was an essential peacetime tool for many research bodies around the United Kingdom, and the war only stimulated greater research requirements. During pre-war planning, it was expected that the reference service – the number of readers – and accessions, with the related work, would decrease, but loans and information services would increase, in both quantity and urgency. Combining these assumptions with the need to reduce personnel, the Library decided to release 40 per cent of the total staff, with individuals being released according to age, experience, physique, and their position. Accessions, and related work, greatly decreased! Publications from enemy and enemy-occupied countries ceased, while the accession of British, American, Dominion, and Allied materials were reduced for reasons of transport difficulties, paper rationing and censorship. The reference service did not decrease! The expected decrease in normal readers did occur, but staff from foreign governments and refugee intelligentsia replaced them. In spite of that, the reference service became less important for the Library. Its most important and dominant functions were the information and loan services. Official records were not kept of the specifics of the information service, but it severely taxed the time and capabilities of the reduced staff. Many requests came from unexpected sources and were often urgent. Many, from the Intelligence Branches and other government departments, were not research-based; rather, they were for geographical and technical information used for planning and operational purposes. The loan service, likewise, was driven by a similar increase in requests from government departments, national institutions, industrial firms, and research organisations. One measure was the expansion of borrowing institutions, which increased from 450 to over 1,000 during the war. This great increase of activity, combined with the drastic reduction of staff, meant that general library work became neglected. In fact, by the end of the war, both Mackintosh and Lancaster Jones, the Keeper of the Library, agreed that too many staff had been released and a more robust staff would have offered better services.[69] The Library was a hive of activity during the Second World War and was the central contribution to the war. The services that the Library provided during the war made it a well-known scientific and technical resource, which continued to be extensively used after the war.

Finally, the Library's addition of a microfilm service should be mentioned. Before the Second World War, microfilm was rare in Britain, but in late 1941 the Science Museum received basic microfilm equipment. Its service was quickly overtaken by the national organisation, Association of Special Libraries and

Information Bureau (ASLIB), which used the Library as its original base until it grew too large and moved across Exhibition Road to the V&A. This was not the end of microfilming at the Museum, as its basic equipment was used 'where speed was not of primary importance' and for creating in-house duplicates.[70] The development of the Museum's microfilm service and ASLIB changed the nature of library services and archiving around the country. For libraries, it reduced space required for duplicate copies and expanded the titles available by eliminating the dangers posed by loaning rare or important works. For archives, it enabled quick and highly portable copies to be produced. Both of these features contributed to the quick and widespread uptake of microfilm throughout the UK, and the Library played a key role in its development.

Reopening the Museum

At the end of the war, the Science Museum was one of the most fortunate National Collections. It was one of only three institutions, along with the Bethnal Green Museum and the National Maritime Museum, that did not sustain a direct hit. Of the museums that were hit, three sustained only minor damage, equivalent to that sustained by the East Block – mainly broken windows, loose ceilings, and superficial damage to frontages.[71] The Museum's greatest loss was the Southern Galleries, which comprised a quarter of the pre-war exhibition space. Yet, the loss was not as dramatic as it may seem. The fire service had condemned the old galleries as unsafe in the 1910s, and as a fire risk since the First World War. So, in late 1940, when the issue of flammable structures became pressing, the decision to remove large portions of the wooden structure was straightforward, and this left the Southern Galleries in a semi-derelict state. Otherwise, most of the work was pretty basic – repairing windows and rooflights, removing the partitions built for the RAF, taking out the linoleum put down during the war, demolishing the air raid shelter built in the basement, and doing a preliminary cleaning and refit of the galleries. In short, the Museum needed manpower and space to restore its exhibitions.[72]

On 14 February 1946, the Science Museum reopened. A total of twelve galleries opened – four for permanent exhibitions and eight for the 'Exhibition of German Aeronautical Developments'. It was a scientific and educational display, which, despite the title, also covered British, American and other international developments. This exhibition had its roots in a display arranged by the Ministry of Aircraft Production for the Cabinet and MPs at RAE Farnborough.[73] Likewise, it followed the precedents set at the Imperial War Museum, where exhibits of captured Axis equipment and advanced Allied technologies were designed to aid the training of Allied military and government personnel.[74] The exhibition was very popular and, like 'Aircraft in War and Peace', it was extended. It was the first time that some items, including the V2 rocket, had been displayed at the Museum. A few of the items, but not the V2, were accessed into the permanent collections.[75] The real importance of the exhibition was that it was the first large-scale public exhibition of the material culture of the Second World War in Britain, and was another clear example of the Museum's ability to construct exhibitions that high-lighted contemporary technology.

The four galleries devoted to the permanent exhibitions were used for a special 'Science Exhibition'. This covered a range of recent developments in a variety of sciences and technologies – including atomic energy, uranium production, X-ray applications and the quartz crystal clock.[76] This was the Museum's first attempt at displaying wartime developments, and it highlighted the problems that it faced during the post-war period, mainly the vast expansion of subjects and material that the Museum could cover.

The Museum opened the majority of the galleries to the public in late 1948. Throughout the period of reopening, the Museum continued a series of con-temporary and historical special exhibitions. After the Science and German Aeronautical exhibits, the 'Naval Mining and Degaussing' exhibit opened in June 1946. It was based on a display of material by the Admiralty Mining Establishment at Havant. Full of working exhibits, it attracted nearly 250,000 visitors in four months. Then, after several historical exhibits, 'Home and Factory Power' opened in October 1947 with the purpose of contributing 'to the solu-tion of a national problem' by educating visitors about the relationship between domestic consumption, industrial consumption, and the production of gas and electricity. Then, in November 1948, the Museum opened 'Science in Building'. It was based on the work of the Building Research Station of the Department of Scientific and Industrial Research, and was used by over sixty school parties, mainly from technical schools. In addition, the Museum held three anniversary exhibitions, all of which focused on showing historical developments and appli-cations to daily life.[77] The range of special exhibitions shows that the Museum's staff were engaged with issues of national importance and of historical concern, with some of the subjects being covered in larger exhibitions. For example, the principles shown in 'Science in Building' were later displayed at the Festival of Britain's 'Live Architecture Exhibition' in Poplar in the East End of London. By using special exhibitions to show subjects that could not be illustrated by the permanent collections, the Science Museum was able to maintain its status as an institution of current scientific and technical training.

The years immediately after the end of the Second World War were a flurry of activity at all of the National Collections. Each museum was trying to, at the minimum, return to its pre-war position. For some, such as the Science Museum, the war had delayed previously approved buildings. In 1938, the Treasury had accepted seven urgent building needs, to be completed between 1939 and 1947, and, in the words of the Standing Commission, the check caused by war 'cou-pled with the serious effects of enemy bombing...has rendered still more serious an already urgent situation.'[78] To direct plans for reconstruction, the Treasury approved a scheme with three tiers: 1) to repair air raid damage; 2) to construct extensions and schemes either begun or planned before the war; and 3) to reorgan-ise the South Kensington site. The Treasury made one very important provision – that repair could be combined with extension and expansion where economical.[79] It was this provision that constrained the development of the Science Museum.

The plan, as originally outlined in the Bell Report, was to demolish the Southern Galleries and replace them with a modern group of three buildings – the Eastern, Centre, and Western Blocks. The war took care of the old galleries, and, because they were not to be repaired, the Museum was pushed down the list of priorities, behind all the museums that sustained significant war damage.

Of the seven scheduled pre-war building needs, several were subsumed in post-war reconstruction, but two of the remaining needs were highlighted as urgent by the Standing Commission. They were the completion of the Entomological Block of the Natural History Museum and the construction of the Centre Block of the Science Museum. These fell to the second tier of the Treasury's reconstruction plan, but the Committee felt they deserved special attention. The Entomological Block had been under construction at the outbreak of war; work had stopped, and its steelwork survived undamaged. The Centre Block had been under consideration since 1912 and was one of the top priorities approved in 1938. By 1948, the condition was 'well-nigh desperate' for the Science Museum, and that was before considering expanding collections to cover the developments of the war.[80] Plans were made for the construction of the Centre Block to begin in 1951.[81] Finally, nearly forty years after the Bell Report, the second of its three phases was to begin.

At the same time, the Festival of Britain office was looking for a place to house the exhibitions of pure science, and the South Kensington museums were discussed. In July 1948, use of the Museum by the Festival was doubtful and some, both at the Museum and at the Ministry of Works, hoped that any proposals would 'blow over'. These hopes were expressed again during late 1948, when South Kensington plans were discussed at a meeting, and W.A. Procter (of the Ministry of Works) declared that plans for the Festival 'would not mature' and that the pre-war plans should go ahead unchanged. Yet, nearly simultaneously with these statements, the Festival plans matured, and it was agreed that the first section of the Centre Block would be constructed to provide space for a number of exhibitions.[82] Originally, these included the 'Exhibition of Science', a planetarium, a Newton House,[83] and a series of models.[84] Each of these would have increased the facilities available at the Science Museum, but drastic budget cuts to the Festival forced organisers to greatly constrain many of the Festival exhibitions. The budget cuts left only the Science Exhibition as part of the finalised programme of the Festival at the Science Museum.

Construction started on the Centre Block in 1949, but only the lower floors were erected to house the Exhibition of Science the Festival of Britain in 1951.[85] The construction of the Centre Block for the Festival did raise some issues. The main issue was whether the 1935 plans were suitable for the Festival, and, more importantly, for the Museum's long-term use. Naturally, the Ministry of Works did not welcome suggestions that required dramatic changes to the old plans, but a few changes were made. Overall, it was expected that the construction would provide 95,000 square feet for exhibition and 49,600 square feet for storage, offices and a restaurant.[86] The Science Exhibition was open to the public from 4 May to 30 September 1951 and used about 35,000 square feet on three floors of the building

(30,000 for the exhibition).[87] Shortly after the closure of the Exhibition of Science, the Ministry of Works notified the Museum that limited money would be available for renovation and that they should take advantage of the materials created by the Festival for their displays in the new space. By economising on redisplay, staff hoped to speed the opening of the new galleries, but further budgetary constraints plagued the project. These included notification that work to renovate the Centre Block could not be scheduled until after 1952–1953.[88] To avoid a three-year delay, the Museum struck a deal with the Natural History Museum to house the Mammals Collection, which had lost its home during the Second World War. In early 1952, because of the agreement, the Treasury approved expenses to make the Centre Block usable.[89] For the Science Museum, the cost of increasing its exhibition space was half of the basement and ground floors of the Centre Block until the Mammals Collection's permanent home was completed. Despite this additional space, exhibition and storage space was still restricted, the planetarium question was still outstanding, and the delays affected the Museum's ability to collect.

Sir Alan Herbert started the post-war push for a planetarium with a parliamentary question to the Minister of Education, Miss Ellen Wilkinson, and later a letter to the *Sunday Times* on 25 February 1946.[90] Throughout 1946, various attempts were made to obtain a planetarium from Europe by any means possible. Many saw reparations as an easy way to acquire a planetarium, but, after investigation into the matter, it was determined that planetaria were not a legitimate target as they had not contributed to Germany's industrial war potential.[91] The Vienna planetarium, which had been put into storage during the war at Schloss Thalheim, was considered and plans were made to offer £2,000 of foodstuffs as payment. After excavations of a small ravine full of rubbish (including everything from the kitchen range to human waste), an assortment of parts were recovered and were determined to be 'simply rusted scrap'.[92] In addition, the Foreign Office investigated the possibility of buying the Hamburg planetarium, but the negotiations stalled over the price.[93] During this period, the purchase of a new Zeiss instrument was impossible because the factory was closed until the early 1950s and the Museum was forced to investigate all other options. Unfortunately, despite its best efforts, the Museum missed the only used instrument that was sold from Europe. In 1947, the Zeiss Model II from the Stockholm planetarium was sold for about £7,000 to the University of North Carolina, where it opened on 10 May 1949 as the Morehead Planetarium. In March 1949, the Museum also made enquiries at the Milan planetarium, which received a quick reply that it would open again in April.[94] Other than Zeiss instruments, the Museum's other options were American-designed instruments, including the Korkosz, which were considered by many to be inferior instruments.[95] The announcement of the construction of the Centre Block started a flurry of press reports and letters to newspapers about planetaria.[96] This flurry of press did little to provide sustained support for the planetarium, and it was quickly overshadowed by the preparations for the Festival of Britain.

The Museum was still caught in the Catch-22 over the planetarium. The Ministry of Works still refused to design the Centre Block to be strong enough to carry the planetarium until the instrument was provided, while the Treasury would not purchase the instrument until the dome was constructed.[97] This debate continued through the early 1950s. Gradually, it became obvious that the planetarium was a major delay to the construction of the Centre Block and that a solution must be found. During early 1953, Sherwood Taylor was able to arrange a £20,000 interest-free loan from the Sir Halley Stewart Trust, but it was far from enough to purchase the instrument.[98] For comparison, the Zeiss instrument purchased by Madame Tussauds in the early 1950s was reported to have cost nearly £100,000.[99] By 1954, the lack of a major financial commitment from the government or a private source to support the planetarium meant that plans for the Centre Block were once again facing delays.

In 1953, the lack of a completed Centre Block was seen as a huge disadvantage to the Science Museum. The expansion of Imperial College posed a serious threat to the Museum's Western Galleries, which housed the Aeronautics Collection. Restrictions on gallery space created the possibility that the Museum could lose significant donations from industry. Beyond that, the lack of storage space hampered the development of the collections. The solution to all of these problems was simple: finish the Centre Block.[100] Yet, gaining approval from the Treasury was not simple.

For the Museum, 1954 was a pivotal year. During the year, three major events occurred. First, the concern over the Aeronautical Collection passed, because arrangements with Imperial College secured the Western Galleries until the Centre Block was completed. Second, Sherwood Taylor divorced the planetarium from the Centre Block.[101] It was a move that upset many of his own staff, especially the Keeper of Astronomy, Henry Calvert. Sherwood Taylor's reasoning was that the separation could guarantee the speedy completion of both projects and it was not an admission of defeat for the planetarium.[102] Finally, in late 1954, the Treasury approved the completion of the Centre Block.[103] The receipt of this approval was met with a renewed surge of interest in plans for the new building and the redevelopment of the older galleries.

This divorce had significant ramifications on gallery allocations and the future expansion of the Museum. After approval of the Centre Block without the planetarium, the question of its appropriate location was opened again. Sherwood Taylor's plan, as approved by the Museum's Advisory Council, was for the planetarium to be built in the basement of the future Western Gallery abutting the basement of the Centre Block. This solved one of the great problems that had plagued the planetarium for years: how to allow separate access at times when the Museum was closed. This new position would, upon completion of the Western Block, provide direct access from the street. Additionally, this move was important because it freed the top storey of the Centre Block from the grip of the planetarium and, by association, the Astronomical Collection.[104] Without the decision, the current display of the Aeronautics Collection would have never been possible. This decision had a larger effect on the actual physical structure of the Museum than any other single choice taken by a Director. This is because it created the space for the Flight Gallery, as it is known today, which has been one of the Museum's signature galleries since it opened in 1963.

While the planetarium was the proximate cause for the delay of the Centre Block, it was not the ultimate cause. The delays experienced by the Science Museum can be traced to two main sources: a lack of independence and consistent leadership. The first of these – and its concomitant lack of administrative and financial autnomy – was the basis for many of the interdepartmental struggles faced by the Museum. The lack of a long-serving Director complicated those interdepartmental conversations because of the inconsistency of the Museum's long-term goals, which changed with each Director's personal agendas, and because of the dearth of continuous professional relationships. Additionally, the absence of a Board of Trustees exposed the Museum staff to the vagaries of interdepartmental relationships and amplified the impact of the change of Directors. These circumstances placed the Museum on an uneven footing with other National Collections and, in many ways, reduced the Museum's influence on the government.

Many people believe that cultural institutions need administrative freedom from the government for a variety of reasons. For most institutions, this involves the creation of a Board of Trustees that acts as a buffer between the government and the institution. One of the earliest examples was the 1753 creation of the British Museum, governed by a Board of Trustees, which reported directly to Parliament. Similarly, in the early twentieth century, when moves were being made to establish the National Maritime Museum, independence from a government department was a high priority, and in 1934 it was created under the control of a Board of Trustees.[105] The Science Museum and the V&A did not achieve similar independence until 1984.[106] Historically, the Museum had been under the control of the Ministry of Education and had to have its approval, and that of the Treasury and the Ministry of Works, for any major project. During this period, the Museum did have an Advisory Council, but it had no real power. While its recommendations were generally accepted by the Ministry of Education, they carried no weight outside that Ministry. This relationship created the basis of the complex situation over the planetarium and the delays in expansion.

Financial independence is also important to the development of museums. While the Science Museum was successful in garnering small donations, especially of items, it was never successful at the level of other National Collections and relied heavily on the Treasury for building funds. In comparison, the British Museum, the Tate, the National Portrait Gallery and the Wallace Collection received donations for new galleries and/or the renovation of old galleries from Joseph Duveen.[107] At the National Maritime Museum, the conversion of the Queen's House was covered by Sir James Caird, at a cost of more than £80,000, and was neither the beginning nor the end of his donations, which included his £300,000 collection and the creation of the Caird Trust.[108] Throughout the period of Duveen's and Caird's donations, which corresponds roughly to the period the Science Museum actively sought a planetarium, it was only able to obtain a £20,000 loan! The inability to find donors compounded the Catch-22 situation and the delays.

An example of the importance of independence can be taken from the National Maritime Museum in the years immediately after its creation. During these years, while the Queen's House was being returned to its original state and the rest of the site was being converted for use by the National Maritime Museum, decisions were taken by the Trustees, chaired by Lord Stanhope, or by the Director, Geoffrey Callendar, without the interference of any government department, and all costs were met by Caird, including landscaping, the cleaning of pictures, and the gesture of thanks to him – the Caird Rotunda.[109] In essence, if the Trustees or Callendar wanted something to happen and Caird was willing to pay, it was done. Because of their independence, their plans did not have to be commented on and approved by the series of departments with which the Science Museum had to negotiate.

The lack of consistent leadership also contributed to the Museum's problems. During the twentieth century, the Science Museum had eleven Directors, and between 1930 and 1960 it had five: Sir Henry Lyons (retired in 1933), Col. Ernest Mackintosh (retired in 1945), Herman Shaw (died in post in 1950), Frank Sherwood Taylor (died in post in 1956) and Terence Morrison-Scott (resigned in 1960). This turnover was high when compared with other National Collections. For instance, the V&A had two, the National Maritime Museum had two, the Tate had two, the Wallace had two, the National Gallery had three, and the British Museum had four directors (but one, Sir John Forsdyke, served for fourteen years). Each Director at the Science Museum had his own agenda but, more importantly, each change disrupted professional (and personal) relationships that could prove important during the interdepartmental negotiations. In addition, because of the absence of Trustees to guide it during periods of transition or to provide support upon the death of a Director, the Museum relied on Civil Servants to make decisions that could greatly influence its future. This resulted in the Museum operating within a diminished sphere of influence.

Again, the contrasts with the National Maritime Museum are significant. The National Maritime Museum's first Director, Sir Geoffrey Callendar, was involved with the project from the early 1920s until his death in 1946. His long involvement resulted in the creation and original layouts of the National Maritime Museum, and he cultivated a large group of contacts within government. Three of the contacts are particularly significant because of their relation to both the National Maritime Museum and the Science Museum. The most senior was James Stanhope, seventh Earl Stanhope. He was the Chairman of both the National Maritime Museum Trust (1927–1934) and the Board of Trustees of the National Maritime Museum (1934–1959). Furthermore, from 1930, he sat on the Board of the National Portrait Gallery and was named the Chairman of the Standing Commission in 1941.[110] Crucially for the Science Museum, he was the President of the Board of Education in 1937–1938.[111] Also important was William Ormsby-Gore, a Conservative and Unionist MP who succeeded his father as Baron Harlech in 1938. On 29 June 1934, in his position as First Commissioner of Works, he moved the second reading of the National Maritime Museum Bill and, because of statements regarding the absence of displays about the 'national maritime adventure', was called the 'accoucheur-in-chief' of the National Maritime Museum by Mackintosh in a letter of complaint.[112] As mentioned earlier, it was his actions that delayed the Centre Block in January 1936. In addition, he was active with many other major collections, including the British Museum and the National Library of Wales, and was the chairman of the Standing Commission in 1949.[113] The final person was Sir Eric de Normann, the chief architect at the Ministry of Works. According to Littlewood and Butler, he was key for the development of the National Maritime Museum and throughout his career helped to move along work at the National Maritime Museum.[114] Each of these men became involved with the National Maritime Museum for different reasons, but one of the continuous features of their involvement was the presence and friendship of Callendar. Additionally, the involvement of these men in competing institutions, especially between the Science Museum and the National Maritime Museum, meant that their personal biases influenced their professional judgements and exaggerated the other disadvantages experienced by the Science Museum.

It is clear that no one individual or circumstance caused the delays that the Museum experienced, but it does not diminish the scale of the delays. Overall, the second stage of the Bell Report was delayed by between thirty and forty years, while the third stage was opened by the Queen in 2000, a full seventy-eight years after it was recommended! Fortunately the Museum was able to expand both its storage and exhibition space outside South Kensington; otherwise its ability to collect would have been restricted much further than it was after the Second World War.

Collecting the war

The Second World War produced a vast array of collectable material. Large items like aircraft and tanks, small items like gas masks and radios, or ephemera like War Office posters and pamphlets could be collected for a variety of reasons. At the end of the First World War, there was a significant gap before collecting began, especially in the case of the Imperial War Museum, which was not opened until 1917. Museums were anxious to avoid the lag in collecting exhibits concerning the Second World War, and collection began much earlier in the war.

In September 1939, the Imperial War Museum made an application to the Treasury to extend the Museum's remit to cover the current war. In October, the Treasury approved the request and the Museum began efforts to acquire material.[115] In April 1940, the House of Lords asked the Standing Commission to consider how war materials should be allocated to the various institutions, mainly the Imperial War Museum, the National Maritime Museum, the Royal United Services Institution, and the Science Museum. The report provided some slightly odd guidance because the Commission gave all museums priority. The Imperial War Museum was given priority for material of a popular or spectacular nature; the National Maritime Museum was given priority for naval material; the Royal United Services Institution was given priority for small personal mementos; and the Science Museum was given priority for technical material.[116] The situation left by this report set the scene for tension between the Imperial War Museum, the National Maritime Museum and the Science Museum.

The report also covered the measures available for the collection of material and recommended the creation of a Disposals Board, on which the Services and the museums would be represented. The board first met in 1945, but earlier steps were taken to preserve materials. The Air Ministry was the first Service to actively collect material, while the Science Museum was the first museum to collect and display objects, as previously discussed. In addition to the three museums mentioned, other collections became interested in material produced during the war: the V&A was interested in design and costumes, and the National Gallery, under the direction of Sir Kenneth Clark, displayed the War Artists' work.[117] During the war, collecting was difficult because of staff limitations and security concerns, but the speed of collecting and availability of items greatly increased after the war.

So, how does one measure a museum's capacity to collect items that represent an event as large and complicated as the Second World War? For the Imperial War Museum, it was simple – collect anything and everything! For the National Maritime Museum, the solution was similar – collect anything naval. At the Science Museum, a more complicated and varied set of criteria was created. The items that the Museum needed to collect were not the sensational pieces, like those collected by the Imperial War Museum, but those that illustrated scientific and technical developments and/or those that were useful in educational displays. This demarcation did not eliminate institutional conflicts, but other factors, such as space limitation at the Science Museum, decreased the occurrence of conflict post-war.

As discussed in an earlier section, Davy, while creating the 'Aircraft in Peace and War' exhibition, was actively collecting items. Of the items used in this exhibition that went into the permanent collections, the majority were models, such as the Armstrong Whitworth Whitley V (Inv. 1940-11) and the Handley Page Hampden I (Inv. 1940-26), but some other items, such as the Bristol Pegasus XVII (Inv. 1940-76), were also collected. Additionally, Davy had, in the 1930s, collected several models of aircraft that saw active service, and so had a lead on the Imperial War Museum's collecting effort. Davy continued his collecting efforts until 'Aircraft in Peace and War' was closed and resumed them as soon as the war ended, when in 1945 he received a captured V1 Rocket (Inv. 1945-77) from the War Office.[118] Over the following years, several models, including one of the Avro Lancaster I (Inv. 1946-206), and engines, including a Daimler-Benz DB 605 A1 (Inv. 1947-181), were acquired. By collecting pieces other than full-scale aircraft, the collection was able to keep pace with technical developments without facing the space concerns. This pattern continued for nearly a decade until planning for the Centre Block display began in the mid-1950s. It was only at this point, when space was available for more full-scale aircraft, that the Museum sought them, including both the Hawker Hurricane (Inv. 1954-660) and the Supermarine Spitfire (Inv. 1954-659). But these iconic aircraft were not the end of the collecting of aeronautical Second World War items; it continued until the 1980s, when the V2 Rocket (Inv. 1982-1264) was finally acquired.[119]

For traditional subjects, like aeronautics, this involved modernising the collections, but with subjects that developed greatly during the war, such as nuclear

Illustration 3.4   V2 rocket arriving at the large object store at Wroughton, near Swindon, c.1982

physics, collections had to be reorganised or created for them. In short, collecting Second World War material was difficult for the Science Museum because it had to be selective, but nonetheless this period is well represented throughout the Museum's galleries and collections. From the model of Bailey bridgework (Inv. 1947-185) to the marine gas turbine from the Admiralty (Inv. 1949-175), the scientific and technical legacy of the war is evident in most galleries. On a question of sheer scale, the Science Museum was never going to be able to compete with either the Imperial War Museum or the National Maritime Museum, but it has been successful in fulfilling its purpose, because it has been able to incorporate many of the important technical advances made during the war into its collections.

Conclusion

The Second World War wrought dramatic changes throughout the world, and this was equally true for museums. The experiences of the war shaped the modern British museum experience by increasing the level of professionalisation, reinforcing the need for atmospheric controls, and pushing conservation reforms. Many of these advances took place because of deficiencies in the previously accepted system, such as the lack of duplicate records at the Science Museum, and because they allowed better preservation of objects. Yet, these advances were tempered by the effects of bombing, which caused the collapse not only of the roofs of museums, but also of the development programmes envisaged for the 1940s and 1950s. This setback was felt by all. While many museums were able to reformat pre-war ideas for the post-war world, the Science Museum was faced with dramatic changes within the fields it illustrated. The Museum was unable to respond, like the Imperial War Museum, by greatly increasing accessions. Instead, it focused on creating temporary educational exhibitions about the developments of science and technology, and, because it focused on presenting, not collecting, contemporary science, the Science Museum was able to maintain its position as a museum of national (and international) importance.

Notes

1. Peter L. Jakab (curator), 'The Wright Brothers: The Invention of the Aerial Age', http://www.nasm.si.edu/wrightbrothers/ (accessed 9 March 2009)
2. David Edgerton, *England and the Aeroplane: An Essay on a Militant and Technological Nation, Science, Technology and Medicine in Modern History* (London: Macmillan Academic, 1991), pp. 2–3.
3. H.G. Wells, *The War in the Air* (London: George Bell and Sons, 1908).
4. Leslie Pollard, *Menace: A Novel of the Near Future* (London: T. Werner Laurie, 1935), pp. 58–9.
5. Standing Commission on Museums and Galleries, *The National Museums and Galleries: The War Years and After, Third Report of the Standing Commission on Museums and Galleries* (London: HMSO, 1948).
6. Various Letters in Victoria & Albert Archives (V&A), ED 84/264.
7. Unpublished manuscript, 'War History of the Science Museum and Library, 1939–1945' by Col. E.E.B. Mackintosh in 1945, SMD, Z 101, p. 1.
8. 11 January 1939 letter in V&A, ED 84/264.
9. 'War History of the Science Museum', p. 2.
10. 'War History of the Science Museum', pp. 2, 18.
11. 15 September 1938 Minute Sheet in V&A, ED 84/264 & 'War History of the Science Museum', pp. 1–3.
12. David Follett, *The Rise of the Science Museum Under Henry Lyons* (London: Science Museum, 1978), pp. 89–94.
13. 'Interim Report of the Royal Commission on National Museums and Galleries [Cmd. 3192]', *PP* (1928), pp. 3–5, 18–19, 43.
14. 'Final Report of the Royal Commission on National Museums and Galleries; Part 2. Conclusions and Recommendations Relating to Individual Institutions [Cmd. 3463]', *PP* (1930), p. 49.
15. David Follett, *The Rise of the Science Museum*, p. 94.
16. Director's Note by Mackintosh dated 1 July 35 in SMD, Nominal 1356.
17. 14 November 1935 Note in SMD, ED 79/38.
18. 23 January 1936 Note in SMD, ED 79/38.
19. David Rooney, *The Events Which Led to the Building of the Science Museum Centre Block, 1912–1951*, Science Museum Papers in the History of Technology No. 7 (London Science Museum, 1997), p. 12.
20. Folder for 'South Kensington Site' dated 1937 in SMD, Z 188.
21. Winston Churchill, *The Second World War: Abridged Edition with an Epilogue on the Years 1945 to 1957* (London: Pimlico, 2002), pp. 112–30.

22. 15 September 1938 Minute Sheet in V&A, ED 84/264.
23. Winston Churchill, *The Second World War: Abridged Edition,* pp. 133–4.
24. Memorandum dated 31 Aug 1939 in V&A, ED 84/264.
25. 'War History of the Science Museum', pp. 20–2.
26. Examples of this behaviour can be found in V&A, ED 84/264; V&A, ED 84/267; and Tate archives, TG/2/7/1/44/2.
27. 31 August 1939 Memo in V&A, ED 84/264.
28. Winston Churchill, *The Second World War: Abridged Edition,* pp. 149–61.
29. 'War History of the Science Museum', p. 2.
30. 'War History of the Science Museum', pp. 9–11.
31. 'War History of the Science Museum', pp. 9, 12, 39, 58, 79.
32. Robert Mackay, *Half the Battle: Civilian Morale in Britain During the Second World War* (Manchester: Manchester University Press, 2002), pp. 54–6; 'War History of the Science Museum', pp. 11–12.
33. Standing Commission on Museums and Galleries, *War Years and After,* p. 13.
34. British Museum Standing Committee Minutes (BMSC), 6 Feb 1940 and 10 Feb 1940.
35. Edward Miller, *That Noble Cabinet: A History of the British Museum* (London: Andre Deutsch Limited, 1973), p. 334.
36. The War Exhibits Register, Department of Prehistory & Europe, British Museum.
37. 15 January 1940 Letter from Davy in SMD, ED 79/43.
38. David Follett, *The Rise of the Science Museum,* pp. 61, 163.
39. December 1939 and January 1940 Letters in SMD, ED 79/43.
40. 16 January 1940 Reply from Vickers in SMD, ED 79/43.
41. Letter to Rolls-Royce from Davy dated 29 May 1940, SMD, Nominal 1453/10.
42. 1 June 1940 Reply from Boulton & Paul in SMD, ED 79/43.
43. SMD, Nominal 1219/17–Dagger and Rapier Engines and Nominal 1453/10–Merlin Engine.
44. SMD, Nominal 1219/17.
45. Science Museum Annual Report for 1940–1951, p. 1.
46. Edward Miller, *That Noble Cabinet,* p. 334.
47. BMSC, 12 July 41 (p. 5708).
48. British Museum War Exhibit Registry.
49. Standing Commission on Museums and Galleries, *War Years and After.*
50. 'War History of the Science Museum', p. 12.
51. John Forsdyke, 'The Museum in War-Time', *The British Museum Quarterly* XV (1952), pp. 1–9.
52. 'War History of the Science Museum', p. 63.
53. 31 July 1940 Report in V&A, ED 84/267.
54. John Ezard, 'How Lords Wrecked War-Time Effort to Save Art', *Guardian,* 12 December 2002.
55. The National Gallery, 'The Gallery during the Second World War', http://www.nationalgallery.org.uk/about/history/war/default.htm (accessed 24 October 2007)
56. John Forsdyke, 'The Museum in War-Time'.
57. SMD, Z 282 and 'War History of the Science Museum', pp. 13–14.
58. Standing Commission on Museums and Galleries, *War Years and After,* p. 26.
59. SMD, Z 282 and 'War History of the Science Museum', p. 16.
60. 5 November 1941 Letter in V&A, ED 84/269.
61. 'War History of the Science Museum', pp. 17–18.
62. The surviving Science Museum lists are inSMD, Z 282.
63. 'War History of the Science Museum', pp. 8, 18.
64. Standing Commission on Museums and Galleries, *War Years and After,* p. 18.
65. V&A, ED 84/263.
66. Tate archives, TGA/2000/1/55/1.

67.   Kew Gardens, '1939–1945: Kew At War', http://www.kew.org/heritage/timeline/1885to1945_war.html (accessed 13 December 2007)
68.   'War History of the Science Museum', pp. 31–3.
69.   'War History of the Science Museum', pp. 59–70.
70.   'War History of the Science Museum', pp. 66–8.
71.   Standing Commission on Museums and Galleries, *War Years and After*.
72.   'War History of the Science Museum', pp. 31, 77–9.
73.   14 February 1946 Director's Note in SMD, ED 79/161.
74.   Standing Commission on Museums and Galleries, *War Years and After*, p. 17.
75.   23 August 1946 Minute Paper in SMD, ED 79/161.
76.   Science Museum Annual Report for 1940–1951, p. 6.
77.   Science Museum Annual Report for 1940–1951, pp. 6–8, 14, 17, 27.
78.   Standing Commission on Museums and Galleries, *War Years and After*, pp. 23–4.
79.   BMSC, 10 July 1943, pp. 5781–2.
80.   Standing Commission on Museums and Galleries, *War Years and After*, p. 26.
81.   26 June 1948 Letter SCM/95B/10/1 in SMD, ED 79/181.
82.   26 June 1948 Letter and 27 September 1948 Notes of Meeting , SMD, ED 79/181.
83.   Also known as a Newtonian-Einstein House, it was described in the Science Museum's papers as a rotating building designed to modify the laws of gravity by centrifugal force and would be used to demonstrate the principles of gravity.
84.   Minute Paper titled 'Planetarium for Festival of Britain 1951' in SMD, Nominal 1356A.
85.   Science Museum Annual Report for 1940–1951, pp. 2–3.
86.   'Note for Director on Return' undated in SMD, ED 79/181.
87.   Memorandum dated 12 Sep 50 SCM/95B/10/9 in SMD, ED 79/181.
88.   15 November 1951 Letter in SMD, ED 79/181.
89.   Science Museum Annual Report for 1952, p. 2.
90.   'Summary of Planetarium History' in SMD, Nominal 1356B.
91.   19 August 1946 Letter in TNA: PRO, FO 943/346 and 20 May 1946 Letter in TNA: PRO, ED 23/1057.
92.   30 July 1946 Letter in TNA: PRO, ED 23/1057.
93.   6 September 1946 Letter in TNA: PRO, FO 943/346.
94.   'Summary of Planetarium History' in SMD, Nominal 1356B.
95.   November 1948 to April 1949 Letters in SMD, Nominal 1356B.
96.   Nominal 1356 Press Cuttings.
97.   10 December 1948 Letter and Reply (SCM/95B/10/3) in SMD, ED 79/181.
98.   25 March 1953 Letter in SMD, Nominal 1356B/6/6.
99.   Anonymous, 'Planetarium Arrives', *The Times*, 7 October 1957, p. 4.
100. Science Museum Annual Report for 1953, pp. 2, 18.
101. 4 December 1954 Letter in TNA: PRO. Work 17/473.
102. Reply to 19 September 1955 Letter in SMD, Nominal 1356/36.
103. Science Museum Annual Report for 1954, p. 1.
104. Paper C for Advisory Council (5 April 1955) in SMD, Nominal 1356.
105. Kevin Littlewood and Beverley Butler, *Of Ships and Stars: Maritime Heritage and the Founding of the National Maritime Museum, Greenwich* (London: The Athlone Press, 1998), pp. 52, 70–2.
106. Anonymous, 'Museums to Get Greater Freedom: Heritage Bill', *The Times*, 25 February 1983, p. 4.
107. Alec Martin, rev. Helen Davies, 'Duveen, Joseph Joel, Baron Duveen (1869–1939)', *Oxford Dictionary of National Biography*, http://ezproxy.ouls.ox.ac.uk:2117/view/article/32945 (accessed 5 March 2009)
108. F.G.G. Carr, rev. Ann Savours, 'Caird, Sir James, of Glenfarquhar, Baronet (1864–1954)', *Oxford Dictionary of National Biography*, http://ezproxy.ouls.ox.ac.uk:2117/view/article/32240 (accessed 5 March 2009)
109. Kevin Littlewood and Beverley Butler, *Of Ships and Stars*, pp. 71–8.

110. Kevin Littlewood and Beverley Butler, *Of Ships and Stars*, pp. 56, 112, and Appendix 1.
111. For a discussion of the conflict of interest in which this placed him, see Chapter 2 of this volume.
112. Kevin Littlewood and Beverley Butler, *Of Ships and Stars*, pp. 71, 107–8.
113. K.E. Robinson, 'Gore, William George Arthur Ormsby, Fourth Baron Harlech (1885–1964)', *Oxford Dictionary of National Biography*, http://ezproxy.ouls.ox.ac.uk:2117/view/article/35330 (accessed 6 March 2009)
114. Kevin Littlewood and Beverley Butler, *Of Ships and Stars*, pp. 60, 120.
115. Personal Communication from Peter Collins, Senior Collections Officer, Imperial War Museum dated 26 August 2008.
116. Standing Commission on Museums and Galleries, *War Years and After*, p.18.
117. Standing Commission on Museums and Galleries, *War Years and After*, pp. 11–18.
118. SMD, T/1945/77 (Part 1)–Technical File: V1 Missile.
119. SMD, T/1982/1264–Technical File: V2 MissileNotes.

4
Ambition and Anxiety: The Science Museum, 1950–1983

Scott Anthony

With hindsight, the Science Museum's post-war record appears impressively assured. Between 1950 and 1983 visitor numbers rose from under a million to over five million a year, the South Kensington site expanded substantially, the Museum's collections grew beyond all recognition, and two new regional centres were successfully launched. But it was a period of considerable anxiety as well as achievement. Each Director of the Museum had to cope with a rapidly evolving range of issues: how should science be presented to the public and with what purpose? To what extent should the Science Museum be concerned with the study of the past? Should a science museum champion new technologies? What is a science museum's educational role and how formalised should it be? The insistent dilemmas of the period were to be as relentless as the rate of scientific progress.

The Festival of Britain, 1951

Staging The Festival of Britain's Exhibition of Science was intended to shake the Science Museum out of its post-war torpor by giving renewed impetus to an array of long-planned developments. For the new Director, Frank Sherwood Taylor, it was a useful forum for stimulating public interest in new methods of display. Borrowing heavily from techniques pioneered by the Palais de la Decouverte in Paris, many of the exhibits on show were models rather than historical objects, and labels became secondary to the linking of objects by an overarching, easily understandable 'story'. 'The exhibition is not being designed for men of science or professional technologists,' explained curator Ian Cox: 'its target audience is that growing section of the general public with no specific scientific training, but nevertheless an active curiosity about scientific affairs.'[1] *Punch*, London Transport and the *London Illustrated News* were among the diverse group of external contributors that provided Cox with everything from animated diagrams to stereoscopic photographs. As well as being a precursor to the 'science centres' of the late twentieth century, the Exhibition of Science also suggests Sherwood Taylor's distinctive approach to display. As he argued:

> An expert in say, musical boxes and their history may not be an expert on how to place musical boxes on public display or choose the captions and type which will provide explanatory labels. We would go further in present circumstances and say that the staff of a Museum who must perforce spend their working days amongst displays and typography conceived in the last century may only too easily be forced into the unconscious acceptance of standards remote from the average accepted by the public.[2]

For Sherwood Taylor, the Festival represented one way of throwing down the gauntlet to the old guard of Keepers. The well-ordered display of the Exhibition of Science stood in contrast to the 'mere accumulation of technological bric-à-brac' that he warned the Museum was in danger of degenerating.[3] 'The History of the Science Museum: Past, Present and Future' accompanied the staging of the Exhibition of Science and provided further elaboration of his vision.

However, the glimpse of the Science Museum's future offered by the Festival was fleeting. The Exhibition of Science was a temporary exhibition, which visitors had to pay to enter, quarantined from the main South Kensington site by its own separate entrance. Rather than impressing or influencing the senior Keepers, the Exhibition of Science merely gave physical form to the fundamental lack of trust and understanding that existed between Sherwood Taylor and his senior staff, further convincing him that he needed greater executive authority to reorientate the Museum.

For the duration of the Festival, the neoclassical south wall of the Museum's East Block was covered by a striking modernist visual display by Brian Peake based on hexagonal arrays of graphite. The Exhibition attracted 213,000 visitors, fewer than the estimates, which had been inflated for political reasons, but still a fifth of the Museum's total annual number of visitors. At a time when Museum staff were wheeling prams around west London parks to gather firewood (as Frank Greenaway remembered it), it provided a welcome burst of gaiety. Michael Frayn described the Festival of Britain as a box of delights, musing, that as with any box of delights, the most delightful thing about it was the packaging.[4] The Exhibition of Science was the Museum's own box of delights and when Peake's board was taken down the elongated process of restoring the Museum resumed.

For the previous Director, Herman Shaw, the quid pro quo of hosting the Festival was that it would kick-start a building programme that had been pending since 1914. The Festival gave the Museum dispensation to bypass the Ministry of Works and employ private contractors to begin work on a new centre block of exhibition galleries. However, the private contractors that built the basement and first floor of the Centre Block for the Festival made no provision for the lighting and heating which permanent museum galleries required. The Centre Block was only upgraded when the Natural History Museum moved its Mammals collection

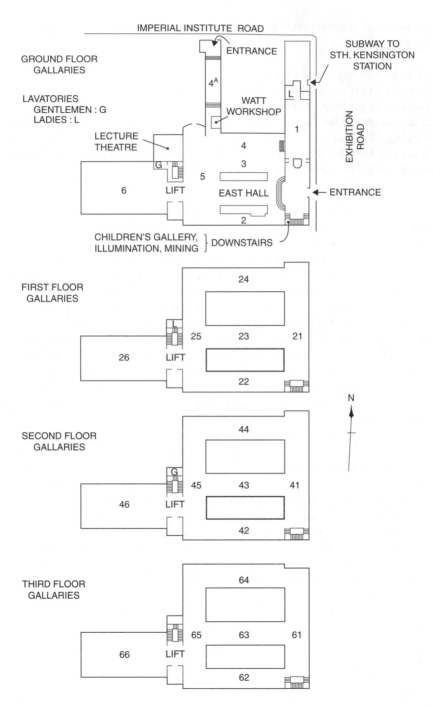

Illustration 4.1   Floor Plan of the Science Museum in the Outline Guide of the Museum published in 1959, with only the East Block in use and thus unaltered since the late 1920s

temporarily into the building. Indeed, the Centre Block was not completed until 1961, and the final Centre Block gallery only opened to the public in 1968.

The Directorship of Sherwood Taylor, 1950–1956

If the Festival of Britain offered the Museum the opportunity to peer into an imagined institutional future, the 'A Hundred Alchemical Books' exhibition of 1952 allowed staff a glimpse inside Sherwood Taylor's mind. 'A Hundred Alchemical Books' was a controversial exhibition – although alchemy could be regarded as an early form of chemistry, it was not a respectable subject for scholars in the immediate post-war period. Moreover, by displaying books (including his own), Sherwood Taylor was staging an exhibition prompted by his research into the history of science and personal interest in the work of Carl Jung. Rather than simply imparting a practical understanding of chemistry, Sherwood Taylor's interest in the history of alchemy was motivated by an unusual combination of principles that included religious conviction. Sherwood Taylor had been one of the founders of the Society for the History of Alchemy and Early Chemistry and promoted the subject by editing the Society's journal *Ambix* from its launch in 1937 until his death, while in books such as *The Fourfold Vision* (1945) Sherwood Taylor attempted to balance science with his Catholic faith. Senior staff at the Science Museum believed such a mix of science and spiritual belief threatened the Museum's reputation as a secular institution. 'A Hundred Alchemical Books' was probably the first exhibition at the Science Museum whose opening was attended by a Dominican friar, Victor White, who was an expert on new practices of psychological therapy.

Despite understanding that his interest in alchemy appeared unusual to his peers, Sherwood Taylor still appears to have been well qualified for the Directorship. In addition to a successful ten-year spell as curator of the Museum for the History of Science in Oxford, he had written both textbooks for children and 'popular' works such as *World of Science*. However, Sherwood Taylor's interest in the history of science had profound implications for how the Museum would curate and display its collections in the future; implications that sensitive, and turf-minded, curators were quick to pick up on. It is worth briefly contrasting the form and emphasis of the Ministry of Education's description of the role and purpose of the Museum with the one developed by Sherwood Taylor. According to the Civil Service:

> The Science Museum, which is the national museum of science and industry, aims at illustrating, by its collections, the history of science and its application to the various branches of technology and industry.[5]

According to Sherwood Taylor:

> Where a sense of wonder and awe is the just and appropriate tribute to the knowledge established in some field of pure science such as astronomy, there that wonder and awe should be evoked. Where admiration is demanded by the achievements of applied science in providing mankind with its present standard

of living, there those achievements should be displayed as part of a historical and continuing process. The object of the Museum may on this view be briefly stated as to exhibit the achievements, wonder, importance and history of science.[6]

Beyond the resentments of the senior staff, Sherwood Taylor's vision was unpopular because it threatened to displace scientists and professional technologists. Sherwood Taylor was scathing about the rigid standards of display in the Museum, believing that they fatigued the eye and thus hindered the Museum's educational mission. He wanted the Science Museum to learn from shop displays, advertising, posters, films and strip cartoons. For instance, Sherwood Taylor purchased Philippe-Jacques de Loutherbourg's 'Coalbrookdale by Night' to illustrate the history of metallurgy. This infuriated Fred Lebeter, the Deputy Keeper of Chemistry (and hence in charge of metallurgy), who complained that the scenes in the 'baroque and romantic' painting (dated 1801) were more typical of the middle of the eighteenth century, and thus the painting 'was of little value in illustrating the actual progress of metallurgy.'[7] Lebeter and his colleagues thus argued that Sherwood Taylor had spent an eighth of the Museum's purchase grant on an artwork irrelevant to the Museum's purpose. By contrast, Sherwood Taylor believed museums could, and should, balance the art of display and the art of instruction. Nonetheless, he wrote a grovelling apology to Lebeter which must be unparalleled in the history of the Science Museum.

Sherwood Taylor's idiosyncratic ideas about the practice, purpose and display of science had been developed in the more academic atmosphere of Oxford. The Science Museum was to prove a less forgiving environment. When Sherwood Taylor was appointed the Museum was in some disarray. The strains of rehabilitating the Museum after the Second World War had seen the previous Director Herman Shaw die in post. 'There is now little control of the museum,' Civil Servants admitted: 'what is mainly needed is an administrator who can control staff.'[8]

Sherwood Taylor got the job because he had the backing of C.P. Snow. Snow's favour meant that, despite having previously declared that administrative ability was crucial, officials at the Ministry of Education appointed an urbane intellectual who, in the judgement of his previous employers, 'might find that Science Museum administration absorbed too large a fraction of his time for his liking.'[9] In recommending Sherwood Taylor, C.P. Snow had dismissed the 'faint doubt that he might be overwhelmed by the administrative burden of the position.'[10] Shortly after arriving at the Science Museum, Sherwood Taylor applied for a junior post at Oxford.[11] Embarrassed, Oxford persuaded him to withdraw his application. Though a heroic figure to the Science Museum's post-war intake, like his predecessor, Sherwood Taylor was to die in post.

Sherwood Taylor did make a lasting impression through the 'Committee on the provision for the needs of children in the Museum', which he set up in 1951. Sherwood Taylor invited Professor Philip Vernon, a pioneer in educational psychology, to investigate children's reaction to the Museum, its effect on them, and its possible benefits. This work resulted in a number of successful innovations. The Museum began to issue schools with 'informal' monthly bulletins, special guides

Illustration 4.2 A child operates a model of a 1924 electric passenger lift in the Children's Gallery, 1963

were employed to develop bespoke tours and public lectures were enlivened with noisy demonstrations. The Committee's final report would profoundly influence the genesis of the new Children's Gallery in the Centre Block.

To some degree, Sherwood Taylor also laid the template for future historians of science at the Museum. With its considerable archival resources and attachment to government, the Science Museum could plausibly have become the central focus for the study of the history of science in Britain. Instead, its post-war practice at the Science Museum assumed a more clubbable character and tended towards the informal and the personal, and the exact status of the Museum's historians of science remained ambiguous.

The Directorship of Terence Morrison-Scott, 1955–1960

When Sherwood Taylor died after a long illness on 5 January 1956, the Museum was in considerable disarray. According to Civil Servants 'practically nothing' had been done to formulate a future policy for the Museum.[12] In a letter to C.P. Snow after the interviews, the Chairman of the Advisory Council, the Earl of Halsbury, stated that:

> The museum has been a rudderless ship for too long, the staff are demoralized and policy decisions long overdue, require to be made without delay due to the Ministry of Works' need to press on with building plans. The selected candidate ought therefore to be of the type who becoming captain of the ship overnight, would make it clear to all that a skilled, experienced touch was on the helm. ... For a long time the Museum has had every type of Director, except a Museum man, from Retired Sappers to Oxford Intellectuals. The consequences of this are part of the legacy with which we have to cope. We want no more amateurs.[13]

Halsbury was a forceful character who had become Chairman of the Advisory Council in 1951. The grandson of a famous Lord Chancellor, Tony Giffard had taken a London University external degree in chemistry after his father forced him to become an accountant.[14] In 1949, after a brilliant career in industrial research at Brown Firth Research Laboratories and Decca Records, Halsbury (he had succeeded his father in 1943) became the first managing director of the National Research Development Corporation, which had been set up to exploit public-sector innovations. He was determined to get the right man, and that person in his view was Terence Morrison-Scott, a relatively junior Assistant Keeper at the Natural History Museum next door. It is clear that Halsbury – who had been his friend since they met at Eton in 1921 – was impressed by both his social and scientific standing. He was an 'Old Etonian with comfortable private means' and a leading member of the exclusive Athenaeum club who, to the surprise of Halsbury, had spurned the obvious careers for someone of his position to take up science and 'then plug[ged] away until he was awarded a D.Sc. for published work.' It is possible that Halsbury's surprise that someone so well connected had stuck with science may have influenced Snow when he wrote his famous Rede lecture on 'the two cultures' two years later.

Despite this powerful backing from Halsbury, Morrison-Scott very nearly did not get the position. His interview was described as 'disappointing'. He admitted he had no interest in education, and that he would look to leave the Science Museum if its financial and political standing were not rapidly improved. By contrast, some members of the selection panel were attracted by the performance of E.C. Baughan. Described as an engaging man of subtlety and depth, Chris Baughan was a Professor of Chemistry at the Royal Military College of Science, Shrivenham, who had studied under Michael Polanyi in Manchester before becoming involved in Operational Research during the Second World War.

Halsbury was having none of this. He declared to Snow that Baughan:

> ...could run any well-run show well, and then impress his stamp on it later, when he had learned all about it. Knowing little of Museums, he would, in his own words, 'lie doggo' for six months and watch. Then he would begin to make his influence felt. The Museum, however, is not a well-run show which will carry on while he lies doggo. It needs a lot of action taking here and now, and I don't think will be good for the staff to watch their new director lying doggo (emphasis in the original).

By contrast, according to Halsbury, Morrison-Scott was 'a completely pukka Museum-man, accepted as such by the Museum world.' Snow found Halsbury's powerful arguments irresistible and Morrison-Scott got the Directorship.

Halsbury admitted to Snow that Morrison-Scott was a 'lusus naturae' who had no bump of reverence and was 'incapable of modifying his forecastle style for the quarter deck of an interviewing board.' This aggressive style was not conspicuously successful at winning over the Keepers or winning support for reforms such as increasing the pay of workshops staff. Although Morrison-Scott successfully petitioned government for the Museum's purchase grant to be raised from £2,000 to £8,000, he was unable to recruit the quality of personnel he wanted. To some extent Morrison-Scott's Directorship of the Science Museum had been seen as essential training for the directorship of the Natural History Museum, for which Morrison-Scott, as a zoologist, was more suited. When his mentor Sir Gavin de Beer retired in 1960, Morrison-Scott was happy to return to his old museum as director.

Clearly the Advisory Council, under the Chairmanship of Halsbury until 1965, was unwilling, following these two apparent failures, to repeat the experiment of appointing an external candidate. David Follett (Director from 1960 to 1973) and Margaret Weston (Director from 1973 to 1986) were internal appoinments and both could lay claim to being the most successful Director that the Science Museum had ever had. Twenty-three years of relative stability followed Morrison-Scott's departure and, paradoxically, it becomes less important to talk about the Directors from henceforth.

The Space Race and the popular appeal of the Museum

At the beginning of the 1960s the press was either ignoring the Museum or disparaging it. 'Its rare excursions into the twentieth century,' wrote Gerald Leach in *The Guardian*, 'are about as relevant as half a dozen abstracts would be scattered on the walls of the National Gallery.'[15] Though British newspaper critics tended to look down on highly interactive North American science centres, they posited the

Deutsches Museum's compromise between old and new as an appropriate model for the Science Museum. Because 12-year-old boys were interested in 'atoms and stars and space, that kind of stuff', what was needed, the media counselled, was a museum of new science rather than of old technology.

Public interest in the Space Race encouraged the Museum to give greater attention to the achievements of modern science. In 1962, Colonel John Glenn's Space Capsule, 'Friendship 7', went on display for two days in a hurriedly prepared Centre Block gallery. The Museum's opening time was extended to 9 p.m. to accommodate the 25,000 people who flocked to see it, police were called in to marshal queues that stretched into Cromwell Road, and Richard Dimbleby interviewed excited visitors for a BBC *Panorama* special. 'Friendship 7' was arguably the Science Museum's first blockbuster attraction. The Museum attempted to repeat the success in 1965 when 'Freedom 7', which made the first US suborbital space flight, was loaned from the Smithsonian Institution and the National Air Museum of the United States. Perhaps it wasn't quite Beatlemania, but the public enthusiasm for similar special exhibitions saw the Museum's attendance figures climb over 1.5 million by 1965. This surge in numbers was given added momentum by the completion of the Centre Block and the subsequent increase in display space. Under Director Follett this extra space, along with the rigorous modernisation and redevelopment of the Museum, saw visitor numbers pass two million by 1970.

Organising the Space Race exhibits apparently owed much to the cultural diplomacy of the Cold War. The aptly titled 'Friendship 7' came to the Museum as part of a worldwide tour sponsored by the United States Information Service. Visitors were handed information and shown films that were prepared in America. Similar motives were at work in 1970, when President Nixon gifted the nation rock samples from the Moon wrapped in a Union Jack with the instruction that they should be displayed to the public at the Science Museum. Indeed, the Science Museum frequently indulged in Cold War cultural relations of its own, building substantial links with the National Technical Museum in Prague during the premiership of Antonín Novotný.

The Museum had actually attempted to address the space sciences before the American loans. Since 1953 the Museum had been pressing government for the funds to install a planetarium on the top floor of the Centre Block. They argued that a planetarium would enable the teaching of astronavigation and would increase national prestige. The government, perhaps not without foundation, labelled the planetarium 'an expensive toy' and dismissed the idea.[16] In 1958, the year after Sputnik's launch, Madame Tussauds opened the Commonwealth's first planetarium and the debate became irrelevant. The top floor of the Centre Block was reassigned to aviation.

To an extent, the Museum's attempts to build a sizeable space collection during the 1960s demonstrated Britain's relative economic and political decline. Crowds would queue to see space exhibits but there were few British exhibits to show them. 'Aerial', the first British satellite, was launched into space the same year that 'Friendship 7' was exhibited at the Museum, but its presence at the Museum was understandably limited to a modest exhibition of satellite instruments. Up-to-date technology was difficult to get hold of and one response to the expense and lack of expertise was for the Museum to make greater use of models. Donations from the BBC or Pinewood Studios further allowed the Museum to begin promoting the science of popular television programmes such as *Doctor Who*. In the winter of 1972–1973, the Daleks proved a massive public draw, adding an extra 20 per cent to visitor numbers. Blockbuster exhibitions did not always illustrate the path of scientific or technical evolution, but they did illustrate why it was exciting. The broader point was that the increasing influence of external expertise and outside sources of funding was exacerbated by the steady growth of exhibition space. The *Daily Telegraph* complained that the sponsors of the Museum's new Iron and Steel Gallery had 'reduced history to a mere introduction and has made the gallery a propaganda show.'[17] While such accusations were exaggerated for dramatic effect,

Illustration 4.3   Doctor Who (played by Jon Pertwee) and the Daleks at the BBC TV's 'Visual Effects' exhibition at the Science Museum, December 1972

there is an obvious correlation between the Museum's excellent atomic energy, gas, oil and industrial chemistry galleries and the expansion of nuclear energy, the discovery of North Sea gas and oil and the continuing industrial might of ICI. Less successful were areas such as computing or the space sciences, where native achievements were more modest and the problems of display more acute.

The professionalisation of the Science Museum

The completion of the Centre Block in 1967 coincided with the creation of Museum Services, the most significant internal reform of David Follett's reign. Prosaically, Museum Services gathered together the Museum's workshops, photographic studio, design office, publishing arm, educational services, press office and front of house. The rationale of Museum Services was simple – ever-increasing visitor numbers required an ever more comprehensive range of services. In Europe, only the Louvre attracted more visitors than the Science Museum, but the South Kensington site did not deign to open a public cloakroom until 1965. More fundamentally, the Museum had only a vague sense of its public appeal. The Advisory Council concluded:

> The staff remains largely ignorant of the age distribution of these visitors, of the reasons that bring them to the Museum, of the collections and methods of presentation that they find most stimulating, of the reasons for the continued decline in the adult's interest in gallery tours and theatre lectures, and of other factors a knowledge of which would facilitate the future development of the Museum.[18]

As well as addressing these gaps in knowledge, Museum Services attempted to build a closer relationship between the Museum and its visitors by running an information service, producing brighter publications, organising a professional press office and developing its educational programmes. By building up a strong internal team of designers it also became possible to more speedily produce new special exhibitions and refurbish existing ones. The acceleration in pace evident during the 1970s owed much to the infrastructure created by Museum Services. Significantly, the department gathered up non-curatorial positions that had sprung up since the war. These jobs generally attracted young people who set themselves against what they perceived as the Museum's over-formal norms. Compared with the V&A and the Natural History Museum, they considered the Science Museum's approach to displays staid and rather second-rate. Thus there was friction between the critically-minded elements of Museum Services and older staff who had considerable technical expertise and were less self-consciously interested in the Museum qua museum. This is a broad brush picture, of course, but the voices from the period are insistent. The expanded 'service' roles at the Museum in the 1960s were filled by a younger generation with an appreciably different outlook. Ironically, when interviewed for this project, they tended to remember the older guard affectionately. There was a reflective appreciation that their considerable practical expertise was an important part of the Museum's success, and an assertion that the Museum was poorer for the loss of their skills.

Museum Services was also an administrative change with some interesting political consequences, because, while it relieved Keepers of mundane duties, it also, to an extent, sidelined them as the new department increasingly served as a nexus between internal logistics, the wider Civil Service and external museum expertise. David Follett had risen to pre-eminence as Secretary of the Museum's Advisory Council; Margaret Weston, his successor, served her 'deputyship' as the first Keeper of Museum Services. Indeed, she attributed much of her subsequent success to the holistic understanding of the Museum's operations that she had learned in what had originally been regarded as a backwater post.

Exploration and the Jubilee celebrations

The Exploration Gallery was the Science Museum's main contribution to 1977's celebrations for Her Majesty's Silver Jubilee. The aim of Exploration was to 'complement the necessarily specialised approach of much formal science teaching by showing the variety of disciplines that contribute to each of its topics.'[19] The practice of Exploration was the artful grouping of objects such as Apollo 10, an eighteenth-century diving suit and a CT scanner. This survey of the frontiers of human knowledge was complemented by an imaginative range of interactive attractions. Exploration both showed how the Museum had adapted to the

Illustration 4.4 Interactive periodic table designed by Robert Wetmore of Oriel Equipe for the new chemistry gallery in 1964

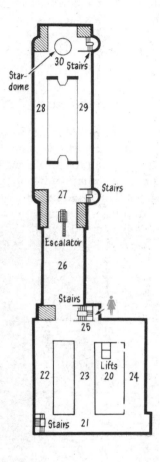

Illustration 4.5 Plan of the first floor of the Science Museum taken from a visitor leaflet published in the early 1980s, showing the Centre Block and the Infill galleries (Gallery 20 on the first floor)

challenge of North American 'science centres' and pointed ahead to LaunchPad. Museologically, Exploration ushered in a period of significant experimentation. Designed with a planned life cycle of three years, it blurred the distinction between permanent gallery and temporary exhibition and hence suggested how the Museum might keep in step with visitor interests. Most importantly, Exploration was a popular triumph.

It is perhaps useful to see 1977 marking the completion of a cycle in the Museum's post-war development, the hurried preparations for the Silver Jubilee celebrations book-ending a process of growth that had begun with the Festival of Britain. During this period methods of display had been revolutionised, the Centre Block galleries had been completed, the Infill galleries were in the process of being added, a new library had been built, and the National Railway Museum at York had been opened. The range and reach of the Museum was to be further extended by the Museum of Photography, Film and Television in Bradford and the donation of the Wellcome Collection.

The flipside of this growth, however, was an increasingly stretched budget and an increasingly stretched staff. Between 1973 and 1987 exhibition space at South Kensington grew by around 15,000 square metres. At the same time, sixty-two posts were lost to manpower cuts imposed by central government. The result was that the relative influence of Keepers and curatorial staff declined and gaps were filled by external contractors and commercial funding. These external influences would grow in significance as the twentieth century drew to a close. The long-term consequence was that the Museum was unable to extend the successful sequence of internal appointments after Margaret Weston's retirement.

Sponsored exhibitions had accounted for just 396 square metres of exhibition space in the years 1973–1977; this number had grown to 4,643 square metres in 1983–1987. Rapid deindustrialisation and abrupt ideological shifts amplified these trends. After 1977, what had been uncontroversial became more contentious, as the funding and governance of museums became an issue of high political debate, a process that culminated in The National Heritage Act of 1983, which removed the Museum from direct government control. The Science Museum had outgrown its immediate post-war vision, but perhaps not quite grown into something else.

The Directorship of Margaret Weston, 1973–1986

Margaret Weston's appointment (as a woman) to the senior staff of the Science Museum in 1955 was revolutionary in itself, but her reign as Director was to prove even more remarkable. Weston describes herself as a natural hands and minds person, which belies her convoluted path through the upper echelons of the Civil Service. Her path from a Gloucestershire grammar school to Oxbridge interrupted by the Second World War, Weston went unexpectedly but enthusiastically into engineering. She was recruited by the Museum from GEC in Birmingham to complete a new Electrical Engineering Gallery. Time pressures were strict and Weston faced a degree of uncooperative disapproval from her colleagues, yet the gallery was a success. According to Margaret Weston, David Follett quickly recognised her talents and pushed her hard. Follet (rather tellingly) sent Margaret Weston to work on the new Children's Gallery, after which she became the first Keeper of Museum Services before rising to Director in 1973.[20]

When Margaret Weston became Director she was able to draw on a considerable fund of knowledge about the Museum and its staff as well as an impressive catalogue of external patrons. When in charge of public relations, Weston had been involved with arranging royal visits and had regular contact with senior industrialists. The heavyweight Board of Trustees that Weston enticed to the Museum in 1983, a Board that featured Sir David Attenborough, Sir John Harvey Jones and Sir Denis Rooke amongst its galaxy of luminaries, was a direct consequence of her public relations expertise. Weston had built firmly on strong pre-existing ties with industry. Indeed, counter-intuitively, it might be that the Science Museum would ultimately suffer most from the period's pressure to work more closely with business. Unlike other comparable institutions the Science Museum had not been insulated from contact with business, thus changes in museum policy would transform its relationship with industry rather than extend it. Arguably this did not work in the Museum's favour, the Museum had always appealed to industry in terms of national duty, prestige and the promotion of scientific education. In the future these relationships would become more contractual as British business culture underwent radical change.

As well as new projects such as LaunchPad, Weston oversaw the completion of large initiatives set in motion by Sir David Follett. When asked at interview how she felt about opening a national railway museum, Margaret Weston replied enthusiastically. Hence Weston spent the first day of her Directorship travelling to York, where she began her reign by announcing the creation of the National Railway Museum.[21]

The first outstation: The National Railway Museum

The launch of the National Railway Museum prefaced many of the political, economic and institutional pressures that the Science Museum would have to adapt to as the twentieth century grew to a close. However, the story of the National Railway Museum had actually begun with Richard Beeching's infamous 1963 report, *The Reshaping of British Railways*. Commissioned by the Conservative government to cut British Rail's losses, Beeching sanctioned the cutting of a third of the nation's 7,000 stations and some 5,000 miles of track. Divesting the railway of responsibility for its museums at Swindon, Clapham and York was part of this drive to turn British Rail into a revenue-generating organisation.

Though opposed to the Beeching cuts in opposition, in office the incoming Labour Prime Minister Harold Wilson both extended them and speeded them up. Wilson wanted British Rail to reinvent itself by launching fast, limited-stopping InterCity services. Thus, in addition to the material need to divest responsibility for the railway museums, the creation of a National Railway Museum bolstered the rationale of successive governments – to emphasise that the preservation of

Britain's railway heritage sat alongside a plan to transform its future. Indeed, as it phased out the steam engines that dominated the network into the mid-1950s, British Rail would sponsor Science Museum exhibitions in South Kensington such as 'Advancing Railway Technology'. As Beeching had it, 'I suppose I'll always be looked upon as the axe man, but it was surgery not mad chopping.'[22]

Although the National Railway Museum was not opened until 1975, the Science Museum began to agitate for the transfer of the British Rail collections in 1964. While there was a high political imperative to divest control of the rail collections, the Museum had to work hard to position itself as the obvious repository for them. The appointment of the railway historian Jack Simmons to the Museum's Advisory Council in 1969 was part of a sustained period of lobbying in which the Museum had faced down suggestions that the rail collection should be split up and used to bolster a proposed museum of land transport. The Museum's argument that 'to do justice to a peculiarly British contribution to civilisation, [the collections] must develop exclusively as a railway museum' went hand-in-hand with a changing conception of its overall role.[23] Its Advisory Council had argued:

> The rate of development of science and technology has in recent years accelerated to such an extent that to form detailed Collections is considered generally to be impractical. The aim of the Museum must therefore be limited to representing this development in outline only, by selecting for preservation only objects whose technical significance is outstanding. Filling in the detail must be left to specialist museums … It seems to us that the ideal way to organise this would be to bring such specialist museums under the care of the Science Museum, to which they would be related as outstations, in the same way as the Victoria and Albert Museum has outstations.[24]

The Science Museum was able to pursue this 'outstations' policy because the National Railway Museum was an enormous success. The National Railway Museum was opened by HRH the Duke of Edinburgh on 27 September 1975 to coincide with the 150th anniversary of the opening of the Stockton & Darlington railway. In the first two months after opening, the National Railway Museum attracted half a million visitors and won the British Tourist Authority's 'Come to Britain' trophy. The success of the National Railway Museum cemented a decade's growth in visitor numbers and palpably lifted institutional self-confidence.

Sherwood Taylor had been anxious to prevent the Museum becoming little more than a collection of 'scientific bric-à-brac'. Journalists urged the Science Museum to 'ruthlessly discard the irrelevant'.[25] This trend was partly reversed by the National Railway Museum as journalists and politicians became more interested in the nation's industrial heritage and aware of its potential to attract tourism. Making use of Civil Service networks, the Science Museum's acquisition of an old Post Office building near Olympia, and a decommissioned Second World War aircraft hanger in Wiltshire allowed Weston to reflect this change in mood. With extra space in which to store everything from early road gritters to aeroplanes, and a purchase grant that jumped from £18,000 in 1975 to £300,000 in

1980, pressure on collection policy dissipated and the Museum looked hungrily outwards.

The impact of deindustrialisation on the Science Museum

'Tower Bridge Observed' was a small exhibition with a wider significance. The decline of London's Docks meant that Tower Bridge's heavily manned battery of hydraulic pumps and engines could be replaced by small electric motors. But 'Tower Bridge Observed' was not an exhibition centred on superlatives of engineering; it was a sentimental attempt to preserve the character of one of London's iconic landmarks. The 1974 exhibition displayed pictures of the bridge with its tower-tops removed during wartime, artefacts donated by the public and a series of specially commissioned paintings by Edna Lumb. 'She manages to make the pumps and boilers, pistons and cogs of Tower Bridge,' gushed the *Arts Review*, 'as compelling and memorable as Van Gogh's sunflowers.'[26]

If the science of the new attracted people to the Museum in the 1960s, 'Tower Bridge Observed' was a relatively benign example of the 1970s resurgence of the science of the old. Deindustrialisation was at the heart of this transformation. The traumatic rapidity with which economic and social change swept across Britain drew divergent responses. At governmental level it was hoped that Britain's industrial heritage could form the basis of a new wave of cultural amenities. The National Railway Museum had been part of this trend. Bradford's active municipal council secured the Science Museum's involvement in the creation of the National Museum of Photography, Film and Television in 1983. Similarly, the Telford Development Corporation offered the Museum a site at Ironbridge for the preservation and display of large-scale machinery from the Industrial Revolution. The Museum planned not only to restore wrought iron machinery but to employ local people to keep iron puddling and other associated skills alive.[27]

The profound impact of deindustrialisation on the theory and practice of British museological approach was exacerbated by central government after the Wright Report of 1973 drew attention to the inadequacies of regional museums. Originally set up in 1971 by the Minister for the Arts, Viscount Eccles, to investigate the managerial and financial efficiency of local authority museums, the Wright Committee on provincial museums and art galleries, chaired by the Civil Servant and leading amateur geologist Claud Wright, identified the need to develop a new breed of more outward-facing museum professionals.[28] Lord Eccles accused regional museums of displaying poorly maintained collections in substandard buildings. 'Millions can afford a car and a television and holidays abroad,' Eccles told delegates at the Museum Association's Annual Conference. 'The wage-earners of today are not asking for culture to be doled out to them as a charity.'[29] According to Eccles, museums had to learn to compete with rival leisure attractions. Thus deindustrialisation was accompanied by the rise of a new kind of social history, which flattened out curatorial expertise to move 'ordinary' people to the centre stage. This tangible 'experience' approach was adopted by pioneering institutions such as the Beamish Museum in County Durham. The heritage boom had begun. It was a strange mixture of conflicting ideological, social and economic prompts.

The Science Museum had been relatively slow to respond to these trends. In this respect its South Kensington location was a marked disadvantage. Nevertheless, in 1963 the Museum had established a Circulation Collection to 'enable smaller museums outside London to present scientific and technological exhibitions which they could not normally devise.'[30] A perennially poor relation in terms of funding and prestige, the Circulation Collection was nevertheless a considerable success. Its first exhibition, on the development of cinema, attracted 21,000 to Doncaster Museum in its five weeks on display and trebled attendances at the provincial Yorkshire museums it toured. By the end of the decade, thirty-eight exhibitions were circulating the UK. Furthermore, the Science Museum's new outstations were able to dovetail their activities with the Circulation Collections – allowing the National Railway Museum, to take an eccentric example, to tour a 'Centenary Express' train round fourteen British stations in celebration of 100 years of railway catering. Leaving the famous British Rail sandwich aside, which most people did, the serious point was that the South Kensington site, its outstations, and the Circulation Collections broadened the Science Museum's reach.

The Science Museum also benefited from the Wright Report's efforts to force the pace of professionalisation in regional museums as the Museum assumed responsibility for administering the Preservation of Technological Material grant. The fund enabled local galleries and museums to purchase everything from motor cars to maps as well as collections of machinery. Regional museums were now to apply to the Science Museum for a share of a £150,000 budget put aside by central government for purchases of technical and scientific artefacts. The weakening of local government control over regional museums enabled the Science Museum to develop a truly national reach, exerting a hitherto unthinkable degree of co-ordination over the national collection of objects related to the history of science and technology.

The end of departmental status

The National Heritage Act of 1983 severed the link between the Science Museum and government established in the aftermath of the Crystal Palace exhibition of 1851. From 1983 onwards the Museum continued to be funded by a state grant, but its future direction was to be determined by a board of trustees rather than a government department, a change that was to profoundly alter the workings of the Museum.

Movement towards trustee status began when James Callaghan's Labour administration was forced to obtain a loan from the International Monetary Fund (IMF) on onerous terms in 1976. Despite being in an expansionary phase and enjoying record visitor numbers, the Science Museum was badly affected, as indiscriminate large cuts in public-sector expenditure were conditions of the loan.

Unofficial official opinion conceded only that trustee museums were better at attracting publicity than departmental museums, but in the context of the IMF's demands their status had obvious benefits. The IMF cut entailed a reduction in the Civil Service. While the Science Museum was forced to reduce its workforce by 11 per cent, museums with trustee status had the option of generating additional revenue or cutting expenditure elsewhere. However, Margaret Weston, the Science Museum Advisory Council and Museum staff remained opposed to the transfer of the Science Museum from departmental control, while the Standing Committee for Museums and Galleries took the opposing view and the Department of Education and Sceince hovered somewhere in between. Education hoped to retain its long-term influence over Science Museum policy, but there was a dawning realisation that it was in the interest of both the Civil Service and the Museum to devolve responsibility for financial management.

In 1978, staffing shortages induced by IMF cuts saw the V&A shut off a third of its display space and close altogether on Fridays. The V&A director, Roy Strong, enlisted heavyweight cultural figures such as Lord Goodman, Lord Gibson, Henry Moore, John Piper and Hugh Casson to protest, and further embarrassed the government by claiming that Shirley Williams, the Secretary of State, had prevented him from making the case for devolution on television. It looked a clear-cut case of a faceless bureaucracy stifling a bold and brilliant curator. Strong's feted Fabergé exhibition had turned a profit of £170,000, but it was a profit that he was unable to reinvest. Inevitably, the media noise created by the controversy sucked the Science Museum reluctantly into the debate. Weston declined to join in the attack on government or to scale back the Museum's operations, to which Strong dismissively responded: 'We are guarding treasures, not fire engines.'[31]

In the short term, a decision on devolution was delayed by unnamed V&A sources who claimed Strong was an egotist who had made a series of bizarre purchasing decisions, Civil Service unions anxious to protect both the career opportunities and the pensions of staff, and Hugh Jenkins's suggestion that Labour should create a new Museums Council run along the same lines as the Arts Council. The election of a new Conservative government then marked a temporary hiatus in the debate as responsibility for the Science Museum shifted from the Department of Education to the newly created Office of Arts and Libraries.

In the early 1970s, Edward Heath's Conservative government had briefly introduced museum charging. This was a politically unpopular move that lowered attendances by 40 per cent and failed to boost finances significantly. When appointed Minister for the Arts by Margaret Thatcher, Norman St John Stevas was anxious not to repeat this mistake. As an official had it:

> The question of charging generally is still a politically sensitive issue. I have been approaching it very gently, allowing some extension of charging where the Museums concerned have asked for it and can clearly justify it. In my judgement it would be wrong to force the pace. The museums know they must do everything they can to supplement what they get from us.[32]

St John Stevas's stance was supported by consultation with the Smithsonian Institution and independent research which suggested that, while organisations such as the Tate and the National Portrait Gallery could generate around a third of their income from charging, and that the V&A had great revenue-generating potential, the Science Museum, with its reliance on visiting school groups, would suffer hardship. Trusteeship was a more flexible way of forcing departmental museums to face their financial responsibilities than the blanket imposition of charges. It was also a politically shrewd manoeuvre, because it devolved the responsibility for the likely future introduction of charging from St John Stevas to the individual museum directors. When Labour and the unions opposed the devolution of departmental museums, it became exponentially more attractive to Margaret Thatcher's Conservative administration. In 1983 the Science Museum's Advisory Council was dissolved and a Board of Trustees was brought into being.

Conclusion

By the close of the twentieth century, the technocratic vision of the post-war world showcased at the Festival of Britain had strayed down some unexpected paths. The Festival of Britain had posited a domesticated future where exaggeratedly over-complicated robots devoted themselves to pleasing but essentially frivolous tasks, whereas the social, political and technological mores of later twentieth-century Britain defined themselves in altogether more aggressive, more immediate and coarser terms. In 1950, the Science Museum was a departmental Museum, staffed by Civil Servants, focused on research, and directed by an Advisory Council. By 1983 the devolved Museum had begun to perceive the general public as a series of niche

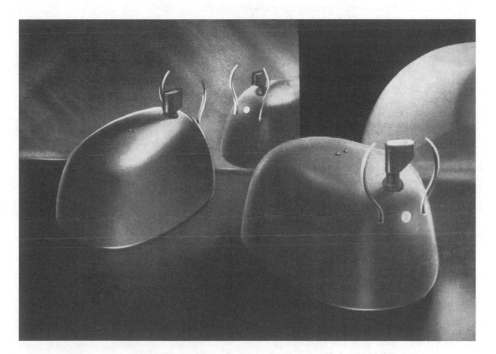

Illustration 4.6   The robotic tortoises displayed at the Festival of Britain, 1951

audiences, its industrial patrons demanded greater influence and its new Board of Trustees began to grapple with the amorphous monster of 'accountability'.

One of the star attractions of the Festival of Britain had been a group of robotic tortoises that steered themselves towards sources of light; this now appeared a charmingly obedient vision of the future.

Notes

1. Quoted in R.G.W. Anderson, 'Circa 1951: The Festival of Britain, The Exhibition of Science, and the Science Museum', in R.G.W. Anderson, P.J.T. Morris and D.A. Robinson (eds) *Chymica Acta: An autobiographical memoir by Frank Greenaway with essays presented to him by his friends* (Huddersfield: Jeremy Mills, 2007), p. 116.
2. Science Museum Annual Report for 1955.
3. Ibid.
4. M. Frayn, 'Festival', in M. Sissons and P. French (eds) *The Age of Austerity 1945–1951* (London: Penguin, 1963), p. 348.
5. TNA: PRO, CSC5/563, Hallett to Howard, 1 June 1950.
6. Science Museum Annual Report for 1955.
7. SMD, Nominal, 8979/1/1, Coalbrookedale by night, 1952.
8. TNA: PRO, CSC5/563, Minutes of meeting to discuss advertising of Director post, 16 May 1950.
9. TNA: PRO, CSC5/563, Parkes to Snow, 17 July 1950.
10. TNA: PRO, CSC5/563, Snow to Parkes and Veale, 12 July 1950.
11. As revealed to me by Tony Simcock, the authority on Sherwood Taylor, and archivist at the Museum for the History of Science, Oxford.
12. TNA: PRO, CSC5/563, Fleming to Padmore, 6 January 1956.
13. TNA: PRO, CSC5/563, Halsbury to Snow, 12 May 1956.
14. David Neave, 'Giffard, John Anthony Hardinge , third earl of Halsbury (1908–2000)', *Oxford Dictionary of National Biography*, Oxford University Press, 2004; online edn, May 2005 [http://www.oxforddnb.com/view/article/73647, accessed 19 Jan 2010]
15. G. Leach, 'The Museum Piece', *The Guardian*, 9 July 1963.
16. Science Museum Annual Report for 1955.
17. A. Michaelis, 'Catching up at the Science Museum', *Daily Telegraph*, 5 July 1963.
18. Science Museum Annual Report for 1968.
19. Science Museum Annual Report for 1976.
20. Personal communication from Margaret Weston, 23 January 2009.
21. Ibid.
22. C. Wolmar, *Fire & Steam. How the Railways Transformed Britain* (London: Atlantic, 2007), p. 286.
23. Science Museum Annual Report for 1969.
24. Science Museum Annual Report for 1964.
25. Leach, 'The Museum Piece'.
26. B. Wallworth, 'Tower Bridge Observed', *Arts Review*, 20 September 1974.
27. Although the Science Museum decided not to pursue their interest in the Shropshire site, it was eventually subsumed into the holdings of the Ironbridge Gorge Museum Trust.
28. For an overview of the Wright Report, see Marion P. Smith, 'The Wright Report on Museums', *London Archaeologist*, Volume 2–05 (1973), pp104–105.
29. D. Walker, 'The Professionalisation and Packaging of Museums', *Times Higher Education Supplement*, 20 July 1973.
30. Science Museum Annual Report for 1963.
31. L. Wilson, 'The Director Regrets...' *Evening News*, 3 February 1978.
32. TNA: PRO, ED 245/112, Monger to Brandes, 15 January 1982.

5

Parallax Error? A Participant's Account of the Science Museum, c.1980–c.2000

Timothy Boon

> This tension, between the wish to preserve and record and the require-
> ment to educate, between the voices of the past and the needs of the pre-
> sent and the future, has formed a continuing thread throughout the life
> of this Museum. – Neil Cossons, 1993[1]

It is generally a mistake, especially for a historian, to assume that the change personally lived through is somehow more dramatic than that of any other era. All
the same, the last thirty years have been marked by significant transformations
in the Science Museum's governance, its public face, its staff structure, and in fact
all the different components that the other contributions to this volume have discussed. A single chapter cannot cover the whole range of this era of the Museum's
history, and, whilst acknowledging the wider scene, do more than note that these
changes occurred in a period of significant change for all national museums.
Instead, this chapter will consider developments in the Science Museum's continuing dialectic between history and science communication – or, to put it differently, collections and interactivity – which was particularly antithetical in this
period.[2]

Science communication and the history of science

If it is possible to discern in the Museum's own history a dialectic between the
history of science and technology on one hand and science communication on
the other, then it is also necessary to note that the two sides of the argument
have taken different forms since 1857. In the nineteenth century, the Science
Department of the South Kensington Museum offered both Smilesian relics to
fuel admiration for engineering genius and scientific instruments for rational
recreation. The interwar period saw sober galleries illustrating the 'lines of development' in the major branches of science and technology complemented by an

introductory gallery soon given the title 'Children's Gallery'. Whilst the Museum was devoted to advances in the practice and application of the physical sciences, the dialectic in play was probably cognitive, that is, between more and less difficult displays. We may consider the commonplace intermingling of 'how it works' exhibits with historical sequences of objects as a tacit expression of this dialectic in action. This was the norm in the post-Second World War galleries and exhibitions up until the 1980s.

The expansionist Directorships of Follett and Weston had several significant influences on the Museum's subsequent history. In the first part of this chapter, I take two major initiatives commencing during their periods in office to exemplify the two sides of the dialectic between history and science communication. First, the transfer of the Wellcome medical collection to the Science Museum on permanent loan in 1976, following lengthy negotiations, shifted the Museum's balance towards the social historical.[3] Separately, the Museum saw, in the opening of the LaunchPad interactive gallery in 1986, the culmination of the long-running trend of supplementing the collections-based galleries with alternative and more educational interventions. This created a new antithesis as a challenge both to the newly social historical Museum and to the older scientific and technological departments. In the second half, I consider how these factors interacted with the reforming spirit of Neil Cossons's transformational decade and a half as Director. Such intellectual differences of emphasis have, since the establishment of the Museum Services Department in 1967, very often not been played out as explicit discussions of principle. Instead, they are visible in changing approaches to the Museum's organisation and the delivery of its products, notably the displays.

Arrival of the Wellcome Collection

For the Wellcome Trust, the transfer of the Wellcome Collection to the Science Museum strengthened a division between their library and artefact collections in favour of a more academic model for the library and iconographic collections at the Wellcome Institute and public display at the Science Museum.[4] But, for the Science Museum, this turned out to be far more than simply the addition of an extra department and a new area of subject interest. This tidal wave of acquisition, documentation, storage, environmentally conditioned display, and to a lesser extent research swept through the Museum. Once the wave receded in the restructurings of the 1990s, it was clear that it had indelibly marked the Museum in a profound way. Partially this was because of scale. The 100,000 objects strong Wellcome Collection doubled the holdings of the Science Museum. Partially it was the interaction with economic and demographic change that was already delivering more graduates into the lowest curatorial grades bringing self-consciously democratic notions of public access and professional aspirations more aligned with the rest of the museum sector. And partially it was the subject matter of medicine, which – building on the tentative 1960s experiments in 'domestic appliances' and agriculture – slowly infused the Science Museum with the spirit of social history.

In his proposal that the Museum acquire the collection, Frank Greenaway, the Keeper of Chemistry, had argued that by 'showing the history of medicine in the same building as the history of physical sciences and engineering the public would be led to see medicine as something rational and accessible, and physical science and technology as having humane associations.'[5] In retrospect, it is the second of these propositions that was borne out, both in the Wellcome Galleries and in the ways that the Science Museum came to display its existing subjects and collections, for example in the reconfigured East Hall and Synopsis Gallery of 1988. 'Glimpses of Medical History' in the Lower Wellcome Gallery, which opened in December 1980, contained a series of reconstructions and dioramas, designed to act as an accessible introduction to the history of medicine. It was stated that by this means:

> the visitor is helped to understand how it might have felt to have been a patient or doctor at other times and in other places, and to appreciate the extraordinary range of activity developed by mankind in trying to understand and treat disease.[6]

This double claim regarding how visitors might respond to a series of visual simulacra marks departures for the Science Museum as a technical museum. The first assertion alludes both to a social historical account of medicine and to ideas of empathy in history teaching.[7] The second is an invitation, not in itself rare in museums, to appreciate human achievement. It is the broad anthropological definition of medicine, matching the range of Wellcome's ambitions, that might have surprised the regular Science Museum visitor in the 1980s.[8] 'The Science and Art of Medicine' in the Upper Wellcome Gallery, opened a year later, was organised in a conventional chronological sequence of showcases and text panels arranged clockwise around the gallery space, linear in its historical account up to the diagnostic revolution of the early nineteenth century, then divided into specialisms for the much larger section on modern medicine, with a small core area devoted to palaeopathology and non-Western medicine. Reading the panel text nearly thirty years on, the gallery's debt to the existing field of the discipline of the history of medicine is clear; this was an academically informed narrative-driven gallery first, and a collections display second. The name Michel Foucault is nowhere mentioned, but there is a whiff of *The Birth of the Clinic*, of Foucauldian epistemic discontinuity, alongside displays that nod to the then-new social history of medicine (displays on medical services, war), the occasional 'great man' (Jenner and Pasteur amongst them) and to liberal social concern (the displays on Third World medicine and public health).

The division of the Wellcome Collections between Euston Road and South Kensington reinforced the tendency for the Trust-funded academic medical history to be written from textual sources. The transfer explicitly outsourced not just the public and display aspects of the artefact collections, but their academic use, to the Science Museum. At the time of the transfer there was a wish to encourage historical work from the collections[9] and this can be seen in the three papers

Illustration 5.1   The Upper Wellcome Gallery, 1990s

on diagnostic instruments published by Christopher Lawrence in this period.[10] At this time, the Science Museum was not known for its historical scholarship; there was a local style of history, a school of engineers' history closely allied to the Newcomen Society. This was certainly concerned with the history of machines and was deferential to the authors' predecessor engineers, who had invented and perfected the devices they discussed in their meetings and published in the Newcomen Society's *Transactions*. The alternative school of historical writing attentive to individual objects was the antiquarian study of scientific instruments, pursued, for example, by the members of the Scientific Instrument Commission and published since 1980 in its Newsletter. Lawrence's papers, distinct in style from his later work in the cultural history of medicine, are closest to this model.

Expectations that significant amounts of academic research into the Wellcome artefact collections would be pursued once they were transferred to the Science Museum were largely stillborn. This was mainly because the vast majority of staff in the department became concerned with the practical management of the collection, either in completing the cataloguing enterprise or in the movement of the reserve collection first from the Wellcome's Enfield store to the Museum's outpost at Hayes, and latterly to Blythe House for orderly storage. Most of the staff employed to undertake this essentially practical work were young and establishing themselves in the museum profession. The main ethos of the department was museological; research was seen as a secondary stage of activity that would follow after collections had been unpacked and laid out in store. It was imagined that collection catalogues, like those produced by art and archaeological museums, would be published.[11] In the 1980s at least two such projects saw some concentrated effort – one of the X-ray collection and one of the pharmacy ceramics – but were overtaken by events well before they were completed.[12] Only Ghislaine Lawrence, in one strand of her work, and Robert Bud stayed true to the ambition to research the material culture of medicine. Lawrence's work on Charles Drew's apparatus for cardiac surgery by profound hypothermia, on the King's Fund hospital bed and on surgical instruments, and Bud's later work on biotechnology and penicillin both showed how trained historical and critical imaginations could weave new stories from and around the artefacts the Museum keeps.[13] The spirit of the medical department at the Museum in the later 1980s was that the foundations for a truly known, fully researched, collection were almost laid, and that the essentially museological work of the curators – in cataloguing and publishing the collections – was perceived to be set for the foreseeable future. It was not to be. Meanwhile, the practical work of the Science Museum's medical department continued, setting precedents in terms of modern museum practice within the institution, which had always been rather separate from the UK museum mainstream. The Wellcome Collection was the first here to be documented to Museum Documentation standards (from 1976), the first to be entered into the computer database (from 1984), the first to be laid out in its entirety at Blythe House, and the first to be displayed in environmentally conditioned galleries.

Supplementing the collections galleries

As other authors in this volume have described, the belief that the Museum's main galleries might benefit from an extra level of explanation beyond the label and the illustration can be traced back at least as far as the Bell Report of 1911.[14] Employment of guide lecturers preceded the opening of the main building in 1928, and by 1949 two staff were employed in this role.[15] Four staff described as 'staff not allocated to [curatorial] departments' provided the broader extra layer in the 1950s. Lectures were delivered in the Lecture Theatre as well as in the galleries.[16] At the same time, there were two film projectionists; the Scientific Films Association collaborated in the public film programme, and frequency of showings increased throughout the 1950s. The Science Museum Annual Report notes that 'the scientific films shown are complementary to the Museum collections and cater for all levels of interest...direct teaching films...are blended with socio-scientific documentaries.'[17] The scientific films programme continued until the late 1980s. The Children's Gallery – defined as it was by audience, not by subject – had an anomalous status amongst the galleries, and was attached to the Electrical Engineering and Communications Department for many years as it was under the charge of the Keeper of that department, Frederick St A. Hartley, who had been its founder.[18] Eventually, responsibility for the Children's Gallery was passed to the Museum Services Department, and latterly more specifically to the Education Service, where it served as a site where the lecturing staff could experiment with educational exhibits, alternatives to the collections-based galleries.

The Advisory Council's committee on the provision for children's interests in the early 1950s had marked an important milestone in the Museum's changing views of its audience.[19] Along with studies by the university-based educational psychologists, James Hemming and Philip Vernon, exploring children's reaction to the Museum, the Schools Inspectorate of the London County Council was also consulted on how teachers could be encouraged 'to make proper use of existing Museum facilities, and how it could guide unaccompanied children, without going so far as to provide teachers in the Museum for them.'[20] The committee determined that 'one of the most useful services the Museum could perform would be to encourage and help school teachers to make best use of it.'[21] As a result of this initiative, one of the guide lecturers was given responsibility for liaison with schools and editorship of a new termly bulletin to schools. At the same time, demonstrations were added to lectures, and special lectures for schools were introduced as an annual event at the end of the summer term.[22] The Education Service grew in influence during the 1960s and 1970s – protected within the Museum Services Department from 1967 – so that, by the early 1980s, three permanent staff were employed to lecture, develop introductory tape-slide programmes, and produce worksheets and other educational materials, with two bookings staff for school groups and a junior office of three Museum Assistants assisting in lectures, developing new demonstration apparatus, running the public film programme and maintaining the Children's Gallery.

But the most significant development to grow from the Education Service in the main period discussed in this chapter was the interactive gallery, LaunchPad. Anthony Wilson, Head of the Education Service, had suggested in 1978 that an 'Experiment Hall' might make good use of the newly acquired Blythe House building in west London, although this was soon discounted as it became clear that the building was not intended for public access.[23] But the seed had been planted; when the Ontario Science Circus visited the Museum in the summer of 1981, the virtues of the North American style of interactive display were forcibly pressed home; Wilson arranged for the exhibition to be evaluated to make the case for an interactive gallery.[24] In the same summer, the Education Service produced the first of its 'Discovery Rooms'. More followed in summer 1982 and December 1983. During LaunchPad's production phase, which started early in 1984, several larger trials of the idea (called 'Test Bed') were run. The gallery finally opened on the ground floor at the front of the Museum in July 1986, during the Directorship of Neil Cossons, but it was emphatically the product of Margaret Weston's regime. It contained a suite of mainly mechanical interactive exhibits built into brightly coloured scaffolding frames. Some have been popular enough to be included in new forms in the latest version of the gallery, on its fourth site, opened in 2007. With the arrival of LaunchPad, the Museum was once more choosing – after more than half a century – to stress interactivity. Following the approach – and copying many of the exhibits – of the North American science centres, including San Francisco's Exploratorium as well as the Ontario Science Centre, the emphasis was placed on separating interactivity from the Museum's other galleries, and focussing on children as a prime audience.[25]

The Science Museum under Neil Cossons

Neil Cossons succeeded Margaret Weston as Director in March 1986, the first external appointment since Terence Morrison-Scott thirty years before. He had the task of realising the potential of the change to arms-length status conferred by the National Heritage Act.[26] This had been passed partially as a solution to long-running problems over the financing of the national museums, and partially so that, in his view, 'by offering opportunity for new thinking, these institutions might be encouraged to escape the incubus of long-standing constraints on the way they had been managed.'[27] With the successful establishment of the Ironbridge Gorge Museum and two years at the National Maritime Museum (NMM) behind him, Cossons was an apostle of museum modernisation. At the NMM, he had chaired a 1984 seminar *The Management of Change in Museums*, where he had argued that 'if museums want to stay in business they have to seriously contemplate circumstances that are changing around them.' He maintained that increasing salaries were absorbing too large a proportion of budgets, and that museums should initiate their own management change rather than waiting for it to be imposed.[28] Willing to introduce charges (from October 1988 at the Science Museum[29]), he was not shy of courting controversy, not least amongst his own staff. As must always be the case, the dramatic changes Cossons introduced applied to a Museum landscape that was in any case being altered beyond recognition by the changes he inherited, not least the Museum's new status, and the counterbalancing developments towards social history and interactivity that I have described in the first half of this chapter.

His intention to be an expansive Director following the lead of Follett and Weston is clear from the early Corporate Plans. In the wake of the National Railway Museum, the National Museum of Photography, Film and Television (now National Media Museum), Infill Block and Wellcome Galleries, in his first decade the Museum began to plan a new aeronautics gallery of hugely increased volume to replace the existing display space,[30] a National Museum of Agriculture in Warwickshire,[31] and an Information Age Museum in Reading with a linked Science Museum gallery.[32] None of these came to fruition, but the public face of the Museum at South Kensington *was* transformed, first by the completion of the long-running East Hall redisplay, incorporating a large-scale audio-visual accompaniment, and, over fifteen years, by the renewal of galleries. Always at the back of Cossons's mind, as with some previous Directors, was the ambition of the Bell Report to extend the Museum westwards all the way to Queen's Gate. All the Museum's galleries were also to be replaced on a rolling fifteen-year cycle, at a total cost of £50 million (£1.6 million per annum).[33] But transformation of the public face of the Museum in this era of Conservative rule and budgetary constraint required dramatic changes behind the scenes too.

Reorganisation

Any director with a desire to revolutionise a national museum in the wake of the National Heritage Act would have had little scope for bringing about change without changing the staff structure, not least because of the significant proportion of expenditure represented by the salaries budget. Cossons had inherited a conventional national museum staff structure of curatorial departments run by Keepers, supported by the Museum Services Department, also headed by a Keeper. In other words, much of the budget was expended in curators' salaries; in terms of its governing ethos and expenditure, the Museum *was* curatorial. Whereas Margaret Weston, as an insider emerging from the powerbase of Museum Services, had made her achievements without structural innovation, Cossons's modernisation was predicated on management change. He announced his intention to take a critical view in an article given the title *Adapt or Die: Dangers of a Dinosaur Mentality* in the generalist BBC magazine *The Listener*. In a Thatcherite tone, the article starts: 'The boundaries of state funding – of the old-style corporatist public sector – are being pushed back, to put spending power, and therefore choice, into the hands of the people.' Increased leisure time and income had created a boom in independent museums; this was the market within which national museums too had to operate; 'the battlefield will be the marketplace and the casualties will be those museums that fail to adapt.' Where the independents were 'vivid, exciting, experiential' and responsive to visitors' needs, 'the larger and well-established museums are in the weakest position: their inflexibility, staff attitudes and near impossibility of recapitalising them at anything like the rate that is needed, all militate against change.'[34]

It became clear that, in the Cossons era, the tension would not be between collections and education, but about the effectiveness of both; about whether more effective management could ensure that national museum practice could be made responsive to the marketplace and visitor preferences, thereby demonstrating the virtues of independence. He refined his argument in *Scholarship or Self-Indulgence*, a talk given at a Royal Society of Arts conference on 'Scholarship in Museums' in 1990. He argued that the growth in the museum sector over the previous decades had been accompanied by the invasion of curatorial space by competing professional groups including conservators, designers, educators, interpreters and managers, the nature of whose work challenged the primacy of the curatorial culture. Particularly in national museums 'where management skills were rare and undervalued,' the growth in staff had been 'unplanned, unmanaged and largely opportunistic.' But, he said, 'a classic process of the specialisation and division of labour was taking place.'[35] We might add that a classic process of professionalisation was also occurring, in which new groups with new values could readily find fault with an outmoded model, at the same time as they established their own foothold in museums.[36] The external publications by Cossons that I have quoted from this period convey an image of the Science Museum as effectively undifferentiated, whereas I have argued that there were discernibly separate (and to some extent antagonistic) cultures of technical museum, social historical medical museum and interactive educational museum. The dialectic I have been discussing is, in these publications, present in arguments about the need to transform the curatorial culture of the Museum. This would be achieved partially by stressing management, and partially by dividing up the very broad curatorial job and transferring aspects to more specialised and skilled professions, either existing within the wider museum profession (conservators) or to be invented (interpreters). The unspoken word is 'elitism'; Cossons argued that 'there is nothing inherently threatening to the scholarly standards of the museum if it chooses to present its collections in a popular manner, in a way that its public understands, finds relevant to its own experience, appealing and stimulating to the imagination. Rather the reverse. ... but scholars do sometimes need expert help in the process of interpretation and they should not be ashamed of that.'[37] 'Scholarship' might have been taken to mean several different things, from connoisseurship via expert cataloguing to social studies of science, but this was a debate that was left to be played out, rather than being explicitly addressed.

Two major restructurings, each intended to engineer a Science Museum more responsive to the needs of the times, occurred in the first ten years of his Directorship. Very soon after his arrival, he set in train the production of a 'Museum Management Plan', whose outline contents were revealed to the staff in May 1987.[38] The Plan reorganised the Museum into five new Divisions of equal status, each headed by a new Assistant Director (AD) post: Collections Management (all the curatorial departments, and curators from Museum Assistants to Keepers), Marketing (all commercial operations[39]), Management Services (finance, personnel, conservation and storage, and buildings and works[40]) and Research and Information Services (where Robert Fox, the assistant director, had responsibility for 'coordinating and strengthening the Museum's scholarly and research output' and running the Science Museum Library). The Public Services Division assumed the former responsibilities of Museum Services, but was also charged with devising and implementing the gallery strategy and was made 'responsible for interpretation, presentation and education,' working with curators. The transfer of these activities, understood to be 'previously the preserve of the curatorial departments,'[41] was one element of this reorganisation that was crucial to Cossons's modernisation of the culture of the Museum. Another was the implied criticism of the Keepers' management of research in the Research Division's brief. A third was the parcelling out to the curatorial Keepers, as secondary responsibilities, of precisely the collections management functions that Cossons held to be poorly done in national museums: conservation, collecting policy, documentation and collections storage. The contemporary documents convey a sense of the pain that the changes were causing; one Museum circular on the reorganisation of Collections Management Division, for example, reported: 'it is envisaged that once the reorganisation is in place a period of adjustment will follow to allow the division to develop the new system of working required of it.'[42]

The consulting firm Touche Ross was commissioned in May 1987 to review the Collections Management Division (at that stage containing the vast majority of the staff); it reported back in the following February. The report's terms of reference included the integration of the four existing curatorial departments into one functional division; a detailed analysis of the role of curatorship and proposing means of monitoring its various aspects; and improving curatorial management. The consultants made fourteen recommendations, which included strengthening the management of the Division by introducing work objectives derived from the Corporate Plan and making the Assistant Director a separate role as the general manager of the Division (and the two northern museums) rather than one of the Keepers doubling-up with their existing job. The result was to alter radically the balance of the Museum's senior management group, where curatorial representation fell from four to one. There was an emphasis on encouraging curators to move into other divisions, and the reallocation of part of the salary budget for recruits to the most junior curatorial grades into a new graduate trainee scheme. Storage and Conservation were to be set up as a 'dedicated activity' within the Division, and headed by a senior curator.[43] In the final version of the structure, fifteen research assistant posts (Curators E) left behind their specialist areas and joined a 'Projects Group', where they could be expected to act as flexible internal consultants to the whole Museum, some doing collections-related research, but the majority conducting operational research or project management roles.[44]

Inevitably, these changes were highly controversial amongst the existing staff. It became a commonplace joke amongst the staff that 'division' was an appropriate noun to describe the separate cultures of the Museum's departments. The early retirement programme that was part of the changes was predictably contentious, with its aim 'to reduce staff numbers to an affordable level over a two year period' (by approximately 10 per cent).[45] Its most prominent casualty was the Keeper who had been Acting Head of Collections Management, Brian Bracegirdle, who took early retirement, became a Museum Fellow and moved to Blythe House to catalogue the microscope collection.[46] Tom Wright, previously Curator of Water Transport, Project Manager of LaunchPad and Keeper of Transport, became the Assistant Director of the Collections Management Division for most of the 1990s.

The gallery development plan

In the older Science Museum, any implicit intellectual tensions between displaying history and representing science and technology were usually not felt by the curators who were responsible for displays. The history presented was essentially whiggish and progressivist; so latest technologies crowned developmental sequences, and interactive exhibits conveyed the principles by which the devices worked. Where time measurement, for example, was on display, the best pieces from the collection would be there, often typologically arranged; the collections *were* the displays. Each of the curatorial departments had its own areas of the

Museum, which were contested by other Keepers whenever new projects were proposed.

An early initiative of Cossons's Directorship was the establishment in February 1987 of a 'gallery planning group' to propose an overall scheme under which the galleries of the Museum would be reorganised according to a new overarching logic.[47] A meeting in August established the principle that the Museum would be organised along the lines of Knowing (third floor), Making (second) and Using (first), with an introductory gallery at the eastern (Exhibition Road) end of each floor.[48] The scheme was not as revolutionary as it might seem on the surface; the 'knowing' theme, for example, was shorthand for science, and the physical sciences had long occupied the eastern end of the third floor. All the same, the sense that a supervening authority could assume the territory of any one of the curatorial departments by, for example, imposing some medical science into the physics space was a foretaste of a Museum much less under the control of Keepers than hitherto. And the scheme was radical enough to provoke some serious resistance from some of the Trustees. Whether it was growing awareness of the difficulty of reaching fundraising targets or Trustee dislike, the scheme was, effectively, stillborn. The former Keeper of Museum Services (who had become Keeper of Physical Sciences), Derek Robinson, was prescient in 1987 when he commented that 'by the end of a c.15 year period, most galleries will have been redisplayed, but... the pace of change is dependent on the availability of financial resources.'[49] All the same, it is possible to perceive its ambitions as present in ghostly form in Food for Thought (Using, 1989); the George III Gallery (Knowing, 1993); Health Matters (Knowing, 1994); and Challenge of Materials (Using with some elements of Making, 1997).

The intended new displays were emphatically intended to be centred on the collections; 'the aim of the Plan is to guide us towards increasing the impact of the Science Museum's pre-eminent collections on the visitor's understanding of the history – and contemporary practice – of science, technology, industry and medicine.'[50] But, at this crucial stage in the development of display ideas, the scheme balanced different interests:

> The essential thematic simplicity of the concept is designed to enable the Museum to provide a new form of access for visitors to its collections. This does not mean that the traditional and orthodox messages that the Museum has purveyed for many years will disappear. On the contrary, the intention is that by juxtaposing messages and collections to amplify each of the major themes, and at points on each floor by presenting important collections in a typological manner, we should be able to achieve much more.[51]

In other words, the scheme was radical *and* conservative, and its displays would be thematic *and* typological. And, in a turning away from the trend of the late Weston era, LaunchPad would not dominate; 'we do not see the Science Museum becoming predominantly an interactive science centre although there will be numerous interactive exhibits throughout the new galleries where this technique of communication is deemed to be the most appropriate.'[52] In struggling with the modernisation of the Museum's display technique, the Plan refined the balance of the dialectic I have been discussing. It presented a 'dual role' of interpreting artefacts using 'ever more sophisticated techniques' and 'other media' whilst staying true to 'our duty to make our unique collections available to the public.'[53]

The failure of the Gallery Plan, which coincided with the necessity of diverting considerable sums of money to replacing the roof of the National Railway Museum Great Hall, led to a reversion to opportunism in seeking new streams of funding.[54] The 1990 Corporate Plan alluded to its shortcomings: 'that appealing intellectual framework has practical disadvantages, chiefly in the length of time required to produce the desired results and the upheaval to the Museum during the implementation period.'[55] Instead it proposed an eminently pragmatic 'multi-museum' of modules: one might be an extended Wellcome Museum, another an education section, and 'there might no longer be galleries on gas, electricity, and nuclear power – but a major section devoted to energy.'[56] Intriguingly, there was discussion of two possibilities, one gesturing to the collections, the other to the present and future of science and technology. A 'Collections Study Base' would be 'visible storage', 'a form of "treasure house" in which the most important objects in the Museum's collections were always accessible.'[57] On the other hand, there was discussion of an innovation centre, provisionally named 'Futures', which would present 'topics of current interest' by means of 'rapid response exhibitions, science films, even plays.' Despite the financial constraints, the Plan proposed development of the Science Museum's west end site. Amusingly, the 1991 Corporate Plan prefaces its description of the new gallery plan with a page-long quotation from the Bell Report.[58]

Interpretation

While LaunchPad in the 1980s marked the first post-war challenge to the notion that galleries would be about collections (even the Children's Gallery displayed a collection), it was Cossons's overall transfer of responsibility of exhibition authorship away from curators, initially to his Public Services Division, that marked the decisive change in the style of exhibitions that visitors would encounter. The 1989 Gallery Plan made the argument, at the conclusion of its section on 'the need for change', that 'many of our staff have the skills to interpret the expertise of diverse communities of scientists, engineers and other communities of scholars, and relate them to objects in the Museum's collections' (that is, curators). But it added a new corollary that 'we shall, however, increasingly come to rely upon our success in developing a new breed of "curator-interpreters" to extend the range of messages we can convey using our unique collection. Towards that purpose an Interpretation Unit within the Public Services Division is to be established in 1989.'[59] The Unit's Head, the advertisement stated, would be appointed to give interpretive support and expertise towards the presentation of collections; support to the Gallery Planning Group, Exhibition teams and Education Service, 'in translating new gallery themes and ideas into comprehensible ideas'; and establishment of an evaluation programme for displays.[60] A physicist from the Open University, Graham Farmelo, arrived at the Museum to take up this post in October 1989. His Unit came to have four subsections, devoted to visitor research (introduced by Jane Bywaters), interactive exhibits, the Museum's audio, visual, electronic and computing exhibit team (AVEC), and a public programmes staff that ran film and lecture programmes. Within the definitional work into the nature of interpretation that the Unit claimed were the two poles of this chapter's dialectic: both 'to improve the effectiveness of the presentation and interpretation of our collections' and a newly conscious determination 'to improve the way in which the Museum communicated scientific and technological ideas to its audiences, notably to non-specialists.'[61]

This unit was to be a consolidation of the principle of organisation trialled in the 'Food for Thought' Gallery project, where the task of creating the gallery was given to a team of erstwhile curators, comparatively junior within the hierarchy of the Museum, reporting to the Public Services Division.[62] These were to be 'the new breed of curator-interpreters,' creating an interpretation from a 'blank sheet', rather than expressing pre-existing curatorial knowledge in display form. Led by Jane Bywaters, formerly a research assistant in the medical curatorial department, the team included Heather Mayfield, who later delivered half the Wellcome Wing project and then became the Museum's Head of Content. This project was a key instance of the change that Cossons intended. To begin with, the project's subject fell outside the Museum's existing curatorial areas; the Museum had agriculture and domestic appliances collections, but had not expressly collected in food technology,[63] and there was no curator of food. The technique, departing from the then normal collections-first approach, was to develop themes and to acquire objects to fit these themes; the objects became secondary to narrative. The project involved removing intellectual aspects of gallery conception and planning from the responsibilities of the Collections Management Division. The ambition to change how the Museum produced new galleries was also seen in the outsourcing of the display's design to the firm Hall-Redman Associates.[64]

An integral part of moving the Museum away from the older model, where galleries were simply displays of specific collections, was to introduce the visitor's voice by proxy into the planning of new exhibitions. In Britain, this had been pioneered as part of the suite of new exhibition techniques by Roger Miles at the Natural History Museum. At the Science Museum, Sandra Bicknell, who had worked on the ill-fated aeronautics gallery project, transferred to become head of visitor studies within the Interpretation Unit in 1989. Here she developed the role of 'audience advocate', a model that had originated in Australia and North America, and which she had encountered during a secondment to the Smithsonian Institution museums in 1989–1990.[65] Here was a systematic response to Cossons's call for the Museum to become more responsive to the public's needs. From this small start, the Science Museum became an internationally respected centre for the application of selected social research techniques to the understanding of visitor attitudes and behaviour.

A second restructuring

After the financial emergency of replacing the roof of the National Railway Museum, Cossons reorganised the Museum's structure once again. He explained the purpose behind the changes as follows: to enhance the Museum's capital programme of building and gallery maintenance and improvement; 'to place greater emphasis on contemporary science and technology'; and 'to maintain and improve scholarship'.[66] This restructuring was explicitly accompanied by a staff reduction programme designed to reduce the salary bill by the loss of 45–55 staff across the three museums and to divert a greater proportion of the budget (£1.4 million per annum) to the Museum's outputs.[67] There was widespread condemnation in both the specialist and the mainstream press, which raged throughout the autumn of 1993.[68] A new Projects Development Division, with Gillian Thomas (formerly Director of Eureka Children's Museum, Halifax) as Assistant Director,[69] replaced the Public Services Division, with the role of providing professional project management to major projects.[70] Research & Information Services became the Science Communication Division, acquiring Interpretation and Education. The Interpretation Unit gained a new mission: 'to promote greater public awareness and understanding of science through effective communication with the Museum's general visitors'.[71] Collections Division dropped the middle term 'Management', which was seen to allude to the practical concerns of conservation, documentation and storage that, from 1992, became the focus of a new subdepartment led by Suzanne Keene.[72] At this point, the experimental 'Projects Group' was dissolved; some of its members became curators and others staffed the new Projects Development Division. Marketing became 'Public Affairs'.

Throughout the 1990s, there was effectively a mixed economy of gallery production at the Museum, in which some were produced using variants of the 'Food for Thought' model, first under the Public Affairs Division and then under Thomas's department. These included the Basement galleries, newly conceived replacements for the Children's Gallery, and 'The Challenge of Materials' (1997), a gallery which started as a curatorial project, but which latterly came to be run by Heather Mayfield within the Projects Development Division using a modified version of the 'Food for Thought' approach. 'George III' (1993) combined the curatorial team of Alan Morton and Jane Wess with the external designer Alan Irvine. Of the others, 'Health Matters' (1994) was similarly transferred to the auspices of the Project Development Division. But, perhaps because of the seniority of Ghislaine Lawrence, the curator in charge, this project never ceased to be curatorial in spirit. 'Health Matters' was a determinedly rigorous response to the issues of science museum display, self-consciously setting out to question and rewrite the conventions of museum exhibitions. At the same time as it sought to use the kind of 'soft' social history visible in the Wellcome galleries, it also gestured to a harder social constructionist account of modern medicine. The gallery was informed by the latest research in the history of biomedicine, but it also aimed to produce for the visitor a highly engaging account of its subject, with dense contextualisation using specially shot and archive films, alternative narratives introduced by secondary objects, and – by the standards of the day – a highly informal verbal style for all panel text.[73]

The Science Communication Division

The engine of displaying contemporary science in the 1990s was what came to be known as the Science Communication Division, under John Durant, who had succeeded Robert Fox.[74] On Durant's watch, from 1992, the Science Museum Library was integrated with Imperial College Library, in fulfilment of one of the original aims of the Department under its previous name. Otherwise, he took the department in a very different direction from that originally intended. He interpreted the research role as establishing an alternative model, into the public understanding of science, not according to the initial aim of driving the quality of curatorial research. Britain was, in the early 1990s, at the height of the 'public understanding of science' (PUS) moment in the wake of the Bodmer Report of 1985. Durant skilfully positioned the Science Museum at the centre of this firmament by establishing the editorial office of the new journal *Public Understanding of Science* in his department. The PUS initiative, often in parallel with the visitor studies work rather than integrated with it, gave a welcome rationale and intellectual underpinning to the Museum's turn to market-responsiveness. The new emphasis of the Division also provided a clarification for the exercise of curatorial research. Under the guidance of Robert Bud in Collections, formal research policies were developed, which accelerated the existing tendency for curators active in research to turn away from the older engineers' history and connoisseurial approaches towards the universities for models of how to research and understand the collections and the cultures that produced them. A small but significant number of middle-ranking curators studied for doctorates, joining the Museum's research culture to the academic mainstream, but maintaining a focus on the material and visual culture of their subjects. This approach produced, amongst much else, Doron Swade's historiographical experiment of building Babbage's Difference Engine No. 2,[75] a scholarly collaboration with the Deutsches Museum and the Smithsonian's National Museum of American History, and the historical chapters of Morton and Wess's catalogue of the George III Collection.[76]

Robert Fox has since written about the missed opportunity in the 1990s of 'what could, and surely should, have been a constructive alliance between the scientific community and the heterogeneous world of historians, sociologists, anthropologists, and policy analysts' involved in PUS.[77] At the Science Museum, the lack of rapprochement was different; the failed union was between two groups with intellectually congenial backgrounds or approaches: the new staff pursuing the PUS agenda (or those undertaking empirical work into visitor attitudes and behaviour), and the curators who had inherited and developed the social historical view associated with the arrival of the Wellcome Collection. Here the Divisions really did divide.[78]

Durant directed his department's efforts more towards adults, whilst seeking to retain the family audience, and towards the Science Box series of small temporary exhibitions on contemporary issues, which ran from March 1992,[79] not towards interactive science displays. In his view, the changes that science museums were undergoing in the mid-1990s resulted, as he said, both from the worldwide success of interactive science centres and, rather aggrandisingly, from the 'pressure on museums of science to contribute more directly to the creation of a scientifically literate society' deriving from the worldwide public understanding of science movement.[80] The scale of his ambitions was fully congruent with Cossons's:

> The museum of science of tomorrow will be a very different sort of place from the museum of science today; and its differentness will reflect in part a redefinition of the nature of the museum itself.[81]

This model of science communication suffused with the spirit of the public understanding of science was more about scientific issues than explaining scientific principles, as was seen in the Science Boxes, with exhibitions on passive smoking and nanotechnology, for example. And it was separate from the object culture of the Collections Division, ultimately to their mutual disadvantage.

The Wellcome Wing and 'Making the Modern World'

The culmination of the separation between the two halves of the Museum's purpose came to public view in June 2000, when the Museum opened 'Making the Modern World' and the Wellcome Wing. This also marked the fulfilment of Cossons's desire, to the extent that it was affordable, to extend westward most of the way to Queen's Gate. The first of these galleries is a consciously 'neoclassical' gallery, showcasing 2,000 of the Museum's objects in a display that emphasises 150 significant 'icons' of science and technology within a highly worked set of contextual displays using the language of objects to illuminate both relevant narratives and the continuity and change of everyday things. Older galleries had often been developed on the naïve model of the rational visitor who was expected – if considered at all – to sequentially absorb every element of the display in turn. 'Making the Modern World', by contrast, was informed by the idea of visitors responding to the visual elements of the display, being drawn by the eye to enjoy what interested or pleased them. No one would see all the displays, but all objects would, over the course of time, be enjoyed by some people.[82] The Wellcome Wing, billed as 'a breathtaking theatre of contemporary science', in addition to an Imax theatre, included at its opening 'Who Am I?', a major biomedical gallery; 'Digitopolis', a gallery about digital technologies; 'Pattern Pod', a mathematical display aimed at preschool children; and 'In Future', a small floor devoted to imaginative multiplayer games on the future implications of scientific and technological change. Probably the most museologically novel in conception of all these innovative shows was the Antenna Rapid strand, which, by way of an arrangement with *Nature* magazine, was able to produce micro-exhibitions on scientific news stories as they hit the press.[83]

Illustration 5.2 'Making the Modern World' Gallery, November 2000

Behind the success of the Wellcome Wing displays was a new level of application of audience research and a broad definition of the kinds of learning the
Museum wanted to encourage. The learning specialists within the exhibition
teams deliberately went beyond the older audience advocate model, establishing
training programmes for content developers, working on 'message documents'
(succinct statements of what the teams wanted to say to visitors), undertaking

Illustration 5.3 'Digitopolis', one of the original Wellcome Wing galleries, looked at the implications of new digital technologies

evaluations, and advising on media and on visitors' existing knowledge and preferred learning styles.[84] These twin gallery openings expressed the dialectic at the heart of the Museum's enterprise in concentrated form: whereas the Wellcome Wing was a highly interactive and fast-paced space, 'Making the Modern World' was deliberately slow and contemplative; the Wellcome Wing was contemporary, and 'Making the Modern World' was historical; and, while the contemporary displays subordinated museum objects to gallery themes, 'Making the Modern World', however much it was informed by the latest scholarly debate, emphasised the objects themselves, and what they can mean to visitors.

Postscript

The Science Museum ventured into the new millennium with the dialectic between 'the voices of the past and the needs of the present and the future' all too apparently separate in the physical form of the new galleries. Behind the scenes, the separation still also took concrete form in the organisational structure that divided its professional staff into a Science Communication Division and a Collections Division. Both soon experienced dramatic contraction, as externally

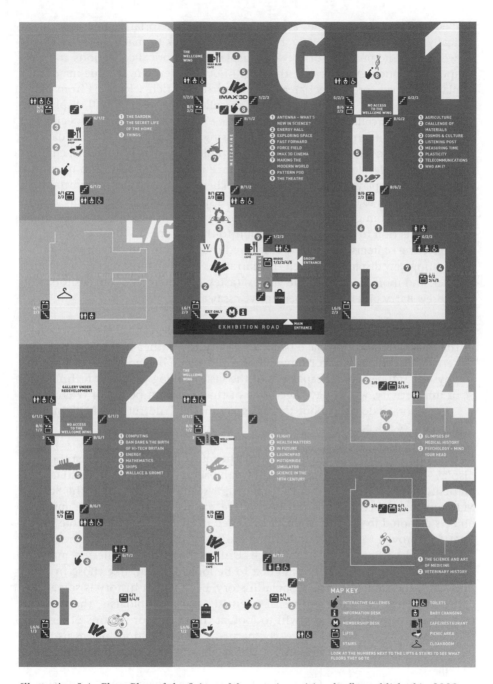

Illustration 5.4 Floor Plan of the Science Museum in a visitor leaflet published in 2009

funded contract posts came to an end and a new directorial regime introduced radically new priorities.[85] Nearly a decade on, we seek to make a new synthesis between past and present – or between displaying historical collections and contemporary science. Our answers, the stuff of our Masterplan, include major new displays of broad core subject areas with diverse and rich selections from many of our collections in 'treasury' galleries, so-called because they will display the 'treasures' of our collections New approaches to the contemporary, especially in recasting the 'Antenna' displays and addressing the major crisis of our day, climate change, also have a high profile. In all projects of significant scale, curators and exhibition developers work with audience researchers seeking to maximise the appeal of displays to target audiences. There is a new spirit of synthesis in the air as we seek to ensure that new galleries speak to both past and present. In the classic Museum galleries of the post-war period, this was a matter of using a specific collection to display the evolution to the present, an ideology on which the collecting of items was, in any case, predicated. In our radical reinvention we are creating new displays in which the narratives are both more diverse and much broader, and more attuned to the values, tastes and interests of our visitors. But, as Andrew Barry observed in a perceptive essay a dozen years ago:

> Although not entirely ignored, however, efforts to rethink the function of the museum collection have been secondary to a rather different set of concerns with questions of consumer choice and visitor behaviour. It has been the physical and perceptual capacities of the visitor, along with new technologies of interactivity, which have been one of the dominant points of reference in contemporary debates about the design, function and future of the modern museum of science.[86]

It is in taking up this challenge to rethink the total collections – not just those on display – that the Science Museum will finally achieve synthesis in its so far permanent dialectic. After LaunchPad opened, it seemed to some that the Science Museum might become one big interactive science centre, with the museum objects demoted to the role of props or exiled in their entirety to remote stores. That fear proved to be unfounded, but we are still grappling with the role of objects, and perhaps always should be. Unlike those in art galleries, many of our objects are not primarily designed to be looked at but, like those in other museums, they are evidence of social history. Even if that history is sometimes concerned with unfamiliar tribes of scientists and technologists, the personal comparison can stimulate the will to look at the object for reasons other than the aesthetic. When we are brave, the nature of our collections places us in the vanguard of Museum practice and of the kinds of stories museums can set out to tell.

Unavoidably, this object-focussed enterprise must largely centre on display in the Museum. Our offer to the visitor will always be a palimpsest of what has gone before, unlike, for example, Mannheim's Landesmuseum für Technik und Arbeit, which was created from nothing in the 1980s and 1990s. The brave modernism of 'knowing, making and using', sweeping away the legacy of the Museum's history, now looks naïve. But, as we live with the inevitable multi-museum, it is essential that we face up to the fact that museums are about objects as well as experiences; we must place effective use of material culture at the heart of our enterprise. That requires us to develop our appetite for novel ways of presenting objects, a preparedness to fail, and the will to experiment again. We need to embrace the multiplicity of meanings that audiences can bring to objects, as well as take away from them. We are fortunate that objects are fickle, enabling different narratives in different contexts for different people; is the V2 Rocket part of the history of rocketry, of ballistics, of war, of Nazism, or the local history of South London? Well, of course, it is all these, and it is important that the institution be confident to accept the multiple meanings that visitors extract from our displays.

It is easier to see the potential of the collections for research. Those versed in the modern manners of science and technology studies often find that the particularity of individual objects makes them a natural focus for historical research. The changing fashions of their display, as this volume shows, also provide crucial evidence for the study of the public culture of science and technology. Serious research into the collections is not optional; it stimulates the Museum's imagination and generates the new, engaging narratives that we need to serve both our audiences and the potential of the collections. But we need to continue to work to make sure that this potential research is enabled, and that its results are fed back into the public-facing enterprise. Equally, it is simply perverse if the Museum turns away from the enthusiasts who come with existing knowledge and passion for objects, whether the rejection is rooted in professional snobbery or in a focus on the numerically greater family audience.[87]

We are fortunate to live in the midst of a media revolution that has opened up the multiple opportunities of websites, blogging, podcasts and the rest. Here, even more than via conventional publication, the total collections will be made accessible.

The division of specialised labour that was accelerated in the 1980s and 1990s has undoubtedly ensured real development in professional standards on both sides of the dialectic that this chapter describes. The 'new spirit of synthesis' I alluded to needs to be nurtured to deliver the goods for audiences and collections. Ultimately, the core elements of our purpose – as Cossons described them, 'the wish to preserve and record and the requirement to educate' – do not have to be in opposition if the collections are valued and the institutional understanding of learning is sophisticated and shared between the Museum's professional groups.[88]

In the end, the scale of change in lived experience is difficult to evaluate against the eras that we know only indirectly. But it is the recent past that makes the conditions of possibility for what we can do in the present and near future. In this sense, the legacy of Neil Cossons is still crucial for the Science Museum today. At the heart of the paradoxes we continue to curate is the collection and what it can stand for. To integrate it fully into our enterprise continues to be the greatest challenge for the staff of the Museum.

Notes

1. Neil Cossons, 'Report of the Director and Accounting Officer', *National Museum of Science & Industry Review* (London: Science Museum, 1993), p. 9.
2. A note on my approach: 'parallax error' refers to my participation in much of what I describe, initially as a very junior member of staff. The examples chosen reflect the developments I have witnessed most closely. A different author might well have produced a very different interpretation, but I have made a historiographical decision to retain the particularity of this insider's view. I have checked and amplified by referring to contemporary documents, but not used oral history to balance my account with others.' Several other participants have read the first draft and helped me to correct errors of fact and emphasis: Sandra Bicknell, Neil Cossons, Graham Farmelo, Derek Robinson. For the record, I started as a Museum Assistant in the Museum's Lecture Service (1982–1985), worked mainly in the medical curatorial department at various grades between 1985 and 2000, and became a senior manager within the curatorial department in 2000. I worked in the teams that produced 'Health Matters' and 'Making the Modern World'. My account focusses on the Science Museum building in South Kensington; space forbids detailed consideration of the National Railway Museum, National Museum of Photography, Film and Television (now National Media Museum), or the stores at Wroughton or Blythe House.
3. See Chapter 7 in this volume.
4. John Symons, *Wellcome Institute for the History of Medicine: A Short History* (London: The Wellcome Trust, 1993), pp. 46, 48–50.
5. Frank Greenaway, 'Future of the Wellcome Collection of the History of Medicine', 13 June 1972, SMD, SCM/1999/0519/001.
6. Science Museum Annual Report for 1980, p.3.
7. See, for example, Richard J. Harris, 'Empathy and History Teaching: An Unresolved Dilemma?', *Prospero* 9 (2003), pp. 31–8. It is debatable whether older dioramas in the agriculture gallery, for example, were also explicitly conceived as social historical. For an account of these, see Jane Insley, 'Little Landscapes: Agriculture, Dioramas and the Science Museum', *Icon: Journal of the International Committee for the History of Technology* 12 (2006), pp. 5–14.
8. Ghislaine Skinner, 'Sir Henry Wellcome's Museum for the Science of History', *Medical History* 30 (1986), pp. 383–418.
9. 'The Science Museum would become the natural centre for research on the physical evidence of the history of medicine and allied sciences... The establishment of a continuing group of research workers is greatly to be desired'; Frank Greenaway, 'Future of the Wellcome Collection of the History of Medicine', 13 June 1972, SMD, SCM/1999/0519/001, 3.
10. For example, Christopher Lawrence, 'Physiological Apparatus in the Wellcome Museum 1. The Marey Sphygmograph', *Medical History* 22 (1978), pp. 196–200.
11. The Wellcome Institute's lead was sometimes seen as a model: J.K. Crellin and J.R. Scott, *Glass and British Pharmacy 1600–1900. A Survey and Guide to the Wellcome Collection of British Glass* (London: Wellcome Institute of the History of Medicine, 1972).
12. Brian Bracegirdle's catalogue of the microscope collection, researched after his retirement in 1990, was published as a limited edition CD-ROM in 2005: Brian Bracegirdle, 'A Catalogue of the Microscopy Collections at the Science Museum, London' (London: Trustees of the National Museum of Science and Industry, 2005).
13. Ghislaine Lawrence, 'Charles Drew's Profound Hypothermia Apparatus (1960)', *Lancet* 358 (9280) (2001), p. 514; Lawrence, 'Hospital Beds by Design: A Socio-Historical Account of the 'King's Fund Bed', 1960–1975', unpublished University of London PhD thesis, 2002; Robert Bud, *The Uses of Life: A History of Biotechnology* (Cambridge: Cambridge University Press, 1993); Robert Bud, *Penicillin: Triumph and Tragedy* (Oxford: Oxford University Press, 2007).

14. David Follett, *The Rise of the Science Museum under Henry Lyons* (London: Science Museum, 1978), pp. 103–5.
15. Science Museum Annual Report for 1953, p. 5. Staff lists.
16. Science Museum Annual Report for 1952, p. 6.
17. Science Museum Annual Report for 1952, p. 6. For SFA, see T. Boon, *Films of Fact: A History of Science in Documentary Films and Television* (London: Wallflower Press, 2008), pp. 115–17.
18. Science Museum Annual Report for 1953, p. 24; 1955, p. 6. Hartley retired in 1957.
19. Described in Chapter 4; see also Science Museum Annual Report for 1955, p. 1.
20. Science Museum Annual Report for 1952, p. 1, my emphasis.
21. They held a conference and sent a questionnaire to 350 schools; three of Vernon's students conducted research in this area. Two concentrated on the effectiveness of the Children's Gallery with existing visitors, and the other looked at the encouragement of school parties. Science Museum Annual Report for 1954, p. 2.
22. Science Museum Annual Report for 1955, pp. 6–10.
23. Aubrey Tully, 'An Informal History of LaunchPad', Chapter from PhD thesis uncompleted at the time of the author's death, 1992, p. 1, SMD, Z 250/3. Tully had been one of the Education Service staff involved in the development of the project.
24. It is worth noting that William O'Dea, the curator who had introduced novel design ideas into the Science Museum, had set the tone of the Ontario Science Centre as its first director.
25. Tully, 'An Informal History of LaunchPad', pp. 2–7.
26. See Chapter 4 in this volume.
27. Neil Cossons, personal communication, September 2009.
28. Neil Cossons, ed., *The Management of Change in Museums: Proceedings of a Seminar Held at the National Maritime Museum, Greenwich, on 22nd November 1984* (London: National Maritime Museum, 1985). Previous Science Museum Directors, including Margaret Weston, had been aware of the tension in efficient expenditure of the Grant in Aid (i.e. government funding) between the need to employ sufficient staff to deliver the Museum's obligations and to retain enough for expenditure on outputs; she aimed at no more than 67 per cent of the total budget on salaries, personal communication from Derek Robinson, 14 Aug 2009.
29. SMO 60/88, SMD, Z 340/2.
30. SMO 6/87, SMD, Z 340/1.
31. See NMSI Corporate Plan 1992–1997, vol. 1, pp. 11–12.
32. See, for example, SMD, SCM/1990/0377 and /0378.
33. Derek Robinson, 'Gallery Development Plan', Memorandum, 2 November 1987.
34. Neil Cossons, 'Adapt or Die: Dangers of a Dinosaur Mentality', *Listener*, 16 April 1987, pp. 18–20. In his view, the Science Museum and other national organisations 'were seen to harbour people who were against change but at the same time unable to articulate an intellectual justification for what they were and why they did what they did. They were thus, by implication, against a more open and audience-focussed approach', Neil Cossons, personal communication, September 2009.
35. Neil Cossons, 'Scholarship or Self-Indulgence?', *RSA Journal* 139 (1991), pp. 184–91, on 185–6. See also his address to the Museums Association in 1982, published as: 'A New Professionalism [Address to Museums Association, 1982]', in Gaynor Kavanagh, *Museum Provision and Professionalism* (London: Routledge, 1994), pp. 231–6.
36. See Harold Perkin, *The Rise of Professional Society* (London: Routledge, 1989), p. 378.
37. 'Scholarship or Self-Indulgence?', p. 187.
38. SMO 35/87, 5 May 1987, SMD, Z 340/1.
39. Assistant Director Mark Pemberton from 1 July 1987.
40. These latter were services recently taken in-house from the Property Services Agency (PSA).

41. Touche Ross Management Consultants, 'Review of Collections Management, 22 Feb 2008', SMD, POL184/185/1.
42. GN 3/90, 9 January 1990. SMD, Z 340/2.
43. Touche Ross Management Consultants, 'Review of Collections Management, 22 Feb 2008', 1, pp. 47–9.
44. See, for example, NMSI Corporate Plan, 1990–1995, draft copy, p. 13, SMD, Pol/237/238, pt. 3.
45. NMSI Corporate Plan, 1990–1995, draft copy, p. 9, SMD, Pol/237/238, pt. 3.
46. GN 3/90, 9 January 1990. SMD, Z 340/2.
47. SMO 14/87, 19 Feb 1987, SMD, Z 340/1.
48. Derek Robinson, 'Gallery Development Plan', Memorandum 2 November 1987, p. 1.
49. Derek Robinson, 'Gallery Development Plan', Memorandum 2 November 1987, p. 2. He had moved into the new post on 1 March (Derek Robinson, personal communication, 14 August 2009).
50. Cossons et al., 'Gallery Development Plan – Thematic Principles', January 1989, p. 3.
51. Cossons et al., 'Gallery Development Plan – Thematic Principles', January 1989, p. 1, my emphasis.
52. Ibid., p. 1.
53. Ibid., pp. 5–6.
54. Corporate Plan 1991–1996, vol. 1, p. 8.
55. NMSI Corporate Plan, 1990–1995, draft copy, p. 35, SMD, Pol/237/238, pt. 3.
56. The old Electrical Power Gallery had been succeeded by the Exploration exhibition in the mid-1970s, but there was clearly a will to redisplay the subject matter; ibid., p. 36.
57. Loc. cit.; the name is given in the following year's report.
58. NMSI Corporate Plan, 1991–1996, vol. 3, iii.
59. Cossons et al., 'Gallery Development Plan – Thematic Principles', January 1989, p. 6.
60. GN 38/89; NMSI Corporate Plan, 1990–1995, draft copy, p. 18, SMD, Pol/237/238, pt. 1.
61. Sandra Bicknell, 'Divisions of Labour: Interpreting Research and Researching Interpretation'. Unpublished paper presented to the Social History Curators Group Annual Study Weekend, 7 July 1994.
62. The initial project team was outlined in the April 1988 circular SMO 19/88, SMD; the gallery opened in the following year. This project is the subject of the ethnographic study, Sharon Macdonald, *Behind the Scenes at the Science Museum* (Oxford: Berg, 2002).
63. With the exception that Alan Morton, Curator of Modern Physics, had begun collecting supermarket laser scanning equipment.
64. In this way, the Museum was able to take advantage of the diversity of approaches available in the growing design sector, whilst seeking to reduce the fixed costs of a permanent in-house design team, Neil Cossons, personal communication, September 2009. This was far from being the first occasion on which external companies had designed Science Museum exhibitions; an immediate precedent was Casson Conder's design for the 1986 Industrial Chemistry Gallery, but here the principal relationship had been with the exhibition's curator, Robert Bud. External designers had been involved with temporary exhibitions as early as Moholy Nagy's design for the 1935 'Empire's Airway' exhibition, and Misha Black contributed to the 1948 'Darkness into Light' lighting exhibition, personal communication from Derek Robinson, 14 August 2009. See also Chapter 8 in this volume.
65. Eilean Hooper-Greenhill, *Museums and Their Visitors* (London, Routledge, 1994, p. 9).
66. GN 38/93, 17 November 1993, SMD, Z 340/2.
67. This restructuring saw the departure of further curators, including John Becklake, long-time Keeper of Engineering.
68. SMD, SCM/1994/0626/1.
69. NMSI Corporate Plan, 1993–1998, vol. 1, p. 5. Thomas left in 1996 to become Chief Executive of the interactive centre, At-Bristol.
70. Terry Suthers, who had been Assistant Director of the Public Services Department since September 1987, left in February 1992; SMO 478/87, SMD, Z 340/1.; GN/3/92, SMD, Z 340/2.

71. Bicknell, 'Divisions of Labour', p. 3.
72. GN 7/92, 26 Feb 1992, SMD, Z 340/2.
73. See Tim Boon, 'Histories, Exhibitions, Collections: Reflections on Medical Curatorship at the Science Museum after "Health Matters"', in Robert Bud, Barney Finn and Helmuth Trischler, eds *Manifesting Medicine: Bodies and Machines* (Reading: Harwood Academic Publishers, 1999).
74. Fox had departed in December 1988, only nine months after he joined the Museum, to take up the Chair in the History of Science at Oxford.
75. Doron Swade, *The Cogwheel Brain* (London: Abacus, 2001).
76. Alan Q. Morton, Jane A. Wess, *Public and Private Science: King George III Collection* (Oxford: Oxford University Press, 1993).
77. Robert Fox, 'History and the Public Understanding of Science: Problems, Practices, and Perspectives', in M. Kokowski, ed. *The Global and the Local: The History of Science and the Cultural Integration of Europe. Proceedings of the 2nd ICESHS* (Cracow: Poland, 2006), p. 175, http://www.2iceshs.cyfronet.pl/2ICESHS_Proceedings/Chapter_8/RE_Fox.pdf (accessed 5 May 2009).
78. See Ghislaine Lawrence, 'Rats, Street Gangs and Culture: Evaluation in Museums', in Gaynor Kavanagh, ed. *Museum Languages: Objects and Texts* (Leicester: Leicester University Press, 1991), pp. 11–32; also see John Durant's discussion of the issues: Durant, 'Science Museums, or Just Museums of Science?', in Susan Pearce, ed. *Exploring Science in Museums* (London: Athlone Press, 1996), pp. 148–61.
79. See, for example, NMSI Corporate Plan 1994–1999, p. 12; John Durant, 'Presenting Contemporary Science: The Science Box Programme', *The National Museum of Science & Industry Review 1993*, 30–2. At this point, curators' temporary exhibition activity was concentrated on the two new 'Technology Futures' cases showcasing new devices, and a selection of gallery upgrades.
80. Durant, 'Science Museums, or Just Museums of Science?'.
81. Durant, 'Science Museums, or Just Museums of Science?'.
82. A model close to that advanced for cultural consumption in general in Michel De Certeau, *The Practice of Everyday Life* (Berkeley: University of California Press, 1984).
83. The Antenna displays were conceived on a magazine model; the Rapids were news stories; Antenna Features were like magazine articles, and followed the style and experience of the Science Boxes. There was also a BBC science news feed, which might be seen as 'stop press' items.
84. Jo Graham and Ben Gammon, 'Putting Learning at the Heart of Exhibition Development: A Case Study of the Wellcome Wing', in E. Scanlon, E. Whitelegg and S. Yates, eds *Communicating Science: Contexts and Channels* (London: Routledge in association with The Open University, 1999).
85. Comparisons with previous eras are difficult to make, because of the deliberate creation of new specialist roles, as I have described. We might note, however, that the sum of curators and content developers c. 2003 was less than half the number of curators in 1980.
86. Andrew Barry, 'On Interactivity: Consumers, Citizens and Culture', in Sharon Macdonald, ed. *The Politics of Display: Museums, Science, Culture* (London: Routledge, 1998), pp. 101, 112.
87. Hilary Geoghegan, 'The Culture of Enthusiasm: Technology, Collecting and Museums', unpublished PhD thesis, University of London, 2008.
88. 'Learning can be defined as the process of active engagement with experience. It is what people do when they want to make sense of the world. It may involve the development or deepening of skills, knowledge, understanding, awareness, values, ideas and feelings, or an increase in the capacity to reflect. Effective learning leads to change, development and the desire to learn more', Jean Franczyk, 'Defining, Planning and Measuring a Life-Enhancing Experience', internal NMSI policy paper, February 2009.

6

Waves of Change: How the Science Museum's Library Rose, Fell and Rose Again

Nicholas Wyatt

The Science Museum Library has always held an anomalous position within the Museum; it has been a library for the Museum and a library for the nation, and it has acted as the main library for Imperial College. This has created tensions as policies have shifted in response to internal and external pressures. This chapter shows how these tensions developed and how the Library responded.

An Educational Library transformed

In 1882 the Science and Art Department organised the Committee on Advice and Reference to investigate the educational collections of the South Kensington Museum. These had begun in 1854 when the Society of Arts organised a special loan exhibition of educational books and apparatus from throughout the world, held in St Martin's Hall, Long Acre. After it closed many of the exhibits were given to the Society of Arts, and in 1857 handed over to the Science and Art Department and displayed as the Educational Museum in temporary buildings adjacent to the new South Kensington Museum. By 1867 the printed collections, now known as the Educational Library, had grown though gifts and purchase into a major resource. In 1883 there were 45,000 works in this library, and it had a thriving readership including large numbers of schoolteachers.

By 1883, the Committee had concluded that a new science library was needed. The seedcorn for the new Science Library would consist of two existing libraries: the South Kensington Museum's Educational Library and the library of the Museum of Practical Geology. It recommended:

> The books relating to science in the Educational Library to be amalgamated with books brought from Jermyn Street, and formed into a Science Library, special facilities being afforded to the Professors of the Normal School to have out on loan such as they require... the books to be housed for the present in the existing Educational Library reading rooms and to be open to the public as at present, so long as there is room after providing for the students of the

Normal School of Science... No additions to be made to other classes of books in the Educational Library...[1]

The books from Jermyn Street came from the Library of the Museum of Practical Geology. This had originated in 1843 when the geologist Sir Henry de la Beche gave his valuable collection of scientific books to the nation. These formed the nucleus of a library used by the museum and the Geological Survey of Great Britain. To meet the demands of the School of Mines, founded in 1851, the library developed into a general library of science. Its collection grew by donation, purchase and exchange until 1883, when 9,730 works no longer deemed useful to the survey were transferred to the new Science Library. These included Agricola's *De Re Metallica* (1556), originally acquired in 1844, Hooke's *Micrographia* (1665) and complete runs of the *Philosophical Transactions of the Royal Society of London* and proceedings of European scientific and engineering institutions.

Scientific material from the Educational Library included early teaching material and a significant number of scientific books such as Guericke's *Experimenta Nova de Vacuo Spatio* (1672). The reconfiguration of the Educational Library into the Science Library took several years. In 1886 a complete set of specifications and patents of the British Patent Office was added. Books on art were transferred to the Art Library, and by 1888 nearly all the educational books had been moved out to form a new departmental library in Whitehall.

In 1891 the first catalogue devoted solely to the Science Library was published, and from then on the collections accumulated and supplementary lists of holdings were issued. Each entry from the printed catalogue was mounted onto card to form a basic card catalogue. During this period reader numbers slumped as educational material was transferred out of the Library.

The focus on collecting became the needs of the Normal School, now renamed the Royal College of Science, as the Librarian, Lionel Fulcher, noted in 1903: 'The aim of the Library at the present time is... to acquire mainly books on subjects now taught at the Royal College of Science, that is pure science, with mining and metallurgy, books on applied science being purchased as far as is considered necessary by the Professors.'[2]

The periodical question

During this period there was a significant expansion of scientific teaching in British universities, with new natural science faculties producing rising numbers of qualified scientists to work in industry and state-funded research facilities. But, while this new scientific culture stimulated research, it was becoming dependent upon the growing infrastructure of scientific publishing. Journals were proliferating and becoming more specialised. In 1894, the President of the Chemical Society, H.E. Armstrong, observed that:

chemical literature is fast becoming unmanageable and uncontrollable from its very vastness. Not only is the number of papers increasing from year to year, but new journals are constantly being established. Something must be done in order to assist chemists to remain in touch with their subject and to retain their hold on literature generally...[3]

The 'periodical question' would vex the Science Museum Library throughout the twentieth century:

> The results of scientific discovery are made known almost invariably in the papers of some periodicals, very rarely in book form. In text-books original matter is out of place or dangerous and no scientific man would publish new work in such a medium. Hence in a Science Library it is of the utmost importance that periodical literature should be adequately represented.[4]

The value of a paper may not be represented by the journal it is in, and Fulcher cites the 'striking' example of Mendel's Law, which had been first published in an 'obscure periodical' in 1866[5] and overlooked until recently. But 'the funds at the disposal of the Library are barely sufficient to meet the growing demands. The average sum spent annually on the Library is about £750, of which the periodicals are estimated to cost about £450.'

The Library was facing other challenges – its staff faced problems in cataloguing and preparing catalogue cards for all its new acquisitions. The Library was

Illustration 6.1   Science Library Reading Room, 1926

becoming overcrowded and in its annual report stated that 'it will be impossible to add many more volumes in its present condition.'

The British Science Guild, founded in 1904, highlighted the importance of scientific and technical information as a commodity that needed to be managed and exploited. For effective research and development there was an urgent need for co-ordinated and timely access to information about the latest research, wherever in the world it was being published.

The Science Library would soon be in a position to address this problem. Its relationship with the Royal College of Science was strengthened when in 1907, after many changes in location within the South Kensington Museum, it moved to the College building in Imperial Institute Road. The books were reclassified and transferred to spacious accommodation. The new reading room opened in 1908, with space for ninety-four readers. The building was linked directly to Southern Galleries occupied by Science Museum collections. The following year, 1909, saw the Science Museum formally gain its independence and the Science Library[6] became its responsibility.

Developing a national role

The Library's collections continued to grow, and by 1912 it held more than 100,000 volumes and was receiving some 570 periodical titles. Its new premises included a book store sufficient for another twenty years of growth. The Library compiled specialised bibliographies and guides to literature; in 1912, for example, it published a list of works on aeronautics to coincide with an aeronautical exhibition in the Museum. Between 1908 and 1914 a valuable collection of historical and modern literature on aeronautics was donated by P.Y. Alexander. In 1913, provision was made to collect portraits and scientists' original manuscripts, and in 1914 sixty-five ships' plans were presented by Harland & Wolff.

Concern about access to periodical literature increased, and the need for an efficient national approach to co-ordinated access to scientific information was thrown into sharp relief by wartime research and a perceived lack of co-ordination of scientific information.

In 1915 the government created the Department of Scientific and Industrial Research (DSIR) 'to promote and organise scientific research with a view especially to its applications to trade and industry.'[7] It was instrumental in the development of research organisations and regional technical libraries and information bureaux. The relationship between scientists and libraries was changing – there was a need to deliver information directly to scientists at their workplace, rather than expecting them to travel to libraries to refer to the literature they needed.

Libraries were to be the new laboratories of technical progress, as the Library Association, the professional body for librarians, proclaimed in 1918: 'arsenals of scientific and technical information will become, nay have become, as necessary as arsenals of war-like materials, and if steps are not to be taken promptly we shall be as little prepared for peace as we were for war.'[8] The chemists of the Faraday

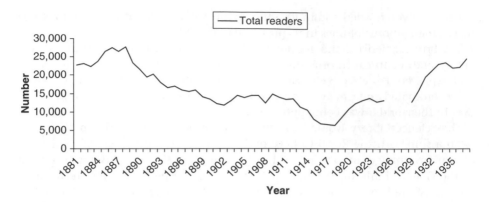

Graph 6.1   Science Library reader numbers 1882–1938

Society 'launched a call for rationalised system of scientific information which would serve both economic and military ends.'[9]

The Library responded to these calls with a combination of vigorous collecting, an extended loan service and a far-reaching programme of indexing the growing numbers of periodicals:

> Efforts are being made to obtain as many of the most important works on pure and applied science which were published during the war, but the high costs of books and of binding is making it difficult to maintain the library at its proper standard, and to make good the leeway of previous years as quickly as possible.[10]

By 1920 the Library was lending material to staff in the Science Museum, Imperial College, other neighbouring institutions and the Geological Museum in Jermyn Street. The numbers of readers were increasing and 12,000 were recorded in 1920. Numbers increased as collecting gathered pace in the following decade to regain the totals achieved in the 1880s.

The grand vision

The Library now held 125,000 volumes and had a staff of fourteen. During the 1920s, under the Director Sir Henry Lyons, concerted efforts were made to increase the Library's holdings, especially in periodicals, in order to make it the central resource for scientific literature in Britain. By 1925 the Library's annual expenditure on books and periodicals had increased from £800 to £1,000, a size-able proportion of the Museum's total purchase grant of £2,500.[11]

Samuel Clement Bradford, who had worked his way up through the ranks since joining the Museum in 1898, became head of the Library upon the retirement of Fulcher in 1925, initially as Deputy Keeper and subsequently as Keeper from 1930 to 1935. Bradford had gained a DSc in chemistry through part-time study and developed his knowledge of librarianship 'on the job'. He initiated a programme of exchanges and presentations, after the British Empire Exhibition of 1924 showed 'with what curious ease one could obtain the gift of enormous numbers of publications from all over the world merely by asking for them.'[12] Staff combed the *World List of Scientific Periodicals* for additions and 1,400 more serial titles were added during 1924.

Illustration 6.2   Samuel Clement Bradford, early 1930s

External lending was extended in 1923 to staff of government departments such as the DSIR, the Admiralty and the Meteorological Office. This was broadened in 1926 to many other institutions after the postal loans service was inaugurated. This enhancement of the service had been approved by the Board of Education, which wanted to gain a better return on the money spent on it, but on the understanding that the new service should be regarded as an experiment and should not entail an increase in the cost of running the Library or in its staffing numbers. The experiment was an immediate success and developed into a major function, with the Board's original condition apparently forgotten (or ignored); in 1926 3,000 loans were recorded, rising to 10,954 in 1930 and 21,000 in 1935.

The significant increase in the size of the collections had an impact upon storage. In 1924 to 1925 rolling stacks were installed, which 'relieved the congestion considerably.' This enabled staff to bring out the complete set of British patent specifications and make them available for reference in the Reading Room. There continued to be financial problems and more funds became necessary for the purchase of material. By 1927 the Library held 195,000 volumes, but, due to a lack of funds, 900 periodicals were cut between 1927 and 1928. The situation had become desperate, Imperial College was complaining and a crisis for the loans service was looming.

The Science Library had developed its own decimal classification system, but Bradford, who was one of the founders of the British Society for International Bibliography, replaced it in 1928 with Universal Decimal Classification (UDC) to arrange the book stock and card index. UDC had been originally developed in 1895 to create a universal bibliography and indexing language and was used to organise subject bibliographies and specialised European libraries. The Science Museum Library became the principal agency in Britain for the dissemination of information about UDC. Bradford believed that by using UDC a World Bibliography could be compiled economically with unskilled labour, so he masterminded the compilation of a vast card index that eventually amounted to nearly four million cards to document the increasing numbers of scientific and technical journal articles. Staff spent considerable time cutting up printed bibliographies and indexes and pasting entries onto cards. But this was a project too far for the Library; the index was little used, took up valuable space and staff time and would eventually be dispersed to other libraries after his death.

Donald Urquhart, who became Assistant Keeper in 1938, described work at the Library critically. It used the British Museum Library cataloguing code to create catalogue cards, which meant that organisations were filed under the name of the city or country in which they were located. The Library also had its own shelving arrangement for periodicals whereby each title was given a unique alphanumeric code; periodicals issued by organisations were allocated a code based upon the city and organisation name. These made it difficult for users to find transactions of societies on the shelves without knowing the catalogue entry. Urquhart called Bradford 'a rather fussy old man' and pointed out that he had no official approval for his aims for creating both a comprehensive index and a comprehensive accessible collection of scientific literature. Bradford had placed a brass plaque on the entrance of the Library stating that this was *The National Science Library* and had headed stationery printed. The Board of Education immediately ordered that name to be changed to Science Museum Library. Bradford relented, but the heading became *The Science Museum Library, The National Library of Science and Technology.*[13]

'A policy of aggrandisement'

In 1924 the Board of Education had appointed Sir Frederic Kenyon to chair the Public Libraries Committee to examine the adequacy of the public library provision in England and Wales and its relation to other libraries, including the Science Library. The subsequent Kenyon Report was published in 1927. Bradford and Lyons lobbied committee members, and were supported by sympathetic scientists and engineers such as Henry Tizard. Their influence is clear in the final report, which recommended that the Science Museum Library should become a central library of science for researchers and the public by developing comprehensive science and technical collections, by creating a detailed subject-matter index and by supplying information and bibliographies to research workers worldwide. The cost was estimated at an additional annual sum of £3,500 for purchases (making a total annual grant of £4,700) with additional funding for staff. The committee stressed that the Library's essential functions should be the provision of publications for reference, abstracting and loan. Their recommendations were endorsed by the Royal Commission on National Museums and Galleries in its final report of 1930.

The acquisition costs for the Library could have been reduced if it had been able to receive British scientific publications by copyright deposit. The Board of Education proposed this to the Royal Commission in 1928, but the idea was vigorously opposed by Sir Frederic Kenyon, who was Director of the British Museum. He argued that it could result in a divided national copyright library. Furthermore, the Science Museum Library loaned its material, whereas the British Museum had always attached great importance to preserving material for posterity by keeping it on site.[14]

The scientific world took a keen interest in these developments, and, at the instigation of the Association of Scientific Workers, a conference of sixteen scientific societies was held in 1930, with discussions led by Alan Pollard of Imperial College, himself a pioneer of subject classification. It concluded that a subject-matter index to existing knowledge was an urgent necessity and that proper provision should be made for the Science Museum Library to carry out the Commission's recommendations efficiently and quickly to avoid the 'increasing bibliographic chaos.' The Commission's findings and the wider scientific community's deputations helped to secure an increase in the Library's purchase grant to £2,700 in 1930 and to £3,000 a year later.

In late 1931 the Science Museum Advisory Council appointed a subcommittee to deal with the Library question, under the chairmanship of the Rector of

Imperial College, Henry Tizard. It concluded that there was a real and pressing need for a national science library, that it would constitute an important link in the chain of industrial development and that the Science Museum Library should receive recognition as the National Library of Science. Its recommendations were similar to those in the Kenyon Report, adding that the Library should lend books and periodicals to individuals through approved institutions, including the research laboratories of industrial firms. To fulfil these functions a purchase grant of £6,000 would be necessary.

The President of the Board of Education, Lord Irwin, promised to review these proposals, but he stressed that in the stringent financial climate at the time it would be impossible to secure further funding for the Library, either for purchases or for staff. The Advisory Council referred the report to the Standing Commission on Museums and Galleries (set up as a permanent body by the Royal Commission in its final report in 1929), which was also considering a Royal Society report concluded that the foundations of a National Science Library existed at the Science Museum.

The Standing Commission's report was passed to the Board of Education, which came to entirely different conclusions. It did not accept that the Science Library should be recognised as a Central Library of Science, believing it would be better to regard it as just one of a number of scientific libraries, the functions of which called for review and coordination. There was no belief in a world index and no justification for spending money on it, and they recommended that the Library should limit itself to an information service based upon its own resources. They were also dubious about lending library stock to industrial firms. The Board strongly believed in a coordinated approach by several libraries to be led by the DSIR.

The Board had become increasingly uneasy with the direction the Science Library was taking. Its Director of Establishments, Robert Wood, noted in his confidential report to his Permanent Secretary that:

> For some years the Science Library has been pursuing a policy of aggrandisement with a view to becoming a comprehensive and all-embracing Science Library covering every field of scientific literature. It has at the same time sought to develop, in addition to its primary function of reference user, an extended loaning service and an information service. This policy has been initiated and followed up with very little, if any, control from the Board who are the responsible Department. It has grown gradually without any reference to the Department.[15]

Wood referred back to the experiment to increase loans and complained that it had been extended without authorisation. He concluded that 'The Librarian is, I think still working in the atmosphere of World Indices, the Library a complete repository of world scientific literature etc. in accordance with the proposals for further development put forward recently by the Advisory Council. The Advisory Council's suggestions were turned down on practically every count by the Standing Commission.'[16]

Wood made further criticisms, especially with respect to the increased staff numbers, with both Lyons and Bradford asking for far more than could be justified: 'One year we are told that a staff of twenty-one will meet ultimate needs, five years later we are told that the figure should be forty-nine.' By 1933 the Library staff complement had actually increased to thirty-two, at which level it remained until the Second World War.

Despite Wood's criticisms, there is little evidence that they slowed down Bradford's momentum. The only effect was that the argument that the Library should be recognised as a National Library of Science was dropped. Nearly 80 per cent of volumes added to the collections were being presented, and these would have been difficult to stop. By the time of Lyons's retirement in 1933 the Library had virtually, but not officially, become the national library for science and technology. Its stock had increased to nearly a quarter of a million with almost 9,000 periodical titles taken. By comparison, the Library of the Deutsches Museum held 152,000 volumes.[17]

'A giant in disguise'

During his tenure Bradford also collected historic scientific books to fill gaps in the Library collections, culminating in a copy of the first edition of Newton's *Principia* (1687) purchased in 1937. Staff continued to go to great lengths to pursue foreign publications, and in 1934 H.P. Spratt 'spent his leave'[18] visiting forty-five libraries and museums in northern Europe and the Soviet Union to elicit further contributions. Bradford promoted his ideas extensively, and in 1935 he gave a paper to the 13th Conference of Documentation, held in Copenhagen, where he concluded that it was 'abundantly clear that the establishment of a central library of science and technology would alleviate the disabilities under which scientific and technical work is carried out at present, while the additional cost of such a library would be small compared with the advantage gained.'[19]

Bradford is still well known for his law concerning the distribution of journal articles within subjects, which he published in 1934.[20] Bradford's Law states that journals in a single field can be divided into three parts, each containing the same number of articles. A core of journals on any subject, relatively few in number, produce approximately one-third of all the articles; a greater number of journals contain the same number of articles as the first and a still greater number contain the same number. Bradford expressed this relationship as $1:n:n^2$. His law of scatter became a cornerstone in the new field of bibliometrics and is still a tool used by librarians to determine core collections for any given subject.

Bradford retired in 1937, the same year that the Library was recognised as the information centre for science and technology by the Institut International de Documentation. His legacy was enormous; he pioneered the statistical study of information and, despite opposition from the Museum's funding department, he developed the Library's collections into a significant resource. His propaganda efforts also laid the foundations for later developments in the national provision for scientific information. By 1939 the Library had the largest collection of scientific periodicals in Europe. Urquhart later admitted: 'The fussy little man I once met was really a giant in disguise.'[21]

Bradford would later assimilate his life's work into his book *Documentation*, published in 1948, the year he died. It became a milestone in the development of documentation and librarianship and had 'a subtle importance that far transcends its primacy. Bradford...was a scientist and a librarian, and thus is himself a symbol of the real unity that brings together under the comprehensive term "bibliographic organisation" the professional interests and activities of both documentalists and librarians – a unity that is given both substance and precision by being founded upon scientific training.'[22]

'A vast Napoleonic information centre'

Bradford's successor Ernest Lancaster Jones became the Library's Keeper in 1935. He had been a curator in the Museum and had written the *Catalogue of the Geodesy and Surveying Collections* as well as papers on applied geophysics. Lancaster Jones played an active role in the Association of Special Libraries and Information Bureaux (ASLIB) and was an active proponent of microfilm technology. During most of the Second World War, the Science Museum was closed, but the Library remained open and escaped being bombed. A large portion of the older stock was evacuated – 20,000 volumes went to Hampshire in 1939 and by 1945 60,000 more had been evacuated. At the outbreak of war Lancaster Jones had expected usage to go down and had laid off 40 per cent of his staff. In fact borrowing increased markedly, concentrated on defence-critical areas such as physics, engineering and chemistry. The number of borrowing institutions increased from 450 before the war to over 1,000 by its end. The Library also took a crucial role in the microfilming of enemy periodicals.

It became clear after the war that the Science Museum Library was unable to provide the detailed documentation services that were becoming necessary. It had neither the staff nor the resources for the indexing, abstracting or translating required. Donald Urquhart, who had joined the Library in 1937 as an Assistant Keeper and was seconded to the Directorate for Scientific Research during the war, returned to the Library to find there was a chronic shortage of staff and external loans were overburdening its work. He had become interested in the problems associated with running a large library with limited manpower. He set about simplifying procedures and sorting out copyright problems, but before he could do much further work Urquhart was offered an interview for a post to take charge of the technical information section of the DSIR's intelligence division. The DSIR had failed to find any suitable candidates after it was advertised and had asked the Museum for suitable candidates. Urquhart was offered the post, which he accepted, and he left in 1948.

The rapid rise in numbers of scientific periodicals was becoming a problem. Loans increased at such a rate after the Second World War that there were complaints about the effects on reference services. There were 79,000 requests in 1947. The introduction of photostatic copying machines in 1948 helped matters, and by 1954 there were 176,000 requests. In 1947 the Library also started a supplementary lending service to augment its loan service. Each member of the scheme had to be willing to lend publications from its collections to other member libraries. The Science Library held a union catalogue of the publications available for loan and continually incorporated entries from the cooperating libraries' accessions lists to keep it up to date.[23]

The Library was still attempting to collect comprehensively, but there were dissenting voices, notably Humphrey Thomas Pledge, who was appointed Keeper in 1945 after the death of Lancaster Jones. Pledge had previously been an assistant master at Wrekin College, Shropshire before being appointed as Assistant Keeper in the Library in 1927, and subsequently wrote *Science since 1500*.[24] In a note for the Subcommittee on Education, written in about 1946, Pledge noted: 'Our task is one of selection, to select at each succeeding period a series of limited objectives in close correlation with the demands of the moment in research and in the nation's aims.' He also noted: 'Let the idea of a vast Napoleonic information centre be abandoned.' Pledge criticised Bradford's initiatives: 'the least used part of the South Kensington's vast collection of references is the UDC cards...they have never been intensively edited; the use of them never justified the great expenditure needed to eliminate the heterogeneity, cryptic quality, multilingualism etc. of the vast array of cards. And, although the UDC made considerable progress, it remained for would-be users, a baffling screen, rather than a crystal-clear lens or mirror, let alone a searchlight.'

The Library's tradition of acquiring rare works in science and engineering was strengthened during the 1950s when Frank Sherwood Taylor, the Science Museum Director, initiated a programme of purchasing antiquarian scientific books to form the core of three book exhibitions held between 1952 and 1954: the first, 'A hundred alchemical books', Sherwood Taylor's own speciality, included items from the Oxford Museum of the History of Science, and was followed by 'Historic books on machines' and 'Historic astronomical books'. Each exhibition was accompanied by a published catalogue written by Museum curators. Notable works that were purchased during this period include Galileo' *Sidereus Nuncius* (1610) and Hieronymus Brunschwig's *Liber de Arte Distillandi Compositis* (1512).

Library staffing

Library staff were employed as part of the Museum's establishment and, as such, were part of the Civil Service. Some were transferred from the Museum or from other Civil Service departments and others were recruited externally. Those who were recruited through Civil Service trawls may not have known they would be working in the Library! For example, Urquhart was interviewed in 1937 for an Assistant Keeper post in the Museum but was not told that it was for the Library until the interview itself. He only accepted the position upon the understanding that he could spend half his official time carrying out research, specifically an X-ray study of the distortion single crystals of metals.[25] Many staff held honours degrees in science or foreign language qualifications; subject knowledge was vital for staff, especially those answering technical enquiries or carrying out cataloguing, classifying and indexing. For those without library experience there was an active training programme, so, for example, in 1957, six Museum Assistants took external librarianship courses.

Unusually for the Science Museum in this period, many of the Library staff were women; in 1949, for example, there were eight women in the Library and six in library accessions, although this was considered to be part of the clerical establishment, not the Library. The employment of women within the Library mirrored their experience within industrial libraries. 'Technical library and information work is significant as perhaps the only field in which both female and male science graduates were routinely recruited across industry, yet clearly they did not occupy an equal position in the prevailing employment ideology of the industrial sector.'[26] Most women were employed in the clerical grades because they were said to have a 'natural flair for routine, clerical type tasks which demanded method, organisation and attention to detail.'[27] But technical libraries could also provide a place for women science graduates to apply their specialist knowledge, so they could write accurate abstracts, select technical books or answer detailed subject enquiries. However, career progression was more problematic, and in the Science Museum Library very few women reached senior levels. Those who did so were Hannah Parker, who, in 1953, was the first woman to become a Deputy Keeper, and Helen Phippen, who started as a senior research assistant in 1955 before becoming an Assistant Keeper in 1965. Marion Gossett, who joined the Library in 1926, left to set up and run a technical library at the Atomic Energy Research Establishment at Harwell in 1949.

A National Lending Library for Science

The call for a new central reference library of science was eventually dropped in the 1950s. In 1944 there had been a proposal to develop a new 'Science Centre' on the South Bank of the Thames which would bring together the Patent Office Library, the Science Museum Library, the Royal Society and learned society libraries in close proximity to the administrative headquarters for the societies and the DSIR. The plan was endorsed in 1947–1948 by the DSIR Panel on Technical Information Services. Its *Report on Scientific and Technical Library Facilities* recommended the establishment of a Central Reference Library (the Science Centre) and a National Lending Library, to be located outside London. The Science Centre idea was received with enthusiasm after the Second World War but was rejected after 1951 as a casualty of public spending caps by the new Conservative administration. Urquhart also objected because he wanted to prioritise the establishment of a national lending library.

A turning point for planning for a national policy for scientific and technical information came in 1948 when the Royal Society, assisted by Urquhart at the DSIR, organised a Scientific Information Conference, which concluded that more effective use could be made of the world's scientific literature by the further development of information services and special libraries. It recommended that the government should give increased support to the Science Museum Library and the Patent Office Library and that they should be made more effective by enabling them to employ adequate staff with special qualifications in librarianship, by increasing opening hours, by extending their collections 'so that either or both, in some form, could hold every publication which contains material of value to science and technology' and that copies of translations should be deposited at the Science Library 'despite restrictions under the present copyright law.'[28]

In reality the conference's vision of planned information provision and its recommendations in other areas were only acted upon incrementally and in a piecemeal approach. The focus was shifting from a model of state-controlled and centralised provision of scientific and technical information to a more pluralistic one based around a mix of provision in public and special libraries and in information bureaux associated with specific regions or industries.

The conference made recommendations relating to the cooperative provision of books, periodicals and related material to avoid duplication and extend access. It suggested that cooperation should commence in local areas, and that regional centres should be linked with national and international systems of interchange.[29] They noted that the proposed scheme of complete coverage would be quite impractical without the cooperation of the national libraries, including the British Museum and the Science Museum Library, and that some modifications to the functions of the latter would be necessary. Greater cooperation would eventually happen, but it would be another factor that would eventually force a radical change to the Library's priorities, the foundation of a truly national lending library for science.

To some in the Museum, including Sherwood Taylor, the Library had grown too big and had become a 'hypertrophied member of the Museum, demanding an undue measure of its attention.'[30] The Museum received no extra funding to run the Library but had to pay for it from its existing government allocation:

> The extensiveness of the present Science Library is not an embarrassment, except administratively; but its value to the Museum as distinguished from the general public, is not commensurate with its cost.[31]

The Royal Society's conference had focussed the attention of the British government on the views of scientists that there was an urgent need for improvements to the provision of scientific and technical information. This initiated further development work by Urquhart at the DSIR and led to the planning and establishment of a central scientific lending library, the National Lending Library for Science and Technology in Boston Spa, Yorkshire. Its nucleus was formed from Science Museum Library holdings. The proposals to move some, or all, of it allowed the Museum to speculate upon the type of library appropriate to the Science Museum:

> If there were no Museum Library, the formation of one to contain some 150 important periodicals and some 15,000 books might well be contemplated, and whatever the outcome on the future of the Science Library, the Museum

must have a library that does not fall below that standard. It should be a specialist library in respect to the history of science and technology, since no other library in London specialises in this field, but should not endeavour to be a complete library of science.[32]

The Museum, however, was not the only user of the Library – it also functioned as the central library for Imperial College. It remained on College premises, and College staff and postgraduate students formed a substantial proportion of users, so any forward planning would need to take their requirements into account.

Urquhart and a reluctant Pledge carried out a detailed survey of the Library's collections and services[33] and found that the Library held 34 per cent of periodicals listed in the third edition of the *World List of Scientific Periodicals*, holdings that were split to accommodate the needs of the Museum, the College and the new library. This was an uncertain time for the Library and its staff:

> The future of the Library, both geographically and administratively, continued to be the subject of close review and gives rise to anxiety. It has occupied much of the time of the senior officers of the Library.[34]

Russian material was first to be moved to Boston Spa in 1957, followed by 7,000 sets of periodicals in 1958. In all, the Science Museum Library contributed 100,000 volumes for the new national library, which opened in 1962.

In spite of these donations, the Science Museum Library remained a significant resource, with 370,000 volumes, including 18,000 periodical titles, 4,500 of which were current. The Library could now concentrate on its local users: the Museum, the public, and staff and postgraduate students of Imperial College. It still acted as a backup for interlibrary lending, but, that aside, it could develop its specialisation in the history of science and technology. But it had other uses, and in 1964 the Reading Room was used to film a scene in 'The Ipcress File'.[35] It was here that the agent Harry Palmer, played by Michael Caine, made contact with the villainous 'Bluejay'.

The jewel of the period

In 1961, John Chaldecott, who had been a curator of physics in the Museum, became Keeper of the Science Museum Library, a post he held until 1976. Yet again, the Library was facing growing space problems, and library stock was now scattered in a number of outlying stores. In 1959 Imperial College had set up its own central lending library for undergraduates, the Lyon Playfair Library, which opened at 180 Queen's Gate. The College also had major plans to develop its site, which required that both libraries vacate their accommodation and move into a new building adjacent to each other. The new Science Museum Library building was designed by the firm Norman & Dawbarn, working to Chaldecott's brief, and after extensive research he was able to incorporate the latest ideas in library design and equipment within a very tight budget.

The Library began to occupy the new building in August 1969, and it was formally opened in the following November by HM The Queen and the HRH The Duke of Edinburgh. To many in the profession, 'The jewel of the period was the long-desired new Science Museum Library.'[36] The new four-storey building had a floor area of 60,000 square feet, had 160 seats for readers, and provided open access to the periodicals of the past twenty-five years and to all twentieth-century scientific and technical monographs. There were two basement floors of closed-access storage with Compactus rolling shelving, so all the Library's holdings could now be brought under one roof. The design allowed for upward expansion to accommodate two more floors and also a lateral expansion to the basement.

The building was full of new equipment; there were pneumatic tubes, personal call-systems, lifts and conveyor belts for the delivery of books to the main enquiry desk. Some of these innovations did not work for long, but others were more successful, such as an IBM 870 Document Writing System and an IBM 82 card sorter. The building included a fully equipped reprographics studio, a microfilming room and a darkroom for photographic work. These were needed urgently – the Library was dealing with 32,000 loan requests each year, and about 30,000 orders for xerographic or photostatic copies, microfiche or microfilm.

The state of the Science Museum Library just before its move is reported in the *Report of the National Libraries Committee* made by Frederick Dainton in 1969. It examined the Library's services, its usage patterns and its stock. Its acquisition policy was centred on the needs of the Science Museum and Imperial College. The report stated: 'It is extremely difficult to make a clear distinction between material acquired primarily for the Museum and that for Imperial College, but the library's exceptionally good coverage of the history of science and technology is intended to satisfy one of the Museum's essential requirements.' The Committee saw the adjacency of the two libraries and their duplication of stock as wasteful and recommended that the Library 'be fully integrated into the Lyon Playfair Library' with certain provisions for continued funding of the special collection in the history of science and technology.

However, when the White Paper appeared in 1971, the Committee modified its recommendations to propose that the Library should not be totally absorbed by the Lyon Playfair Library, but that it should be developed as a reference library of the history of science and technology and for the provision of specialist services for the Science Museum and other neighbouring museums. Moreover, it suggested that general reference requirements, and a large part of the existing stock, should be taken over by the Lyon Playfair Library. However, there were proposals for rehousing the Museum Library in a future west block of the Museum, so these recommendations were not followed. The Library was defended by Chaldecott, whose skill and determination as a negotiator helped to ensure that it remained under the Museum's management. It was thought best to keep the libraries' separate identities and roles while maintaining close cooperation. The Library, therefore, continued to collect as comprehensively as possible in its historical speciality, while maintaining some coverage of non-historical material.

Expanding collections

During the 1970s the pictorial collection and the archives collection were established in the Library. The pictorial collection centralised the rich collections of pictorial material in the Museum, including prints, oil paintings, coins and medals, but was transferred back to the Museum in 1992. The archives collection was built up from existing archive material held in Museum collections supplemented by donations or purchased items, and became a rich source for material on a wide diversity of subjects, from ships' plans to the papers of Charles Babbage. In 1974 two further collections were added: the Bidder Collection, which included both archival and printed material on railways and other engineering works; and the Simmons collection, which included documentation and photographs on British windmills and watermills.

Chaldecott retired in 1976 and Lance Day was appointed Keeper of the Library, a post he held until 1987. Day had started his career in the Library as a research assistant in the early 1950s, but had transferred to the Museum, where he became Assistant Keeper for Chemistry and Keeper for Communications and Electrical Engineering before his return to the Library. During his tenure the Library continued its series of published bibliographies[37] and Lance Day published two guides to the historical collections in academic journals.[38] The Library's historical collections were also strengthened significantly, as important collections were acquired or created from existing stock.

From 1978 to 1980 a significant personal collection of antiquarian scientific books, the Honeyman Collection, was sold by Sotheby's. The Library was awarded a special grant to bid for items and filled many gaps in its collection, including Harvey's *Exercitatio Anatomica de Motu Cordis* ... (1628); Kepler's *Harmonices Mundi* (1619); and a copy of Einstein's *Über die spezielle und allgemeine Relativitätstheorie* (1917), signed by the author. In 1979 the Museum published a catalogue of books printed before 1641, with a supplement issued in 1982 to include material purchased from this collection.[39]

Two further collections were moved from the Museum to the Library in the 1980s: the Watt collection and printed items from the Penn-Gaskell Collection of aeronautica. The Watt Collection was part of the personal library of James Watt, and included a wide range of works on medicine and literature as well as science and technology, such as *The Works of Shakespear / collated and corrected ... by Mr. Pope* (1766) and Abraham Bennet's *New Experiments on Electricity* ... (1789). The Penn-Gaskell books include a copy of Gaston Tissandier's *Histoire des Ballons et des Aeronautes Célèbres* (1887), signed by the author, and Cyrano de Bergerac's *Comical History of the States of and Empires of the Moon and Sun* (1687).

In 1976 the Wellcome Collection was deposited in the Museum, so the Library widened its collecting to include the history of medicine, although it only collected selectively. It could never compete with the resources available to the Wellcome Library, with its comprehensive collections in the history of medicine. The coverage of veterinary history was expanded considerably in 1987 by the purchase, with the aid of grants from a number of bodies, of a major collection in this field, the Comben Collection. This comprised some 900 printed books and pamphlets from the sixteenth to the twentieth century. Five years later a detailed catalogue of the collection was published.[40]

During the early 1980s Library staff recognised the value of trade literature as an information resource, and resources were directed into collecting, maintaining and making it accessible. Collections from around the Museum were transferred to the Library and supplemented by material scattered throughout the Library. International exhibition publications were also centralised.

Compared with other museums, the Science Museum was relatively late in acquiring a computer to administer its collections. The Museum catalogued its objects using a combination of ledgers, index cards and registered files, although it did use input cards to document its Pictorial and Wellcome collections. The Library had also been generating automated printouts of periodicals lists since the 1960s, but its main catalogue remained on cards. In the early 1980s Library staff successfully argued for the computerisation of both its catalogue and museum object data. The Museum acquired a Prime minicomputer, which was located in the Library building. Its value was recognised immediately by Museum administration, which used it to manage finances. The Adlib software package was bought for the cataloguing of objects and library material, and records could be linked to each other despite being on separate databases. In 1984 Library cataloguers started cataloguing new stock using Adlib.

Reorganisation and renewal

Lance Day's retirement in 1987 coincided with a reorganisation of the Museum under its new Director, Neil Cossons. Staff structures changed and the titles of Keeper and Assistant Keeper disappeared. For a short time, the historian of science Robert Fox took charge of the Library as part of his role as the Assistant Director of Research and Information Services. However, he left in 1988 to become Professor of the History of Science at Oxford, and was replaced by John Durant, who joined the Museum in 1989 as Assistant Director of the new Science Communication Division. Durant also became a visiting Professor of the Public Understanding of Science at Imperial College, so collecting policy was again widened and the Public Understanding of Science became a priority. This covered a wide range of subjects including science education, social aspects of science and science communication. A new post, Head of Library and Information Services, was created and filled by Leonard Will, a professional librarian and keen exponent of computerised information systems.

The original recommendations of the Dainton Committee were adopted in 1992, and the two neighbouring libraries became 'The Imperial College & Science Museum Libraries'. This involved moving, integrating and reorganising collections and services, and significant building work. However, the Science Museum Library remained a separate entity with its own budget, staff and management. Staff numbers were reduced by a third in 1993 and Leonard Will took early retirement. He was replaced as Head of the Library (until her retirement in 2003) by Pauline Dingley, who had managed the merger of the Library's catalogue with the College's Libertas computer catalogue.

The Science Museum Library provided specialist services in the new Science and Technology Studies Collection, which incorporated the History of Science collection and popular books on science and technology. Here, Library staff continued to serve the needs of the Science Museum, Imperial College, the international scholarly community and the general public. This arrangement continued until the early 2000s, when the Library was reviewed yet again. After a lengthy options-analysis an agreement was signed with Imperial College, and 85 per cent of the Library's collections were moved to new facilities at the Science Museum's large object store at Wroughton, near Swindon, leaving the heavily used Science and Technology Studies Collection in London.

The Science Museum Library is now a large specialised library focussing primarily on the history of the physical sciences, industry and engineering. It holds about 500,000 volumes on its two sites – collections of international importance – and is one of the larger libraries of its type in Europe. By comparison, it is approximately 130 per cent the size of the Library of the Conservatoire National des Arts et Métiers in Paris, 55 per cent the size of the Deutsches Museum Library and a third the size of the Smithsonian Institution Libraries, the largest museum library in the world.

Conclusion

The Library has had a complex and tumultuous history as it has reacted to internal and external pressures. Its origins lay not in the needs of the Museum, but in those of scientific and technical education, and in many ways it has grown independently of the Museum. For a time it assumed a national role in the absence of any government policy for scientific information, but Bradford's aspirations were overconfident; the Museum received no extra funding for the Library and it had neither the space, the staffing nor the government support for so ambitious a project. Bradford had the backing of Henry Lyons, but later Directors would not be persuaded by this grandiose vision. However, the Library could never be reduced to the 'ideal' size for the Museum; the needs of Imperial College also had to be considered, so, even after the establishment of the national lending library for science, it shrank by only 21 per cent. For most of its history it has occupied College property, and only recently has it settled on land actually owned by the Museum, albeit far from its original location. Staff and accommodation have changed, but its collections have grown and have remained the core of its activities and a key resource for the Museum and wider world. The Library has renewed itself on many occasions, a process which is still underway today – some things never change.

Notes

1. *Catalogue of the Science Library in the South Kensington Museum* (London: HMSO, 1891), p. v.
2. L.W. Fulcher, 'Science Library: Report on Periodicals', 3.iii.[19]03, Science Museum Library.

3. Quoted in Dave Muddiman, 'Science, Industry and State: Scientific and Technical Information in Early-Twentieth Century Britain', in Alistair Black, Dave Muddiman and Helen Platt, eds *The Early Information Society: Information Management before the Computer* (London: Ashgate, 2007), pp. 55–78, on p. 57.
4. L.W. Fulcher, 'Science Library: Report on Periodicals'.
5. Gregor Mendel, 'Versuche über Pflanzen-Hybriden', *Verhandlungen des naturforschenden Vereines in Brünn* 4 (1866), pp. 3–47.
6. The name of the Library remained 'The Science Library', but it was also often referred to as the Science Museum Library, a name which gradually took precedence.
7. Donald Urquhart, *Mr Boston Spa* (Leeds: Wood Garth, 1990), p. 8.
8. Library Association Record (1918), on p. 110.
9. Alistair Black and Dave Muddiman, 'Reconsidering the Chronology of the Information Age', in Alistair Black, Dave Muddiman and Helen Platt, eds *The Early Information Society: Information Management before the Computer* (London: Ashgate, 2007), pp. 237–43, on p. 240.
10. Advisory Council Annual Report for 1920, p. 7.
11. Noted in Charles R. Richards, *The Industrial Museum* (New York: Macmillan, 1925), p. 17.
12. H.T. Pledge, 'The Science Library', in R. Irwin and R. Staveely, eds *The Libraries of London*, 2nd edn (London: Library Association, 1961), p. 49.
13. Marion Gossett, 'S.C. Bradford, Keeper of the Science Museum Library, 1925–1937', *Journal of Documentation* 33 (1977), pp. 173–6, on p. 174.
14. P.R. Harris, *A History of the British Museum Library 1753–1973*[Can't understand note] (London: British Library, 1998), p. 526.
15. Quoted in David Follett, *The Rise of the Science Museum under Henry Lyons* (London: Science Museum, 1978), p. 134.
16. Quoted in Follett, *Rise of the Science Museum*, p. 135.
17. H.P. Spratt, 'Technical Science Libraries', in *The Year's work in Librarianship* 6. (London: Library Association, 1933), pp. 114–34, on p. 119.
18. Advisory Council Annual Report for 1934, p. 49.
19. S.C. Bradford, 'The Organisation of a Library Service in Science and Technology', *British Society for International Bibliography Publication* 2 (September 1935).
20. S.C. Bradford, 'Sources of Information on Specific Subjects', *Engineering* 137 (26 January 1934), pp. 85–6.
21. Quoted in Robert Wedgeworth, *World Encyclopedia of Library and Information services*, 3rd edn (Chicago: American Library Association, 1993), p. 142.
22. J.H. Shera and M.E. Egan, 'A Review of the Present State of Librarianship and Documentation', in S.C. Bradford, ed. *Documentation*, 2nd edn (London, Crosby Lockwood, 1953), pp. 11–5, on pp. 11–12.
23. H.J. Parker, 'Science Museum Library', *Libri*, 3 (1954), pp. 326–36.
24. H.T. Pledge, *Science since 1500: A Short History of Mathematics, Physics, Chemistry, Biology* (London: HMSO, 1939).
25. Urquhart, *Mr Boston Spa*, p. 11.
26. Helen Plant, 'Women's Employment in Industrial Libraries and Information Bureaux in Britain, c. 1918–1960', in Alistair Black, Dave Muddiman and Helen Plant, eds *The Early Information Society: Information Management before the Computer* (London: Ashgate, 2007), pp. 219–34, on p. 221.
27. Plant, 'Women's Employment', 22.
28. *The Royal Society Scientific Information Conference, 21 June–2 July 1948: Report and Papers Submitted* (London: Royal Society, 1948), pp. 201–2, 207.
29. 'The Co-operative Provision of Books, Periodicals and Related Material', *Library Association Record* (December 1949), p. 383.
30. Advisory Council Annual Report for 1952, p. 41.
31. Ibid, p. 42.

32. Ibid, pp. 41–2.
33. Donald Urquhart, 'A Domesday Book of Scientific Periodicals', *Journal of Documentation* 12 (June 1956), pp. 114–15; Donald Urquhart, 'Use of Scientific Periodicals', in *Proceedings of the International Conference on Scientific Information, Washington, D.C., November 16–21, 1958* (Washington, DC: National Academy of Sciences, 1959).
34. Advisory Council Annual Report for 1955, p. 29.
35. 'Ipcress File'. Directed by Len Deighton (London, Network, 1965). Chapter 4, The Science Museum Library. Video:DVD.
36. H.A. Whatley, ed., *British Librarianship and Information Science, 1966–1970* (London: Library Association, 1972), p. 522.
37. S.A. Jayawardene, *Reference Books for the Historian of Science: A Handlist* (London: Science Museum, 1982).
38. Lance Day, 'Resources for the History of Science in the Science Museum Library', *BJHS* 18 (1985), pp. 72–6; 'Resources for the Study of the History of Technology in the Science Museum Library', *Iatul Quarterly* 3 (1989), pp. 122–39.
39. Judit A. Brody, *A Catalogue of Books Printed before 1641 in the Science Museum Library* (London: Science Museum, 1979); *Supplement Comprising Acquisitions to End of 1981* (London: Science Museum, 1982).
40. Pauline O. Dingley, *Historic Books on Veterinary Science and Animal Husbandry: The Comben Collection in the Science Museum Library* (London: HMSO, 1992).

7
'A Worthy and Suitable House': The Science Museum Buildings and the Temporality of Space

David Rooney

Introduction

Our great museums are concrete symbols; they are our civilisation made manifest. They contain, and become, our achievements, the tangible image we choose to project to our world, and they are our time machines: our culture preserved and thrown forward to some temporally distant other place. Even before the Science Museum as we know it today was built, the symbolic value of the South Kensington museums was strongly articulated; the role of the museum building as beacon of our techno-scientific human culture shining into other cultures was clear. In his powerful early novel *The Time Machine*, H.G. Wells describes the Time Traveller's forward flight to the year 802,701, to a time when human civilisation as he knew it had ended, replaced by a bleak polarisation between the technocratic subterranean Morlocks and the decadent, indolent Eloi. Symbolic at the heart of the novel is the 'Palace of Green Porcelain', the Time Traveller's only concrete link to his own culture, home of dinosaurs and fossils, machines and matches: 'Clearly we stood among the ruins of some latter-day South Kensington!'[1] And yet a ruin it was: the museum that was originally man's construction against nature was now being sent inexorably back to the soil by the forces of time.

This tension was voiced by Wells in 1895, eight years after he had finished studying at South Kensington's Royal College of Science, and it sits at the heart of this essay: how much are the Science Museum's buildings and permanent galleries the result of explicit plans by their makers and users, and how much have they been shaped by external forces? Science Museum Directors and staff have always known the power of bricks and mortar in establishing, presenting and maintaining subjects and status. It's about politics and people as much as artefacts and education; local rivalries alongside global ideals. The Science Museum buildings have all been long in the planning and detailed in the design. Yet the results have rarely matched the original intentions as the real world has collided with the theoretical over time. If the buildings shape the visit, whose vision is being articulated, and how long ago was it relevant?

A visit in 1936: 'Electric Illumination' in the East Block

On 15 December 1936, the entertainment correspondent from *The Times* called in at the Science Museum for the press launch of a new temporary exhibition entitled 'Electric Illumination'. It was held just inside the main entrance on Exhibition Road, and it was hard to miss. Bright, brash and brilliant, the exhibition was a deliberate attempt to popularise the latest products of industry and technology in an interactive, hands-on display that promised, according to Lord Rutherford, the nuclear physicist who gave the opening speech, to be 'of great interest not only to scientists, but to every man, woman and child.'[2]

The reporter enjoyed himself amongst the displays. He observed scale models of London's Bush House and St Paul's Cathedral, showing floodlighting schemes at the push of a button. He examined the lighting of an aerodrome required for night flying. He saw the effects of coloured lights on flowers, textiles and pictures, and decided which sort of lamps provided the most pleasing 'look' in a domestic room. Shop windows vied for his attention with historical sequences of lighting technology, and all these bright lights had been curated for his enjoyment and education by William O'Dea, an ambitious young Assistant Keeper with a background in electrical engineering five years into his tenure at the Museum, and the staff at the Electric Lamp Manufacturers' Association (ELMA), a trade cartel controlling most of the British lamp supply network. Their main man was Walter Jones, a senior manager well known and respected for his publicity work. What kind of museum had this reporter come to see in 1936, and how did its physical nature – its buildings, its layout, its display techniques – affect his view of its subjects and objects?

In 1910, Hugh Bell's committee had looked at, among other things:

> the special characteristics which should be possessed by the new buildings which it is hoped will shortly be erected on the South Kensington site to house [the collections of the Science and Geological Museums], so as to enable the latter to be classified and exhibited in the manner most fitted to accomplish the purposes they are intended to fulfil.[3]

A year later, Bell reported that the Science Museum 'ought also to be a worthy and suitable house for the preservation of appliances which hold honoured place in the progress of Science or in the history of invention.'[4] The form of this worthy and suitable house was described by Bell in 1912. The Science Museum would be built in three distinct phases. First, and immediately, the eastern block should be built. A central block would follow soon after, when lessons learned from the first phase could be incorporated. Finally, a western building would complete the Museum when demand for space grew. John Liffen has argued elsewhere that Bell's assessments of the Science Museum's collections 'are mostly excellent examples of the blindingly obvious,' and perhaps the same could be said for his analysis of building requirements.[5] But, whilst his master plan was not particularly controversial, Bell did not foresee the delay that would be caused by external forces in the timescale of the building work.

Bell believed that the first two stages would be built by 1922. Work on the East Block began in 1913 but was halted by the war in 1915, and in 1920 Director Francis Ogilvie wrote in frustration, 'Archaeology, Fine Arts, The Industrial Arts and Natural History are all housed in palaces. Physical and Mechanical Science and the applications of science in industry are still in the wilderness!'[6] Two years later work restarted, but only because of external pressures for space, not owing to the force of argument for a palace for science: the Imperial War Museum was looking for a new home for its collections, then housed in the Crystal Palace under a lease that was to end in 1924, and it needed the Science Museum's western galleries. Indeed, a visitor to the Museum a decade or more after Ogilvie's lament had a stark choice of accommodation. The grand eastern halls, opened eventually in 1928, were certainly impressive. A grand central atrium rising to third floor level was surrounded by sidelit galleries; a western wing provided a stack of long, high-ceilinged cuboids; and, to the north, bright halls took light through long rows of windows. But the rest of the Museum's collections were still housed in buildings erected for the 1862 International Exhibition, which Lyons in 1933 pointed out 'were condemned as unsuitable and dangerous some 40 or 45 years ago.'[7] Illustration 7.1, taken from the Science Museum Annual Report for 1914, shows the site of the Museum as it was from 1913 to the mid-1920s.

Whether housed in dramatic or dangerous conditions, special exhibitions at the Science Museum had long been on the agenda. As early as 1926, before the East Block was fully open, space was offered as an experiment to both the National Physical Laboratory and the Department of Scientific and Industrial Research to exhibit the fruits of their labour, since 'Trade Exhibitions lasting only for a week or ten days do not give sufficient time for the Research Associations of the various Industries to show effectively the results which they have obtained by their research work.' The results were 'most successful', in Director Henry Lyons's opinion, as 'they attract many visitors, they give publicity to the organisation which is exhibiting, and they give the impression that there is always something new for visitors to see in the Museum.'[8] The 1936 Electric Illumination exhibition throws light on aspects of the physicality of the Museum and its relationship with the content on show. Seeking Director Ernest Mackintosh's authorisation of his scheme in February 1936, O'Dea proposed a lively, interactive exhibition. Seeing first a historical section, the visitor would then move through a series of 'experimental' demonstrations of lighting techniques before reaching the centrepiece, two full-sized demonstration rooms showcasing the latest lighting technology. On emerging from the demonstration rooms the visitor checked out the cutting edge in vapour tube discharge lighting – with the fluorescent lamp set to be launched on the public a couple of years later[9] – and then saw demonstrations of street lighting and floodlighting. It was an ambitious scheme, and very timely. Lighting was a key concern to architects in the 1930s. On the one hand, good natural lighting was considered vitally important – indeed, the long ranks of tall windows at the front of the Science Museum were incorporated to give the best lighting to the glass-topped display cases displayed inside. Since Britain first advanced its clocks for summer time, in 1916, daylight was something to be 'saved'. On the other

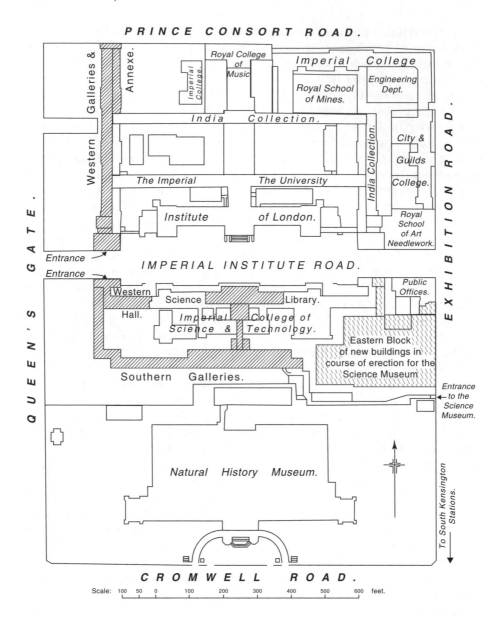

Illustration 7.1   Plan of Science Museum site from the Annual Report for 1914, showing the East Block under construction (opened fully in 1928) and the southern and western galleries which were legacies of the 1862 International Exhibition

hand, electric lighting was of growing interest in the building trade as incandescent lighting was maturing and the new electric discharge lamps (including fluorescent tubes) promised efficiency, increased brightness and new colours.

Moreover, lighting was raised in the museum context as an enabler of (or barrier to) learning. *The Electrical Age* in April 1937 described 'endless walking though endless badly-lit galleries to see a depressing collection of Egyptian mummies which meant little to one, or Chinese antiquities which meant even less. Those were early impressions of a museum, and inevitably a museum became in the logical youthful mind a place to be AVOIDED AT ALL COSTS.' Yet, as one reviewer of the Electric Illumination exhibit explained, new approaches and new displays were confounding that outdated view. 'Ten-year-old James finds as much fun as the older or technical person can find instruction, in turning switches on and off to see the various effects.' Here, lighting as both museum subject and museum architectural approach proved a saviour of young minds and the education of the masses. More than this: the exhibition was to be a life-saver. 'Every year hundreds of people are killed, thousands of persons' eyesight is ruined, and thousands of pounds are wasted by inadequate or unsuitable artificial lighting. This is the state of affairs that the Electric Illumination Exhibition ... seeks to remove.'[10]

Illustration 7.2   Science Museum special exhibition gallery, entrance hall (Gallery 1), 1928, showing the recently arrived King George III Collection of scientific instruments

One reason why O'Dea and his collaborators wanted to run this temporary exhibition in the Museum's front hall was to do with publicity and spatiality: the rhetorical power of place. The end-goal was a permanent gallery on electric lighting, but both O'Dea at the Museum and Jones at ELMA knew that this installation, which was destined for the Museum's basement, would get little press attention, whereas a temporary show in the main front hall would act as a high-profile 'trailer' that was guaranteed publicity and visitors.[11] Publicity arrangements even extended to special notepaper for temporary exhibitions. 'This is a great asset,' O'Dea wrote to Jones, 'in drawing attention to the importance of the exhibition particularly for the press etc. who thereby differentiate it from not very noteworthy special arrangements of existing collections which are quite common to other museums.'[12] This was conceived as a popular show, having 'all the elements to interest the public, and the fact that the visitor is given such exceptional powers of control should appeal strongly to the younger generation.'[13] Illustration 7.2 shows the front hall in 1928 on the opening of the new museum building, showing an in-house exhibition of scientific apparatus. Illustration 7.3 shows the Electric Illumination exhibition in the same space eight years later. It is clear that the new focus on publicity, industry support and a nascent use of display lighting

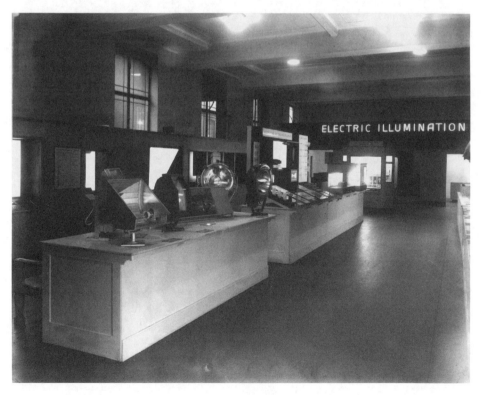

Illustration 7.3   'Electric Illumination' exhibition in the entrance hall, 1936, showing novel lighting techniques

techniques – note the use of electric lighting in the exhibition compared with the earlier picture – resulted in a more visually arresting – and one might therefore assume more engaging – exhibition, likely to pull in the crowds.

It seemed to work. Some 250,000 people visited the show over its five-month run (averaging about 1,600 per day, 45 per cent of the total visitors to the Museum in that period), and nearly 150,000 people were logged (automatically) using one of its interactive light-level-testing exhibits. The Museum had provided a physical space where people could engage (with their hands and senses as well as their brains) with scientific concepts, and this was crucial. 'The Museum had been asked to preserve the results of these tests as a basis from which important conclusions were to be drawn – an example of how the Science Museum can fulfil a useful function where mass statistics are required as the material of scientific investigation.'[14] A strong argument made for the relevance of a physical space for public science. Yet the Museum's new building, designed in the 1910s but not opened until 1928, had been made for daylight and glass cases. The 1936 exhibition demanded darkness and drama. The building, so new, so hard-won, had proved wanting, but it was modified to suit the new requirements, daylight shaded, artificial light ushered in. Buildings adapt; people like O'Dea shape them, because fresh from the construction site they are rarely what their users need any more.

A visit in 1951: The 'Exhibition of Science' in the Centre Block

On 4 May 1951, visitors began to arrive at the Science Museum for a special opening event of the 'Exhibition of Science', part of the Festival of Britain. It was held in a new building at the west of the Museum's existing premises and was reached from Museum Lane, the private roadway dividing the Science Museum's site from that of the Natural History Museum next door. Running until the end of September, the exhibition attracted over 200,000 visitors and contained many important exhibits. The visitor figures were lower than predicted (as with many subsidiary exhibitions that year, perhaps overshadowed by the main focus on London's South Bank), but, if only a partial success for science education, it was a triumph of politics for the Science Museum's management, who, in a finely played game of brinkmanship, finally saw work begin on the second phase of buildings for the Museum's collections.

Lyons had seen the results of the intervention of the Great War into the plans for the Museum's first building and hoped this delay would provide impetus for the immediate development of the central block, planned by Bell to be built by the early 1920s. The fundamental argument for a second block was to allow the representation of current practice in science and industry, subjects which were not displayed or even collected at that time because of the lack of space. The Bell Committee and the Museum's officers had consistently stressed this reasoning through the 1910s and 1920s, but it was never enough to win funds from an over-stretched government, and progress was slow. A subtle shift of focus occurred in 1931, when Lyons visited the Deutsches Museum in Munich with the chief architect of the Office of Works, Richard Allison. The two men wanted to learn lessons from their continental counterpart.[15] Lyons concluded that the Bell approach was now outdated; pragmatism resulting from the inordinate delays in completing the East Block had led to a museum far removed from that envisaged in 1912. Galleries had been occupied one at a time whenever they had become available, whereas in Munich like subjects had been grouped logically together. In South Kensington, the layout was random and illogical. To solve this problem, the Centre Block would no longer be a new building for new stuff. What was needed was a long-term strategic plan for the entire Science Museum site: a master plan for the eventual layout of the entire collections, which would be followed whenever funds became available. This was an argument for the museum as three-dimensional encyclopaedia; a taxonomy of scientific and industrial subjects manifest in physical layout. But the trouble with taxonomies is that they reflect their period and change with the times. It might have worked if the whole Science Museum building could have been constructed in a short period during which the taxonomy could have remained in vogue. But the real world wasn't like that. Two years after the Munich visit, Lyons retired, with his hopes, it seemed, frustrated. The new Director, Ernest Mackintosh, seemed to draw closer to securing funds when, in 1937, new accommodation requirements for the Royal College of Art in South Kensington prompted a government rethink of the whole area, and under these proposals the Science Museum's central block got a high priority. But local politics were soon to be eclipsed by the global storm that was about to hit. Two years later the Second World War broke out and all plans ground to a halt.

With peace declared in 1945, Herman Shaw took over from Mackintosh and attention turned once more to the Centre Block plans. Here, finally, there was some light at the end of a dark tunnel, for the aftermath of the ghastly war provided the impetus to extend the Science Museum. Plans were afoot to raise the nation's spirits with a Festival of Britain, to be held in 1951, and one key part of the show was to be an exhibition of science. Organisers needed space to house it, and O'Dea recognised a golden opportunity: 'In 1947 I began a campaign (which was enthusiastically furthered by Shaw) to get the Centre Block built for the Festival of Britain.'[16] With so little time, the Ministry of Works tried to block the idea, one official, for instance, telling Shaw in 1948: 'I am hoping the 1951 proposals will blow over!'[17] But Shaw did not share his pessimism and, with O'Dea pulling the strings, manoeuvred his way through.

In August that year, Shaw wrote to the ministry to offer 35,000 square feet for the exhibition in the four-storey western extension of the East Block, still unfinished since its occupation in the 1920s (more on this presently). A further 200,000 square feet could be offered if *part* of the Centre Block was begun and completed by the end of 1950. Even this compromise had been rejected out of hand as impossibly quick by the Ministry of Works, but Shaw proposed in his letter that the problem could be solved 'even at this eleventh hour' if external contractors were employed. 'I have been making some enquiries through one of the Museum Keepers who has personal contacts with a large firm of public works contractors. The enquiry has been strictly unofficial with no intention whatever of invading the province of the Ministry of Works.' The ministry was

over a barrel. The education minister backed Shaw's proposals personally, his officials stating that 'we have satisfied ourselves that Shaw and the Keeper concerned
have solid grounds for their assertion.' The Ministry of Works squirmed, Herbert
Morrison (the Festival supremo) charmed, the education lobby flattered and the
result – a compromise commitment to build the basement and ground floors of
the new Centre Block by the end of 1950, covered with a temporary roof – became
inevitable. In 1949, work started.

Alongside the partial construction of the much-needed Centre Block, the exhibition also demanded use of the East Block's western wing. This building had never
been finished: in 1921 it had been occupied unfinished to relieve pressure on display space, and was given a temporary cladding.[18] Now was the time to complete it,
and O'Dea, the power behind the decision to house the 'Exhibition of Science' at the
Science Museum, had plans for this too. Building on his experience of curating the
electric lighting exhibition and gallery in the 1930s, O'Dea ushered in a new paradigm of museum display, based on electric lighting, which is still valid today and,
as such, was profoundly influential in changes to the physical nature of museum
buildings worldwide.[19] Now that the Centre Block was under way, he was planning
what would eventually go in the East Block western wing, and, instead of the 1920s
vogue for long rows of windows to admit maximum daylight, O'Dea argued that
'this first-floor space should have no windows. Now that efficient fluorescent lighting is available, the museum planner should be less ready to submit to the inelastic

Illustration 7.4 Photograph of architect's model of entrance to the 'Exhibition of Science'
between the Science Museum and the Geological Museum, March 1950. File Work 17/388,
The National Archives, Kew, and reproduced by kind permission

Illustration 7.5   Photograph of William Thomas O'Dea (1905–1981) taken in September 1966, just before he left the Science Museum to become Director General of the Ontario Science Centre which was under construction in Toronto

and restrictive tyranny of daylight...it is too variable in intensity and direction for many types of display and casts shadows that can be most discomfiting.'[20] Instead, he specified a plain brick windowless curtain wall, to give him free rein with artificial lighting, which he used to great effect in the Agriculture Gallery that followed.

Therefore, the Centre Block had been started and the eastern building finally completed. O'Dea, the bright spark in museum lighting, had manoeuvred his way through global and local politics to get his world view stamped indelibly into the bricks and mortar of the Science Museum. It was a world away from Bell's imagined 1912 museum, and even Lyons's and Mackintosh's 1930s holistic master plans, but it was rooted indelibly in all three. Buildings change slowly, and so do the people who build them, it seems. In March 1950, proposals for the entrance forecourt design of the 'Exhibition of Science' were sent to the Ministry of Works by the Festival architect. Illustration 7.4 shows a model of the treatment: this was an atomic world. The entrenched and bitter Ministry of Works had always said there was not enough time. One senior official saw the photo and said, 'Personally, I think this sort of treatment quite deplorable but there is so little time left that we can hardly put the Festival Architects to the trouble of redesigning.'[21] Twelve months later the exhibition was built, O'Dea's star was high in the heavens and the first visitors to the 'Exhibition of Science' started trickling down the newly decorated Museum Lane to glimpse the future.[22]

A visit in 1963: 'Aeronautics' and 'Sailing Ships' in the Centre Block

On 10 July 1963, the press crowded into the third floor of the Science Museum's newest building to see the opening of the Aeronautics Gallery, hot on the heels of the Sailing Ships Gallery on the floor below revealed five months previously. Its top-floor location must have seemed like an ascent to the skies for some; a newly installed suite of escalators ferried visitors from the entrance hall up to the aeroplanes in their lofty hangar of suspended animation. After an earlier preview of the gallery, the *Sunday Times* exclaimed that: 'a war against boredom is being waged at the Science Museum...Instead of row upon row of glass cases, planes are suspended in mock flight from the roof of a hangar...and ships and boats are displayed in the form of real ocean-going liners.' There was one star of the show. 'The man who is waging the war is a 58-year-old Lancastrian keeper at the Museum, Mr W.T. O'Dea. He considers the traditional museum in Britain "awful".'[23]

As the 'Exhibition of Science' brought new hope to the Science Museum, all eyes focussed on planning the content of the rest of the Centre Block. Shaw died in 1950 and the new Director, Frank Sherwood Taylor, inherited a set of plans half-formed and embroiled in internal strife. In August 1951, a fresh set of proposals for gallery layouts in the new wing was sent to Museum Keepers for comment, and O'Dea had decided to launch a broadside attack on a planning situation he felt had lost its way. 'There is so much that seems open to criticism that I have paused to think again broadly about the whole business,' he said.[24] His note outlined the role of circumstance and personal politics in the ultimate form of the Museum. The Keepers wanted to devise a coherent scheme for display, but they kept a close eye on the space allocated to their own collections. Size was everything. But Shaw had faced an additional, external, challenge. At the end of the

Second World War, only a proportion of his staff had returned and a great deal of space had been lost to other, more pressing, demands. He had eventually got fed up with the internecine warfare among his Keepers and started to develop the space allocation plans himself. O'Dea recalled, 'I believe matters developed under Shaw's personal eye and that he was in the middle of complicated calculations when he died suddenly.'

It was complicated indeed. 'It was all a question of timing,' said O'Dea, for, besides the internal politics, dealing with external bodies (such as the chemical or electrical industries) for exhibit assistance left the Museum at the mercy of late exhibits and commercial jockeying for the best spots. Added to the mix were practical display complexities dictated partly by O'Dea himself in demanding the exclusion of light in certain galleries. He summed up the frustration and the politics. 'I have observed that the logic of the arrangements under considera-tion appears to be dwindling, under the influence of various natural restrictions (such as inappropriate gallery sizes or conditions) and prejudicial factors (such as the alleged commitments with the electrical and chemical and gas industries of which only the latter has got to the stage of producing the goods!) to insignificant proportions. Could we therefore start again with unbiased minds, or less biased minds, and ... examine a few fundamental issues?' He observed that the loser in all this politicking and internal jockeying for space allocation was the visitor. 'Too much regard, I think, is being paid to exact percentages; to the artificial barriers (to the public) between departmental responsibilities; to the placation of puta-tively generous industries by allowing them to regard space booked for them that time has later disclosed to be unsuited to their subject.'

Having solicited opinions from his Keepers and received the robust response from O'Dea and others, Sherwood Taylor sketched out his own draft plan for a newly reorganised Science Museum with a complete central block. If its realisa-tion was all about timing, as O'Dea had suggested, its form was (as Mackintosh had envisaged in 1931) an attempt to map the material culture of science and industry onto a three-dimensional building. He ordered the space by broad clas-sificatory groups: the third floor would house air transport, physics and com-munications. Next came water transport, chemistry and its applications (such as photography, glass and pottery). The first floor was a split between 'Earth and heavens' and 'industries', and land transport and power occupied the double-height ground floor galleries. The basement took up mining and metallurgy alongside its long-standing Children's Gallery. But it was by no means pure. Sherwood Taylor had a set of 'fixed positions' that dictated the basic layout: mining was entrenched in the basement where a demonstration mine had been constructed; the beam engines built into the East Hall's ground floor (with special foundations through to the basement) could not go anywhere else; the ground floor of the Centre Block was the only place big enough for the full-sized road and rail transport exhibits, and so on. The problem with a science museum is the nature of the objects. They just don't like to be captured, and force-fitting them to a matrix of departments, each with its own Keeper (and agenda), coupled with their own physicality and that of the building, makes the job of the Director a very tricky one indeed. Sherwood Taylor's sketch layouts of 1955 were issued as he himself lay gravely ill, sent with a set of instructions and apologies that he had not completed his designs. Like Shaw before him, he died before he could complete a rational plan agreed by all, and on his untimely death in January 1956, aged just 58, Sherwood Taylor's plan was thrown back up in the air.

Few aspects of the layout caused more argument than the top floor of the Centre Block. It was prime real estate, and there were two candidates for occupation: Aeronautics (O'Dea) and Astronomy (Henry Calvert). Calvert wanted astronomy displays with a staircase to a rooftop planetarium, and an astronomical observatory at the west end of the existing building roof. O'Dea, on the other hand, demanded a hangar-style, column-free space to display full-sized aeroplanes. Under this option, the planetarium, if built, would go in a ground floor building in the car park, reached by a staircase from a first-floor astronomy gallery. Sherwood Taylor had suggested that the aeronautics scheme would have the advantages of 'keeping departments pretty well together.'[25] Calvert had tartly retorted, 'I advise very strongly against an arrangement which puts an Astronomical Observatory on the roof, an Astronomical Planetarium in the basement at the other end of the building and the Astronomical Instruments somewhere in between.'[26] The Director was not convinced. After Sherwood Taylor's death in January 1956, Calvert immediately lobbied his successor, Terence Morrison-Scott, explaining, 'I wrote a memorandum to the late Director about the proposed plan, but he was not able to give it consideration because of ill-health.' His trump card was to invoke intermuseum rivalry. 'I have several splendid large telescopes of great public interest ... but if the late Director's proposed plan is adopted there will be little prospect of showing them and I should suggest that they be sent to Greenwich where the National Maritime Museum is forming an astronomical museum in the old observatory.'[27]

But it was not enough. The argument between astronomy and aeroplanes was, in the end, resolved by an outside pressure. After the war, the aeronautics collection languished in the old Western Galleries on the site of the Imperial College, which wanted to demolish the old buildings for a new department of aeronautics (appropriately enough). Replacement space was urgently needed for the Science Museum's aeroplanes, and it was this impetus that finally convinced the Ministry of Works to recommence the building of the Centre Block. Circumstances had transcended internal politics, the top floor of the Centre Block was given to the pragmatic aeronauts against the idealistic astronomers, and politically astute O'Dea had ensured that his own collections were the first to be housed in the new block and occupied the best gallery space. On Morrison-Scott's departure in 1960 to run the Natural History Museum, the Centre Block was finally under construction, and was handed over to the Museum for fitting-out in September 1961.

Morrison-Scott's replacement was, like O'Dea, a Science Museum man through and through. David Follett, Keeper of Electrical Engineering and Communication, had joined in 1937 just as O'Dea's lighting exhibition had come to an end, and in 1960 his first project was the realisation of O'Dea's latest work. The new Aeronautics

Gallery, at 18,000 square feet, was 50 per cent bigger than that which currently housed the aircraft, and designed with a cantilevered roof suitable for the suspension of some twenty full-sized aeroplanes. A central walkway raised visitors nine feet above ground to inspect the exhibits close up, with space for showcases containing smaller artefacts housed underneath. This gallery, together with its marine counterpart on the floor below, was O'Dea's most dramatic intervention yet in museum displays. He spent much time expounding his views on museum design to visiting journalists as the galleries took shape. To Raoul Engel, writing in *Discovery*, he showed the almost-complete Sailing Ships Gallery: 'This conception would probably have shocked my Victorian predecessor half to death; to even *contemplate* the need to please the public was heresy! Well, then, we're heretics; we've frankly set out to please and stimulate non-specialists; we want them to enjoy themselves, as well as to learn.'[28] Engel was impressed, and his advance viewing of the aeronautics display upstairs cemented his admiration for 'the forceful and imaginative keeper in charge, who strongly influenced their design.' The best bits of the central block had been given to O'Dea, the man who had played such a significant role in shaping the Museum's buildings since the 1930s. The Astronomy Gallery finally opened in May 1966, on the first floor, with no planetarium and a long way from its observatory. Forcefulness and imagination had favoured the engineer, but O'Dea had wanted the Director's job back in 1956 and in 1960, being turned down both times. In the autumn of 1966, three long decades after 'Electric Illumination', aged 61, he called it quits and headed overseas.

A visit in 1981: The Wellcome Museum of the History of Medicine in the Infill Block

December 1981 saw visitors to the Science Museum confronted with exhibits of a type never before seen at the South Kensington attraction. Those who made their way up to the fifth floor of a newly built block of galleries were faced with a very visceral science: human nature, red in (false) tooth and (prosthetic) claw. Princess Alexandra had cut the ribbon on the new Wellcome Museum of the History of Medicine, an air-conditioned and purpose-built suite of galleries housing parts of the most remarkable single collection of artefacts ever amassed.

Henry Wellcome – patron of science and medicine research, philanthropist, entrepreneur and businessman – was also a collector, who conceived in the late nineteenth century the idea of a museum 'for the purpose of demonstrating by means of objects... the actuality of every notable step in the evolution and progress from the first germ of life up to the fully developed man of today.'[29] The Wellcome Historical Medical Museum opened in London's Wigmore Street, north of Oxford Street, in 1914 following a temporary exhibit the previous year, and by the time of his death in 1936 Wellcome had amassed a collection of something like one million artefacts – five times more than the Louvre.[30] Yet few, and certainly not the general public, had the opportunity to see Wellcome's collection in his lifetime. His museum was for specialists and researchers, attracting only a few dozen visitors each month. Wellcome was not particularly keen on widespread

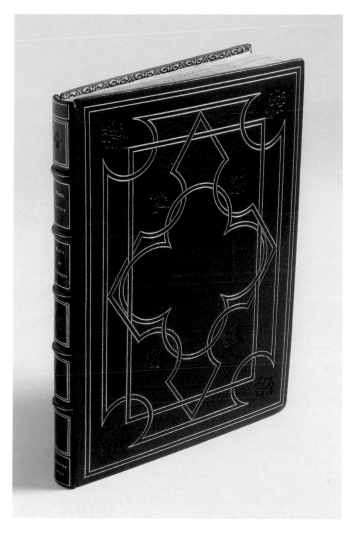

Plate 1 Martin Cortes, *Breve compendio de la sphera y de la arte de navigar*, 1551

This beautifully bound book can stand as a metaphor for the history of the Science Museum Library. The *Breve compendio de la sphera y de la arte de navigar* (Short compendium of the world and of the art of navigation) deals with three great themes: the composition of the world and the universal principles of the art of navigation; the movements of the sun and the moon and the effects they produce; and the construction and use of instruments and the rules of the art of navigation. The Library has had its own difficult course to navigate through uncharted political waters, it has had to observe and react to the shifting priorities of government and the scientific community and it has developed tools and techniques to record the ever-increasing numbers of scientific publications. The Library purchased this important book along with many other major works of science and medicine at the auction of the Honeyman Collection in 1979. This book also represents the international nature of the Library's holdings, outstanding collections that contain the written and printed record of 500 years of science and engineering.

Plate 2　Francis Thompson's engine, 1791

This engine is the oldest survivor of the type first developed by Thomas Newcomen around 1712 to survive complete and essentially unaltered. Its arrival in the Science Museum relied as much on luck as it did on judgement. The Museum's East Block, the construction of which began in 1913, was intended from the start to incorporate a number of 'engine pits' – spaces to house large steam engines, a number of which formed the nucleus of the Motive Power collection. However, the collection lacked an atmospheric steam engine and Henry. Dickinson was tasked with making good the deficiency. By the end of 1914 he had completed a survey of candidate engines but thereafter his plans were suspended until 1919 when a friend visiting Pentrich Colliery, Derbyshire, warned him that the engine there was derelict and in danger of being scrapped. Dickinson oversaw the purchase, dismantling and re-erection of the 105 tons of iron, stone and timber comprising the engine and large portions of its house in the Museum's new East Hall. It remains there today, symbolising the substitution of mineral energy for wind, water and muscle power thereby laying the foundations for Britain's industrial revolution, but also for the greenhouse effect three centuries later.

Plate 3   Coalbrookdale by night

This painting of the Darbys' famous Bedlam iron furnace on the River Severn by the French artist Philippe Jacques de Loutherbourg (1740–1812) is currently displayed in 'Making the Modern World' Gallery. It is one of the most requested objects for loan in the Museum's collections and is widely reproduced as an icon of the Industrial Revolution. Yet it was one of the most controversial acquisitions ever made by the Science Museum. In April 1952, the new Director Frank Sherwood Taylor was alerted to the sale of the painting by A. S. Crosley of the Newcomen Society and after viewing it at the London art dealers, Pictura, immediately bought it for £250. This snap purchase brought him into conflict with Fred Lebeter the Deputy Keeper of Chemistry in charge of metallurgy, because Lebeter regarded the painting as inaccurate and thus worthless from a historical point of view. With his background in science popularisation, Taylor saw the value of this 'striking and important picture' of the Industrial Revolution 'to fire the imagination of the spectator' in contrast to the 'aridity' of the existing metallurgical collections. It is interesting to note that this painting was acquired when dioramas were at their peak at the Science Museum since de Loutherbourg was a pioneer of dioramas with his miniature theatre, the 'Eidophusikon'.

Plate 4   Stephenson's 'Rocket' locomotive, 1829

The 'Rocket' illustrates both the change and the continuity characteristic of the Museum's collections. The Museum has continued to acquire objects that have been repeatedly reinterpreted in terms of changing concepts of science, technology and history. The much-modified 1829 locomotive was donated to the Patent Office Museum in 1862. A few years after the publication of Samuel Smiles's *Self-Help*, it seemed to symbolise the ingenuity of the self-made British engineer, Robert Stephenson who had designed it originally. By the time of the 1876 loan exhibition of scientific apparatus in South Kensington, the 'Rocket' was being shown as a masterpiece of 'applied mechanics', within which locomotives were shown merely as a subsection of 'prime movers'. After the takeover of the Patent Office Museum, in 1884 it was incorporated within the 'Machinery and Inventions' division of the Science Collections of the South Kensington Museum. Subsequently, it was included in the Transport Department of the Science Museum as part of the development of land transport and, from the 1960s to the 1980s, exhibited in the shadow of the much larger and more impressive then-modern Deltic diesel locomotive. Currently the locomotive is shown in the Science Museum's 'Making the Modern World' Gallery as an icon of the Industrial Revolution.

Plate 5 Five-needle electric telegraph, 1849

The Cooke and Wheatstone five-needle electric telegraph is an example of how an object acquired many years ago as an examplar of its technology has mistakenly come to be regarded as the original artefact itself. Two five-needle instruments were among several designs demonstrated by the inventors, William Fothergill Cooke and Charles Wheatstone, to the directors of the London and Birmingham Railway during 1837. Though the demonstrations were successful the directors declined to adopt the system. The two instruments were purchased back by the inventors and taken to Wheatstone's laboratories in King's College London. One of these, on loan to the Science Museum since 1963 but held in store, has recently been identified as an original 1837 instrument. The five-needle instrument on display at the Museum since 1876 has for many years been described as one of the 1837 originals, but several anachronistic details make this impossible. Recent research has instead identified it as one of several working models made in 1849 for demonstration at a patent infringement trial. First placed on show at South Kensington in 1876 in the 'Loan Collection of Scientific Instruments', it is currently displayed in the 'Making the Modern World' Gallery.

Plate 6   Mauve dye crystals, c.1863

When 18-year-old William Henry Perkin discovered an artificial purple dye at Easter 1856, he began an industrial revolution which led to a myriad of industrial products based on synthetic organic chemistry. Known as mauve from 1859, his dye was soon replaced by other artificial dyes. By the fiftieth anniversary of his discovery, Perkin was long retired, and the British dye industry was overshadowed by its German counterpart. British chemists led by Raphael Meldola used the anniversary to assert British priority in this field and lament the subsequent failures. The Science Museum first exhibited mauve in the 'Seventy-five Years of Progress in Technical Industry' exhibition held in the summer of 1932. The section on dyes including mauve proved popular and was retained as the 'Manufacture of Artificial Dyes' (during the autumn). The jar of crystals lent by Perkin's son Arthur George Perkin was returned before his death in 1937 and has disappeared. The jar shown here was donated by Annie Florence Perkin, daughter of W. H. Perkin Senior, for the 'Centenary of the Chemical Society' exhibition (1947). Recent chemical analysis demonstrates that this sample (mauveine acetate) probably dates from 1863–4 when Perkin was researching the structure of mauve, and has also led to the discovery of new isomers. The jar is not on display at present.

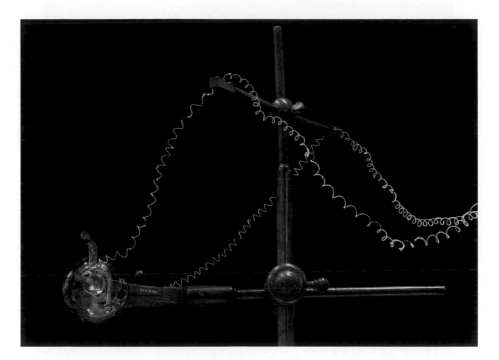

Plate 7   Russell Reynolds' X-ray Set, 1896

This X-ray set is one of the oldest anywhere in the world - a remarkable survival from the months immediately after Wilhelm Röntgen announced his discovery of a 'new type of ray'. While still at school and working with his general practitioner father, Russell Reynolds was enthused by the first British report of the discovery of X-rays in January 1896, and set out to build his own X-ray apparatus. It was possible in those days to purchase items of physical apparatus from the many firms trading in the Hatton Garden (Holborn) district of London. Reynolds completed the apparatus within a year. But by donating this set to the Science Museum in 1938, Reynolds was not merely preserving an early X-ray machine. Like so many other donors, by giving the Museum his own apparatus, Reynolds was making a claim to be considered a pioneer of radiology, the profession to which he devoted his life. This example demonstrates the status that the Museum had acquired by this date, as inclusion of apparatus within its collections could be seen as connoting fame and significance.

Plate 8   Ford Model T motor-car, 1916

The acquisition of the Ford Model T in 1997 for the 'Making the Modern World' Gallery typifies the changing criteria which have governed acquisitions and objects on display, for it has two stories. The obvious one is as a sound, utilitarian design that motorised America. But its significance now is seen less in its particular features than as a symbol of Henry Ford's production line techniques and as rhetorical focus of the worldwide admiration of 'Fordism'. The moving production line is the perfect realisation of the Enlightenment project 'to turn men into machines', for one of its lesser known ambitions was to extract the knowledge of the secretive craftsman and put the workplace under the direction of 'a philosopher'. Adam Ferguson wrote that 'mechanical arts ... succeed best under a total suppression of sentiment and reason. ... Manufactures prosper where ... the workshop may ... be considered as an engine, the parts of which are men.' This could be a perfect description of Ford's Highland Park factory. Intriguingly, Ford and his methods impressed both Hitler and Stalin.

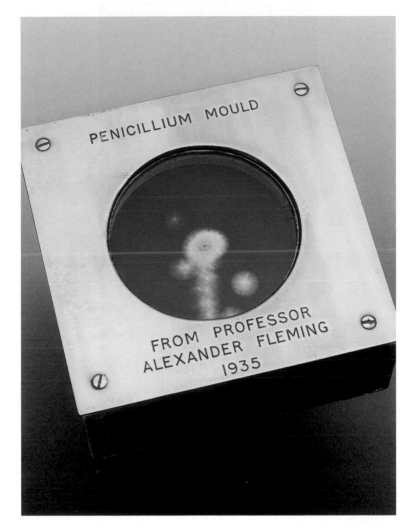

Plate 9 Penicillium mould sample, c.1935

Penicillin is one of the true icons of scientific creativity and of the good that can follow from scientific research. Over the previous sixty years the Science Museum has
acquired many artefacts relating to the development of penicillin as a medicine at Oxford
University early in the 1940s. This sample of Penicillium mould represents the slow transformation of penicillin from a curious phenomenon into a life-saving medicine. In 1935
Alexander Fleming first learned about the benefits of the newly discovered Prontosil
sulphonamide drug. He ruminated to a friend that he had discovered something potentially much more powerful yet, but had found no interest. He then gave this sample to
the friend. It was purchased at auction by the Museum in 1997. It is currently displayed
in the Science Museum's 'Making the Modern World' Gallery as an icon of that period
of 'Defiant Modernism' in which Britain turned from the Empire to Science as a source
of greatness.

Plate 10   V2 Rocket Engine, 1944

This V2 (or more accurately, A4) rocket engine symbolises, as do all of the surviving engines, many things: its debut in the Second World War as the propulsive means of the first long-range ballistic missiles; the innovative technologies the engine employed, not least the propellant pumping and inertial guidance systems (which in turn paved the way for the invulnerable missiles of the Cold War); the barbarous treatment of its slave manufacturers, not to mention the death and destruction that over three thousand V2 rockets brought to England; but, for most of its display time in the Science Museum, its role in the development of the first space launch vehicles. This particular engine was acquired in 1949 from the Ministry of Supply's Rocket Propulsion Department at Westcott, Buckinghamshire. When put on display in 1963 as part of the new Aeronautics Gallery on the third floor of the Museum it was slotted into a propulsion narrative in which various rocket engines were displayed near to the air breathing machines of aircraft. But in 1986 the engine was moved down to the new Exploration of Space Gallery on the ground floor where, with its painted pipe work, it illustrated the arterial routes by which fuel and oxidant were pumped from tank to combustion chamber in the early space rocket. In 2000, the engine was moved a few metres to its present location, in what became the Exploring Space Gallery (2006), where minimal interpretation evinces object as (technocratic) icon of the Space Age.

Plate 11 Pilot ACE computer, 1950

The Pilot ACE (Automatic Computing Engine) was one of the first electronic stored-program computers in Britain. Created at the National Physical Laboratory in Teddington, Middlesex, the machine embodies the original ideas of the genius mathematician Alan Turing (1912–1954) and his conceptual discovery of the general-purpose computer. Pilot ACE ran its first program in May 1950, and for the next six years it was employed in advanced scientific and engineering work, including aircraft wing flutter, X-ray crystallography and calculating bomb trajectories. Its development was intended to launch a national computing industry, but in reality the Pilot ACE only had a short working life. When it was brought into the Science Museum in 1956 the *Daily Mirror* reported 'The first robot brain has been sacked. It is now so old and tired that it makes mistakes' (28th June 1956). Pilot ACE was placed on display in the Museum's computing gallery, to reflect the fundamental shift of computers from calculating machines to programmable machines that could be applied to solve a range of tasks. Today its display in the 'Making the Modern World' Gallery illustrates the level of confidence of British post-war technology.

Plate 12   DNA model, 1953

The explanation of how our inherited characteristics are determined by the chemical structure and ordering of our DNA could count as one of the major scientific developments of
the second half of the 20th century. The breakthrough work to find the structure of DNA
published in 1953 and recounted by James Watson in his later book *The Double Helix* has
become one of the great public tales of science. At the heart of the story is the building
of a model which would demonstrate that a double helix structure could meet the exact-
ing criteria set by X-ray evidence gathered at King's College London. Within weeks of the
first excited realisation, Francis Crick and James Watson at Cambridge built a large model
to demonstrate the validity of their hypothesis and were famously photographed next to
it. That model was subsequently dismantled but most of the components were preserved.
They were reassembled in the 1970s to form this model which has been a treasure of the
Science Museum ever since. Currently, it is displayed in the 'Making the Modern World'
Gallery at the Science Museum.

Plate 13   Apollo 10 Command Module, 1969

The Apollo 10 command module, call sign 'Charlie Brown' (after the cartoon character), touched down at the Science Museum in 1976. The Smithsonian Institution, its owner, had been touring it around mainland Europe but was now keen that it be displayed in a more stable environment. Well established relations with the Smithsonian's National Air Museum – from which the National Air and Space Museum (NASM) was to be born later the same year – enabled the Science Museum to negotiate a three year loan of the spacecraft. The Museum swiftly exhibited it as a 'taster' for its forthcoming Exploration Gallery in which it would stand near to a full-scale replica of the Apollo 11 lunar module. The NASM extended the spacecraft's loan and when the Exploration of Space Gallery opened in 1986 'Charlie Brown' remained in position with an expanded interpretation of its role in the Apollo 10 mission to the Moon. In 2000 it was moved into the nearby 'Making the Modern World' Gallery for display as one of the Museum's most significant historical artefacts. With the passage of years 'Charlie Brown' appears to gain in popularity with the Museum's visitors and in the Summer of 2009, with the coincidence of the Museum's Centenary and Apollo 10's 40th, the NASM agreed to the Museum opening the spacecraft's hatch so that visitors could look into the specially lit crew compartment.

Plates of the Museum of the Future:

Plate 14   Artist's impression of 'The Beacon', a proposed new Exhibition Road façade for the Science Museum

Plate 15 Artist's impression of 'SkySpace', a proposed new Cosmology Gallery on the Third Floor

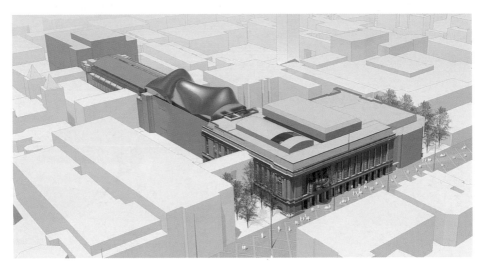

Plate 16 Artist's impression of the golden roof of the proposed new 'SkySpace' Cosmology Gallery.

public access. 'A great many people visit museums simply as stragglers,' he said in 1929, with those mainstream museums 'arranged for popular entertainment, to gratify those who wish to view strange and curious objects,' whereas his (research) museum was for 'intellectual people', 'those genuinely concerned and interested in the subjects represented there who attend entirely for beneficial information.'[31] But even amongst specialists the museum was old-fashioned and unpopular.[32] Over the forty years following Wellcome's death in 1936, parts of the collection were dispersed by the Trustees of his legacy to museums around the world, although the museum itself and its associated library remained open and active in new Wellcome Trust buildings in the Euston Road.[33]

Yet the spirit of Wellcome's distaste for 'stragglers' seeking 'gratification' over 'strange and curious objects' lived on. In 1966, a formalised Trust policy stated that the museum and library should 'provide facilities and materials for senior students of medical history,' with its space in the Wellcome building being used only for 'the promotion of research and education in medical history, and its use for subjects of peripheral interest, merely because the material is available, is contrary to the Trustees' policy.'[34] The Trust was concentrating its resources on establishing university departments for the history of medicine, and, following discussions at a 'symposium of experts' in 1967, it was starting to become unclear how, or indeed whether, the artefact museum could fit into university teaching. One attendee at the symposium, and a founder member of the History of Medicine Panel established in its wake in 1968 to advise the Trust, was Frank Greenaway, Keeper of Chemistry at the Science Museum, and from his earliest involvement in 1967 Greenaway had plans for the collection. The Wellcome museum enjoyed a refurbishment carried out between 1969 and 1972, but its place in the sun was to be short-lived. The Trustees had become increasingly concerned that the museum was being steered by its director, Frederick Poynter, towards a populism that ran against the terms of Wellcome's will, which could become vastly expensive in staff and collection costs, and increasingly hungry for space – and all of this for the public, not for advanced research, which by then was focussed on the historic library. They were worried about their museum collection and started looking around for places into which it could better fit.

In January 1972, the history advisory panel at the Wellcome Trust presented a forward strategy for the museum collections, but two months before, Greenaway had written separately to the Trust proposing that the collection be transferred to South Kensington, to enable the Science Museum to cover biological and medical history not then represented. The Science Museum had the space, staff and skills to make the most of Wellcome's legacy, but now the rhetoric had to shift. The restrictive academic focus was out; public engagement was in. The Wellcome museum only attracted some 20,000 visitors each year, explained the Science Museum in a briefing note, 'far fewer than the interest of the subject merits', and in January 1972 the Trustees resolved that the museum collection 'should be transferred to the care of the Science Museum with a view to its being kept up-to-date and exhibited to the public on a scale which would not be possible with the Trustees' financial resources.'[35]

The opportunity afforded by the Wellcome collection nevertheless raised museological questions about the visitor's experience of science and technology in the Science Museum's displays. Should the history of medicine be displayed at all? 'It is widely accepted that the origins of modern physics and chemistry are intimately connected with the origins of modern medicine,' reasoned the Museum. Medical science and technology was a major gap in national museums, and this was an argument for seeing both medical and physical science as part of everyday life too. 'By showing the history of medicine in the same building as the history of physical sciences and engineering the public would be led to see medicine as something rational and accessible, and physical science and technology as having humane associations.' Against the proposal, though, were practical considerations. Fifteen thousand square feet of exhibition space would be needed, and this was not readily to hand. Storage and reserve collection space would be needed, although the Wellcome Trust was prepared to assist with this. Money would be available from the Trust, too, to fund historical research into the collections, and their establishment of specialised university departments would provide the stock of expert staff who could be recruited to work with the new subject. The idea seemed, on balance, a good one. The only serious problem was lack of room in the South Kensington buildings. But, at exactly the same time, the Museum was in the middle of talks about extending the East Block to create a substantial tranche of new, purpose-built exhibition space.[36]

Plans to extend the Museum following the completion of the east and central blocks went back as far as the second part of the Bell Report in 1912, but, given the protracted completion of the first two buildings, the final phase was not considered until the early 1970s, once the Centre Block had been filled and the eastern galleries redisplayed. The main plan was to extend westwards to Queen's Gate, reorienting the Museum's main entrance onto that end to provide for the first time a spacious and properly served introduction to the Museum's subjects.[37] It hasn't happened yet; in the late 1990s the Wellcome Wing was built on the west end of the Centre Block, occupying about half of the remaining site and housing modern science and technology exhibitions, and the Wellcome Wolfson Building fronting Queen's Gate was opened in November 2003: a small office block housing a centre for public science engagement. The main entrance remains the cramped east frontage on Exhibition Road. But the 1970s did see a building extension project that substantially increased exhibition space in the old east building – the infill of a light well with a structurally independent building tied through to the 1920s galleries, with two extra floors emerging above roof level, the top floor of which was cantilevered over the existing East Block roof, providing a substantial amount of column-free space.[38]

Plans for the Infill Block began around 1968 when David Follett discussed options for an East Block extension with the superintending architect at the Ministry of Works, who had experience of filling in courtyards at the National Gallery. This was an ambitious and creative scheme, yet one that would have been unthinkable thirty years previously. As the 1930s attempts to seek funding for the Centre Block had rumbled on, one newspaper reported the President of the Board of Education, Lord Stanhope, suggesting that 'the Science Museum had better abandon hopes of lateral expansion and grow upwards.' The Museum had been aghast, noting that it was 'contrary to modern practice in museum buildings throughout the world. Beyond two, or at most three, storeys people will not go, and such upper galleries remain "unvisited mausoleums".'[39] Nevertheless, by the 1960s museum fashions had moved upwards; the Centre Block possessed its four storeys plus basement and was interconnected by a set of escalators, and some of its most popular exhibits were housed on the top floor. For Follett, both lateral and vertical expansion was welcome, and setting the foundations for the new block was one of his final acts at the Science Museum before retiring in 1973.

Follett was succeeded by Margaret Weston, a long-time Science Museum curator who had joined in 1955 as Sherwood Taylor was struggling with the internal warfare of the Centre Block project. Now the Museum was grappling with the implications of the most significant addition to its collections and scope in its history: taking over 'the greatest collection of its kind ever made'.[40] The Infill Block provided the answer to the accommodation problem and in 1974 Weston allocated the large top floor, plus parts of the floor below, to the Wellcome Museum. Building work began on site in 1975, the handover of the collection was announced to the press in 1976 and legally approved in 1977, and the new block was handed over soon after.[41] The new Wellcome Museum of the History of Medicine gained its permanent presence at the Science Museum in December 1981 with the public opening of the 'Science and Art of Medicine' gallery. *Times* reporter Mel Lewis found the contents of the 537 showcases in the new gallery 'fascinating, if grim',[42] whereas a colleague, Agnes Whitiker, had reported the previous year that its sister gallery, 'Glimpses of Medical History', would be ideal for 'teenagers interested in medicine'.[43] The criticism levelled at high-rise museums as 'mausoleums' in the 1930s must have seemed somewhat ironic in this lofty suite of medical galleries; the sole artefact chosen by Lewis to illustrate the review was a human skull.

Conclusion

In 1896, Louis Sullivan, the American modernist architect, said of buildings, 'Form ever follows function.' Twenty-eight years later, Winston Churchill said at an Architectural Association awards dinner, 'We make our buildings and afterwards they make us.' Stewart Brand, building historian, engineer and inventor, thought both these insights stopped short. Instead, he suggested, 'First we shape our buildings, then they shape us, then we shape them again – ad infinitum. Function reforms form, perpetually.'[44] He was taking a whole-life view of buildings, rather than seeing them as finished, unchanging and unproblematic on the day the ribbon is cut.

With this in mind, we have seen O'Dea respond to the constraints and opportunities of the Science Museum's East Block in his 1936 'Electric Illumination' exhibition, which was an attempt to get valuable publicity and visitor figures in the big front hall in advance of a permanent show destined for the bowels of the basement. The East Block had been specified twenty-three years previously, and times had changed. O'Dea reshaped its bright, windowed halls to embrace new display techniques and colonise space for his subject. Fast-forward fifteen years to 1951, and O'Dea's new gallery on agriculture opened, ushering in a new paradigm of museum display and lighting technique that is still fresh today, and which caused the old East Block to be reshaped in a way that would never have been dreamed of, let alone sanctioned, in its planning days of the 1910s and 1920s: with a windowless wall to exclude daylight. Function reforming form, perpetually, with doubtless a little bit of post-war austerity thrown in for good measure. We have seen the consequences of the politics of space, both global and local: victims and victors of circumstance, deals done, territory fought over. O'Dea played the game and won, at least for bricks and mortar, as his lobbying finally led to the building of the Centre Block, for which he was rewarded with prime real estate. And money talks, of course: shifting resource priorities in the Euston Road shaped a South Kensington building extension and brought the history of medicine to the Science Museum.

Amidst all of this, the visitors visit, trying to find their way, to make sense of the buildings and layouts, stories and histories at the Science Museum. Stewart Brand also said: 'When we deal with buildings we deal with decisions taken long ago for remote reasons. We argue with anonymous predecessors and lose. The best we can hope for is compromise with the *fait accompli* of the building.'[45] A worthy and suitable house would always be a tall order.

Notes

1. H.G. Wells, *The Time Machine* (London: W. Heinemann, 1895).
2. This *et seq.* from *The Times*, 16 December 1936, p. 12.
3. Terms of reference of the Bell Committee, as cited by David Follett, *The Rise of the Science Museum under Sir Henry Lyons* (London: Science Museum, 1978), p. 19.
4. *Report of the Departmental Committee on the Science Museum and the Geological Museum*, Part I (London, 1911), p. 5.
5. John Liffen, 'The development of collecting policy within the South Kensington Museum and the Science Museum', unpublished briefing paper, April 2002.
6. Francis Ogilvie, Memorandum, 24 January 1920, SMD, ED 79/3.
7. H.G. Lyons, 'A Memorandum on the Development of the Science Museum from 1920 to 1933', August 1933, SMD, Z 186.
8. H.G. Lyons, 'A Memorandum on the Development of the Science Museum'.
9. See W.E. Bijker, *Of Bicycles, Bakelites and Bulbs: Towards a Theory of Sociotechnical Change* (Massachusetts: MIT, 1995), chapter 4, for a history of fluorescent lighting.
10. *The Electrical Review*, 18 December 1936, p. 865.
11. Science Museum Minute ScM 5497/1/1, SMD, ED 79/47.
12. Letter, O'Dea to Jones, 22 April 1936, SMD, ED 79/47.
13. *Light and Lighting*, January 1937, p. 22.
14. Advisory Council Annual Report for 1938, p. 9.
15. SMD, Z 188.
16. Confidential memorandum to Director and Keepers, 29 May 1951, SMD, Z 183/1.
17. Letter, Procter to Shaw, 26 July 1948, SMD, ED 79/181.
18. Memorandum from the Director, 13 February 1950, SMD, ED 79/181.
19. See, for example, his obituary in *The Times*, 24 November 1981, p. 12.

20. W.T. O'Dea, 'The Science Museum's Agricultural Gallery', *Museums Journal* (1952), pp. 99–301, on p. 299.
21. Memorandum, from Eric de Normann to R. Auriol Barker, 22 April 1950, in TNA: PRO, Work 17/388.
22. Notes of meeting, 27 July 1948, and subsequent Festival / Science Museum correspondence, in TNA: PRO, Work 17/388.
23. *Sunday Times*, 3 February 1963, in SMD, Z 183/2.
24. Confidential memorandum by William O'Dea to Director and Keepers, 29 May 1951, SMD, Z 183/1. Subsequent correspondence cited is in this file too.
25. Memorandum, from Sherwood Taylor to Calvert, 20 November 1954, SMD, Z 183/1.
26. Memorandum, from Calvert to Sherwood Taylor, 24 November 1954, SMD, Z 183/1.
27. Memorandum, from Calvert to Morrison-Scott, 23 June 1956, SMD, Z 183/1.
28. *Discovery* (March 1963), SMD, Z 183/2.
29. Ghislaine Lawrence, 'Wellcome's Museum for the Science of History', in Ken Arnold and Danielle Olsen, eds *Medicine Man* (London: British Museum, 2003), pp. 51–71, on p. 59.
30. Lawrence, 'Wellcome's Museum', p. 51.
31. Henry Wellcome, in *Great Britain, Royal Commission on National Museums and Galleries, 'Oral Evidence, Memoranda and Appendices to the Final Report'* (London: HMSO, 1929), quoted in Ghislaine Skinner, 'Sir Henry Wellcome's Museum for the Science of History', *Medical History* 30 (1986), pp. 383–418, on p. 400.
32. See Lawrence, 'Wellcome's Museum', pp. 65–9, for an examination of the museum's users.
33. For the dispersal of parts of Wellcome's collection, see Georgina Russell, 'The Wellcome Historical Museum's Dispersal of Non-Medical Material, 1936–1983', *Museums Journal* 86, (1986), Supplement 1986, pp. S3-S15.
34. Quoted in B.A. Bembridge and A. Rupert Hall, *Physic and Philanthropy: A History of the Wellcome Trust 1936–1986* (Cambridge: Cambridge University Press, 1986), p. 134. See this work, pp. 134–49, for the subsequent treatment of this aspect of the Wellcome collection history.
35. Trustee Minute, 10 January 1972, quoted in Bembridge and Hall, *Physic and Philanthropy*, p. 143.
36. Set of briefing notes, 1972, SMD, Z 191.
37. Memorandum, Follett to Keepers, 4 January 1973, SMD, Z 158.
38. Ralph Mill, 'The Science Museum East Infill Scheme', *The Structural Engineer* (February 1978), pp. 29–35.
39. *Birmingham Post*, 10 May 1939, in TNA: PRO, ED 23/883.
40. Set of briefing notes, 1972, SMD, Z 191.
41. Briefing note, Lance Day, 22 July 1977, SMD, Z 207/15.
42. *The Times*, 18 December 1981, p. 21.
43. *The Times*, 29 November 1980, p. 5.
44. Stewart Brand, *How Buildings Learn: What Happens After They're Built* (London: Phoenix, 1994), p. 3. The two quotations are cited here too.
45. Brand, *How Buildings Learn*, p. 2.

8
Exhibiting Science: Changing Conceptions of Science Museum Display

Andrew Nahum

An enduring language

In 1910, the future novelist and aircraft designer, Nevil Shute, discovered the Science Museum. By bunking off school and paying a penny excess fare on the London Underground, the 11-year-old discovered:

> a wonderland of mechanical models in glass cases, among examples of the real thing. There was the actual original locomotive, Stephenson's Rocket, and dozens of scale models in glass cases, some of which would go by compressed air when you pressed a button. There were working models of steam hammers, and looms, and motor cars, and beam engines, and above all, there were aeroplanes. For ten days I browsed in this wonderland ...[1]

Shute's subsequent technical career as a designer on the Vickers airship R100, and then as a founder of the aircraft company Airspeed, could be seen as a direct fulfilment of the explicit agenda for education and technical exposition envisaged for the Science Museum from the very outset. However, an intriguing element in this encounter with the Museum (which parallels that of many who subsequently made careers in engineering and science) is that the displays Shute would have seen were very little different in style, content and general approach from those put in place by the first generation of curators or 'Museum officers' who first populated new galleries with machines, cases, models and labels some fifty years before.

The visual language, and what we might now call the narrative content, of these austere yet charming displays reflected the ideological ambitions of the Museum's founders, advisers and curators, but also their views on the very nature of their subjects and on what a museum of science and industry itself was. In itself, the extraordinary longevity, or perhaps rigidity, of this display form that they constructed is remarkable. But even the next forty years after Shute's boyhood visit saw little change in technique beyond some tentative but suggestive work in the 1930s, though there was then a growing current of new practice from the 1950s.

Today, curators and interpreters are aware that they are, in a sense, publishing or 'producing' a creative product (the use of these words here is meant to point up

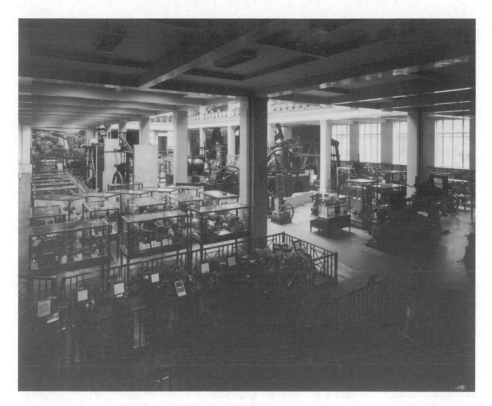

Illustration 8.1   The East Hall in the 1950s still demonstrating a display language that Nevil Shute would have recognised before the First World War

the parallels with the worlds of print and television). As with these other creative forms, the narrative or 'messages' for permanent galleries or special exhibitions are likely to be clearly articulated well in advance, while many other interpretive and technical choices will be consciously made. These particularly include design as an adjunct to narrative, applied both 'locally', to graphics, illustration, exhibits and cases, and also holistically (and if it can be afforded), to the control or reworking of the whole space with a regard, perhaps, to circulation – the way visitors encounter exhibits in sequence or consume the exhibit – but also the way it is intended to be perceived, and its ambiance.

By contrast, these questions of narrative, structure and appearance were hardly considered in the Science Museum from the 1850s to the 1950s. Rather, it was the purpose and function of the Museum which occupied internal discussion, together with issues concerning the representation of different subjects and the balance between them. It is as though the principal actors felt that, if these ruling principles for the Museum were clear, the desired outcome, in terms of public uptake of information and understanding, would naturally follow. Issues concerning what we might now call the 'language' of display seem rarely to have arisen. Indeed, the technique of communication must have seemed entirely unproblematic, being answered quite simply by the provision of galleries and cases that were 'well laid out' and furnished with 'suitable' labels.

Typical of this thinking is the introduction to the *Catalogue of Machinery* for 1890.[2] The introductory note observes:

> this collection...has been formed by direction of the Lords of the Committee of Council on Education with the view of affording in the best possible manner information and instruction on the immense variety of machinery employed in the manufactures of this country....It is desired by their Lordships that the Collection should be made as complete as possible, so as to include examples of all good and new machines.

The 'Collection' in this case, and at this time, of course means the collection on display. Clearly, therefore, this expressed desire for completeness rather precluded selection of exhibits in the interest of framing or refining a story or approach. The Museum, as willed by the Committee, is thus not a vehicle for comment on, or analysis of, the world of mechanism and industry, but was seen as an idealised microcosm – a compressed representation of the landscape of current machines, prefaced by the steps by which these had been reached.

Writing in 1922, Director Henry Lyons still maintained an identical outlook. The historical exhibits were 'to represent the steps by which progress has been made.'[3] In parallel with this, Museum officers also aspired to a taxonomic approach that would make the coverage of science and technology as complete as possible. In such a scheme the nearest parallel to the developing visible displays of the Science Museum were natural history collections. There, collections of monkeys, bats or butterflies were celebrated for their completeness, and certainly the same spirit operated in the Science Museum. Thus, in 1937, a 'spatial analysis' of the Museum still aspired to this taxonomic and evolutionary approach in a discussion of the areas to be given to each subject, noting the space needed for Astronomy 'to fill gaps in the historical series.' The pumping machinery collection was 'fairly fully developed, especially historically', but, though the reciprocating steam engine 'was not likely to be greatly developed in the future', the (actually rather extensive) historical series had 'a few gaps'. In addition, 'many important modern types of boiler...are not represented' while 'there are gaps in the historical series of locomotives which should be filled.'[4]

Dioramas and the new pictorialism

By the late 1920s, while the traditional display technique remained pre-eminent, there were signs of new approaches and influences. Ghislaine Lawrence, in an insightful paper, has argued that, at this time, retail window dressing and shop display techniques became an admitted influence on Science Museum exhibits. Thus in 1929 'a painted backdrop was placed in a case containing...aircraft models...depicting appropriate sky and cloud backgrounds executed from technically accurate paintings....The event was noteworthy enough to appear in the Museum's Annual Report.'[5] Lawrence sees, in this (still quite rare) move to deploy backboards in nineteenth-century cases, originally constructed for all-round viewing, and in the infusion of retail display techniques in the 1930s, an increasing 'pictorialisation' of display, inspired perhaps by the 1902 Education Act, which provided for children to visit museums in school hours. This was complemented by calls in the *Museums Journal* to 'educate rather than collect' and to replace the 'purposeless mumble-jumble hoard of previous years' with a systematic, logical and full explanatory display.'[6]

In response to this mood, perhaps, a fuller and more highly worked form of pic-torialisation was introduced from 1931, with specially commissioned dioramas, as deployed in the 'Introductory Galleries' of December 1931 (quickly renamed the Children's Gallery). The production of these intricate and often charming model views with composed perspectives became an intriguing art form in itself. Sometimes these settings were used as backdrops for existing scale models in the collections but, in the most developed form of diorama art, real models were unnecessary, for the scene was composed as a complete artwork and the land-scapes were populated with new, specially executed machine models and human figures, scaled by the ruling perspective.[7]

A recent study by Jane Insley suggests that the idea of using dioramas appears to have originated with the extraordinary effect they had in brightening the gal-leries of the Imperial Institute. Its director, Sir William Furze, had noticed the impact of peopled models at the British Empire Exhibition at Wembley in 1924–1926, and had invited the artists responsible to undertake scenic model-making for the displays in a rather unprepossessing exhibition space. Over the next three decades, the studio run by Raphael Roussel and other independent artists sup-plied the Science Museum with series of dioramas to illustrate many new galler-ies. This tradition, starting with the Children's Gallery, continued intermittently until the 1980s, with the gallery, today called 'Glimpses of Medical History', rely-ing almost exclusively on the technique and deploying historic models formerly at the Wellcome Trust's museum in Wigmore Street in the 1930s as well as many new pieces and room settings.[8] However, by the mid-1980s the use of dioramas had declined rapidly. This reflected, in part, the passionate preference of the new Director, Neil Cossons, for 'the real thing' wherever possible, but also a growing sense among curators and exhibition developers that dioramas were now some-what '*passé*', though no real research-based evidence was adduced to indicate that they had lost their power to charm. Indeed, Insley has asserted that from their inception visitors 'appreciated them deeply' and speculates that 'part of the appeal [is] the lure of the brightly lit miniature in a darkened room.'[9]

Clearly there is something intriguing and quite mysterious in the encounter with a model which we empathise with but do not fully understand. Ludmilla Jordanova has suggested that 'the idealisation present within "model" indicates clearly a kind of longing that is implicit... in models as material objects.'[10] More simplistically, perhaps, it seems to this author that our viewing a model, whether as child or as adult, allows us to 'own' the scene briefly in a way in which we are powerless to do in the real world.

This elusive appeal of models applies also to models of machinery or, more recently, to molecular structures. It reflects, too, the way that both formal education and research have privileged the two-dimensional, whether as texts, drawings or images, over the three-dimensional representation. Such sources, and particularly engineering drawings, require considerable fluency in the conventions that they use to communicate 3-D information to the reader. Models are direct and short-circuit this, allowing a broader audience to comprehend a complex mechanism or arrangement.[11]

The invention of the Children's Gallery, with its dioramas and other new display ideas, reflected a sense, we surmise, that time had moved on and that the serried ranks of cases, no matter how 'well laid out', were obscure to many visitors. They remained, though, the ruling style for the majority of the Museum's spaces for decades to come. However, it could perhaps be argued that the Science Museum of this time was not as devoid of narrative and analysis as these passages imply, particularly because there were a large number of special temporary exhibitions, but the narrative scope of industry-focussed exhibits such as 'The Rubber Industry' must have been highly restricted, while the major post-war show on Local Government Services, mentioned by Lawrence, was surely pushing the medium of temporary exhibitions to the limit of what it could usefully do.

The new diorama technique, meanwhile, remained fairly incidental to the main landscape of the Museum, being most visible, as we have noted, in the Children's Gallery until it was deployed on a major scale in the Agriculture Gallery, which opened in the new Centre Block in 1951; a project which also explored a new and much more explicit way of describing the operations and even glimpses of the working life associated with a particular industry. Previous museum practice had emphasised the need for large window areas, as well as showcases glazed on all sides, but also on top. As David Rooney has also pointed out in the preceding chapter in this volume, as a former electrical engineer William O'Dea understood the potential of the new electric light fittings and saw the use of controlled lighting as an advantage, arguing 'now that efficient fluorescent lighting is available, the museum planner should be less ready to submit to the inelastic and restrictive tyranny of daylight.'[12] Accurately directed local (theatrical) lighting, of course, resonated well with the tentative moves to 'pictorialisation' that Lawrence has identified in pre-war practice, which was now given free rein with some twenty specially commissioned dioramas giving an insight into the seasons of the farmer's year and into all the important processes on the farm. Today, it is still marvellous to see the survival, some sixty years later, of such scenes. A tiny tractor is frozen in the act of crop-spraying, with cotton-wool tufts on the spray boom artfully evoking the chemical mist, while close by three other button-operated tractors ceaselessly plough, harrow and roll their land. These 'little landscapes', it must be said, though for a while unfashionable and even embarrassing to a later generation of curators, still retain a deep and visible power to enchant children and even adult visitors.

Dioramas were not the only innovative technique deployed in the gallery. O'Dea visited Sweden with Welbury Kendall, the architect of the new Centre Block, to study 'what were commonly held to be the most advanced examples of

Illustration 8.2   The 1951 Agriculture Gallery, reflecting a both a new awareness of design and of contemporary international trends

museum technique.' He found the standard high but was not impressed by all he found, deciding, he wrote, 'to break new ground'. It is certainly possible, though, to see echoes of the Scandinavian genre of folk museums in the modelled full-size historic ploughman (now removed) and in the light varnished wood finishes for cases and walls. O'Dea promoted his new ideas in a brochure aimed at external organisations (although it may have also been circulated internally), which was intended to be an intimidating document, deliberately 'made an awkward size and the two pages...dry-mounted on boards so stiff that they could not easily be torn up or even got rid of.'[13] This brochure demonstrates that what particularly distinguished this gallery, apart from a commitment to a fully worked and fresh design, was a very overt statement about its informative intentions and narrative structure, and these clearly permeated the planning and policy documents. The intentions also survived to become expressed in the gallery text itself and the selection and sequencing of exhibits. The gallery:

> was to be organised in sections to show the development of methods of tillage with dioramas to show mediaeval farming, horse ploughing, steam plough, tractor cultivation: sowing, reaping, threshing, binding, milling etc. (mainly historical) and...the work of the modern farm (arranged by seasons). This section should be completely new and only the best up-to-date practice should be included.

O'Dea continued:

> The new section is on the first floor of the Museum and it will be windowless in order that the methods of display may incorporate recent developments in artificial lighting technique. All the display cases will be internally lit and the space between the top of the case and the ceiling along the long walls of the gallery will be provided with an illuminated cyclorama background against which will be shown a number of scenes representing agricultural processes such as ploughing, sowing, reaping, threshing and winnowing. This background should give a light and spacious atmosphere to the gallery which will be accentuated by the recessing of the ceiling to provide the recesses from which the floor exhibits may be spotlit. The furniture, specially designed, will have a light natural wood finish.[14]

The considered use of design was certainly not due to O'Dea's sole influence. Design techniques had been seeping into the museum world from the 1920s, while the association of prominent architects and designers with the 1951 Festival of Britain made it evident that the bar had been raised and that design was now an almost expected component of modern display. However, O'Dea deserves full credit for realising this and also for making the gallery the first in the Museum, arguably, to be conceived and executed with a clear and explicit narrative intention and 'content scheme' as well as a total architectural design treatment of the gallery's internal space and furniture.

Nevertheless, the Agriculture Gallery must have been seen as an exceptional event, in terms of its comprehensive visual language and strong implementation of a design scheme, and we can surmise that it was accomplished because O'Dea was able to enlist the services of the Centre Block architect to consider also this important new gallery within it. Elsewhere, the impact of design remained limited. The records and photographs of the Museum support this impression and the Museum's annual report for 1950, which describes a Drawing Office populated only by 'draughtsmen', shows how little resource could normally have been applied to design, since much of their work related to mechanical design of exhibits, drawing models, mechanisms and redrawing artefacts thought to be of value for instruction and for actual display. By 1961, the annual report noted the increasing attention paid in recent years 'to the rapidly growing art of display technique,' leading to the creation of an 'exhibition studio', which came to be called the Design Office in 1967.[15] This new design consciousness was directed more towards graphics and layout, rather than the broader appearance of whole galleries, with the Director Terence Morrison-Scott arguing particularly that label length had to be reduced. 'Nowadays...the public is accustomed to pithy, tabloid exposition.... All [this] may be regretted, but it must be recognised.'[16] Predictably some passion was aroused by this diktat, but hardly as much as by the rather later proposal to unify typefaces throughout the Museum. This brought Michael Preston, as Exhibition Officer, into confrontation with a 'curatorial body addicted to serifs'. He argued that Times Roman was 'only originally intended for newsprint with thin inks' and lacked 'the visual quality [of] Adrian Frutiger's Univers; a sturdy workmanlike elegance of the 1960s.'[17]

Some twelve years after the new Agriculture Gallery was opened, O'Dea revealed another innovative approach with the Aeronautics Gallery, which opened in July 1963. The location and size of an aeronautical display had not been without its earlier tussles, with Frederick Hartley, the Keeper of the Mining and Electrical Engineering Department, arguing in 1950 that 'we should strongly resist giving [aeronautics] more space than it properly merits, because of its popular appeal and political influence.'[18]

However, the energetic O'Dea secured a prime spot on the top floor of the new west block, with a purpose-designed hangar-type arched roof. Here, the main innovation was the display of twenty-two full-size aircraft suspended 'as if in flight.' If this seems commonplace now as a display technique, it was certainly a major step at the time on the scale employed. Indeed, during the earlier years of aviation the Museum had shied away from acquiring full-size aircraft and considered that essential developments could be conveyed by representative models and some specially commissioned examples of structure. Aero engines, however, were acquired keenly, and from an early date, as it was judged that these represented the area of aeronautics that was most technically challenging and most subject to improvement – an approach that gave the Museum one of the world's best collections in the field. Indeed, when the Museum was offered a complete Blériot XI, claimed to be the first aircraft to fly the Irish Sea in 1912, it was accepted only as a 'donor' source for a truncated display of its cockpit section and controls, deemed the most technically interesting parts, the rest being scrapped. The first complete aircraft only arrived in 1913 (Colonel Cody's winning entrant for the 1912 military aeroplane trials). However, by 1919 the power of aircraft to interest visitors had become evident when Alcock and Brown's Vickers Vimy arrived, refurbished after the first non-stop crossing of the Atlantic, and was said, anecdotally, to have actually doubled, for a spell, the daily admissions to the Museum. Aircraft thereafter arrived steadily, giving the Museum an important collection of nationally and internationally significant machines.

The display approach for the new Aeronautics Gallery was novel enough for the *Sunday Times* to trail the gallery, in February 1963, under the headline: 'A war against boredom'. It revealed that 'the man who is waging this war is a 58-year-old Lancastrian keeper at the Museum, Mr W. T. O'Dea. ... Instead of row upon row of glass cases planes are suspended in mock flight from the roof of a hangar.'[19]

But though O'Dea, as David Rooney has pointed out in the previous chapter of this volume, considered the traditional museum in Britain to be 'awful', beyond the main innovation of suspending aircraft the gallery was, in fact, very traditional. Glass cases still abounded, not so much in rows, but sprinkled liberally around, some tucked under the elevated walkway – in itself a nice invention – and even under the wheels of Cody's aeroplane. The aircraft themselves were arranged more or less in chronological order with the aero engines alongside, although these engines were, in fact, mustered taxonomically, arranged into groups according to the principle of operation. This actually produced a rough, imperfect chronological series which did not map onto the one established by the aircraft – a dissonance that, today, curators might not be happy with. Perhaps the most evident surviving traditional element was the way exhibits were labelled. These individual 'object labels' pretty much represented the entire text of the gallery, which retailed, in effect, a fairly comprehensive history of aviation, but only to

Illustration 8.3   The new Aeronautics Gallery which opened in 1963, showed a new drama in the display of suspended aircraft but a resolutely taxonomic approach to the arrangement of aero engines

those who were prepared to read most of this small text, which was largely specific to each object. What the Aeronautics Gallery did do, however, was to increase the stakes and demonstrate that spectacle, on a large and quite expensive scale, was now a legitimate part of the experience that the Science Museum offered.

At about the same time, O'Dea also developed the huge Shipping Gallery, which relied overwhelmingly on models ranging from vanishing British coastal craft, hypothetical reconstructions of antique boats, and traditional craft from around the world, as well as warships, yachts and ocean liners. Here, the organisational principles for the gallery seem more obscure than in his other projects, perhaps due to the profusion of material, though he again relied heavily on the diorama technique, as with the Agriculture Gallery. However, in most cases these were not totally composed dioramas but landscaped settings to enhance the context of existing boat models, though this still required a high degree of painting and modelling skill for both backgrounds and human figures.[20]

Nevertheless, the essential 'founding spirit' of the Science Museum – as seen in the earliest statements of purpose and reinforced by Lyons in the 1920s – remained substantially intact. Even the progressive and technically impressive Aeronautics Gallery was seen in this spirit, and Lyons's intentions found an almost perfect echo in the notes prepared by David Follett (though apparently not used) as a brief for the Minister's speech at the opening of the Aeronautics Gallery. It was still 'the basic job of the Museum to collect and preserve [objects] for posterity' and then use them to show 'development...from the earliest times up to the present day.'[21] There was little room for exploration or selection implied by this mission, or any licence to explore the reciprocal interplay between societies, their science and their inventions. But perhaps the most surprising assertion that Follett wished to make, particularly in the context of a new and dramatic extension of display technique, using some of the Museum's most popular and attractive exhibits, was the assertion that 'unlike the exhibits at, for example, the National Gallery, or the Victoria and Albert Museum – paintings, sculpture, mosaics...and so on, most scientific objects have no aesthetic appeal; they are of interest only in the context of their origin and role.'[22] Indeed, Lyons himself had written, more than thirty years earlier, that 'while no one will dispute the utility of technical museums, it must be admitted that they fall far short of art museums in the attractiveness of the objects which they contain.'[23] Perhaps Follett's remark was intended as a corrective to the perceived glamour of aircraft, but it seems to us now an almost wilful refusal to accept the potential that technical and scientific exhibits can, have by their presence, to elicit interest or engage viewers in the subjects they represent.

Nevertheless, Follett's austere sense of the aesthetic status of the Museum's exhibits cannot have been shared by all curators, though they may not have had the type of opportunity that the Aeronautics Gallery offered to O'Dea. A little later, Keith Geddes, as Assistant Keeper of Telecommunications, certainly appreciated the artefacts of his own specialism, their iconography and their place in the wider visual and social landscape. This was evidenced by the temporary exhibition on the history of television that he assembled, with colleague Eryl

Davies, placing each television set in a room furnished from the appropriate era and showing an appropriate programme – probably the first time the Museum had tried something so explicitly theatrical on this scale.

Narrative and design

Keith Geddes also brought something of this approach to the new Telecommunications Gallery, opened in 1983 as a celebration of the centenary of the telecommunications giant, Standard Telephones and Cables (STC). This proposal arose largely through contacts between senior executives in STC and Director Margaret Weston, an electrical engineer herself – actively assisted by Derek Robinson as Keeper of Museum Services. Discussions started in 1979, looking ahead to the STC centenary in 1983, in which it came to be understood that an exhibition at the Science Museum should form part of the celebrations, and that STC would contribute £200,000 towards this. But, despite the groundwork done by Weston and Robinson, the news was not welcomed unequivocally in the Museum and few curators today would greet the news of such industrial largesse with the insouciance of Keith Geddes, who commented: 'I do not see how we can decently do other than go along with the proposal, but I must admit to finding it difficult to contemplate it with much enthusiasm, even if I assume that someone else does all the work.'[24]

Geddes's apprehensions seem to have centred on the emphasis STC wished to place on future innovation, and on the scale of the exhibit. 'STC's desire [is] for a forward-looking show' but one also charting 'man's attempts to communicate throughout history from cave paintings to optical fibre.' For Geddes, 'the problem [was] likely to be to convince STC that grandiosity is counter productive and that they should settle for a more modest exhibition than this one promises to be. ... A recital of wonders to come soon palls.'[25] By the following year Geddes had become reconciled to the impending task, but the interest of the project for this chapter was the way Geddes and Davies set out to frame the story and coverage of the exhibition. Geddes noted that a timeline approach was not the only one possible and he included, in discussion papers, Eryl Davies's idea that 'the exhibition's main dimension' need not be the 'march of time, but the progress of a message from originator to recipient' dealing with history within each link.[26]

In negotiations with STC it soon came to be agreed that the exhibition would still celebrate the company's centenary, but would have an extended life as the Museum's new Telecommunications Gallery, a considerable diplomatic feat for Weston and Robinson and an enlightened gesture from STC. This took care of one of Geddes's objections: that the new STC-funded gallery would duplicate some of the areas covered in the existing Gallery 66, which covered telecommunications and electronics display. This latter display was, Geddes admitted, 'pretty boring' but contained some interesting material from other makers which, presumably, would have been omitted in a pure celebration of STC history.

Despite Geddes's earlier cavils, and perhaps encouraged by this wider scope, he addressed the new gallery, together with Davies, with his usual enthusiasm.

Illustration 8.4 The STC Gallery (later renamed the Telecommunications Gallery) marked the emergence of a new narrative style, although the use of graphic panels was still tentative

Interestingly, the 'content scheme', as it might now be called, came to be a combination of the historical timeline proposed by Geddes, with 'progress of a message from originator to recipient' as suggested by Davies. The 'march of time' idea was represented as a general historical introduction in a linear gallery area, which then branched into three separate subsidiary galleries dealing with message origination and transmitters, with the technology for broadcasting and carrying the message, and with the reception equipment.

These subgroups were announced with a title over each portal, but it was necessary to raise one's eyes to see it and the logic of that part of the display must have been quite opaque to most visitors. After a few years, in fact, these elements were removed and the introductory area has stood alone as the Museum's telecommunications offer. This is dominated by two tableaux showing a ship's wireless room (c. 1910) and an office 'then and now' relying on the Pepper's ghost technique.[27] The ship's wireless room was based on a reconstruction created by Marconi Marine in 1950, which was copied by Gerald Garratt and placed in the 1971 Telecommunications Gallery. Sadly, though, the life-size station for the wireless operator in a Lancaster bomber, animated with background sound and transmissions – a really virtuoso piece of Museum workshop display technique – was removed several years ago.

From the point of view of this chapter, however, the particular interest of the gallery was that it was conceived with a definite thematic structure. It seems to stand between the emerging ambitions of the Agriculture Gallery to be a 'narrative environment' (although the narrative in that case can be expressed as the rather simple proposition 'what goes on at the farm') and the kind of approach that might be adopted today. Today, although the introduction plus triple theme layout might be proposed at the planning stage, ongoing discussions during development would probably reveal its flaws, as this idea is too likely to be obscure or invisible to visitors. Exhibit developers are tougher on themselves now and subject their ideas to a robust process of internal critiquing and discussion.

What is also significant and interesting in the display is the use of graphic panels bearing pictures and text. This magazine-type graphical format was becoming increasingly standard in museums, and we could speculate that Geddes, as a former BBC engineer, was more interested in, and comfortable with, using a structured narrative medium than many of his museum contemporaries. These panels devised by Geddes and Davies, however, are still somewhat tentative, in that they do not clearly 'rule' the display, ordering it and signalling its divisions and content areas. Rather, they have a kind of interstitial role, being of moderate size, inserted quite demurely between cases and commenting on the contents as a type of 'super label'. That said, they still speak of the authors' perhaps quirky interpretive approach. For example, the panel entitled 'The Indian Summer of the Electromagnet' (and another headed by 'Waving not Drowning') must rank among the earliest attempts in the Science Museum to link poetic metaphors with technological developments, and are still relished by those with an interest in the arcana of exhibition craft.

One project that clearly adopted explicit narrative, and even used a kind of 'industrial theatre', was the Chemical Industry Gallery, which opened in 1986. This was a bold attempt to take on the problem of presenting a subject which seemed, to many, hard to interpret and potentially dull. The 'conceit' conceived by curator Robert Bud and architects Casson Mann was to envelop the visitor in a convincing evocation of a chemical works in which the industrial structure and perforated steel walkways acted as a backdrop to real exhibits and to the narrative and explication. Here, the graphic panels ordered the visit and were not optional additions to the cases and tableaux – they were essential (to most people) for any comprehension of the obscure equipment on display and for glimpses of the chemical transformations involved in production. The gallery also marked the emergence of the desire, if it can be afforded, to create exhibitions which are designed as total architectural spaces in support of the narrative or 'spirit' of the exhibition.

These trends, which now seem inevitable, were of course only partially visible at the time. For example, when the Aeronautics Gallery was reworked in the mid-1990s, there was no instruction or institutional 'steer' to move from an object label-based display to one with a more overt panel-based narrative, though it seemed both natural and necessary to do this. Neither was there any particular standard procedure or model to follow that had been unequivocally endorsed within the Museum. The strategy adopted included strengthening the chronology and the clarity of the gallery by some relocations, bringing in new exhibits (four new aircraft and a Boeing 747 'slice'), replacing the existing unruly casework with a single unified wall run, and composing a new narrative synoptic history of flight with a clear chapter structure.

However, narrative, or 'message', is not always necessarily at the service of a large pre-existing display collection, which was almost certainly acquired in a different age and for reasons which may no longer seem to be of concern. Thus the new exhibition philosophy developing in some other centres, such as the Natural History Museum, in the period, rightly stressed the importance of defining and refining the intentions and message of the project, but also revealed the tension that can develop between the narrative and the collection. In the 'strong' version of this approach, exhibits were to be few in number and ruthlessly selected, since they were seen merely as three-dimensional illustrations to a defined educative and informational mission. Some specialist museum designers and interpreters, who embraced this philosophy more quickly than many curators, were fond of challenging them with the question 'couldn't the story of pocket watches (or aero engines, or steam power) be better told with *a few well chosen examples?*'

Although such challenges could be a useful test, they perhaps masked an assumption that curators found their collections more interesting than the public, and that, if unrestrained, might be expected to overload the exhibition with a finely graduated series showing the micro-evolution of their special subject. But against this is the odd fact that audiences do appear to find 'unreconstructed' collections fascinating, as shown by a great appetite for visits to museum stores, or by the success of sites like Snowshill Manor in Gloucestershire, which offers an extraordinarily eclectic collection garnered by an antiquarian autodidact. Perhaps the redisplay of aero engines, albeit in a condensed and more interpreted form and periodised by a historical commentary, helped to make the point that collections, in themselves, can be eloquent. In this particular case, the long series, with its extraordinary variations in mechanical architecture from the dawn of flight to now, says something powerful about the sheer weight of human ingenuity that has been devoted to this problem.

The different gallery projects touched on here are certainly not intended to portray a trajectory from 'bad' to 'good' or to point inevitably towards a particular style of exhibition. In each period the Museum's offer has reflected expectations, reading habits and the conventions of other media to remain credible. Neither is the framework of the type of narrative that has emerged in recent years fixed. The institutional milieu in which each new gallery is framed, and the expectations invested in it, inevitably shift in each period, and with successive Directors. 'Making the Modern World', which opened in 2000, was, in a sense, the historical element in a broader new offer that the Wellcome Wing was designed to bring. It would form an object-rich historical antechamber and convey the visitor from 1750 to 'now', in preparation for the Wellcome Wing's topical and fast-changing exhibits on contemporary science and technology.

The gallery reflected, in part, Neil Cossons's express desire to see the gallery make an argument for the cultural value of our scientific and industrial artefacts and that it should rest on the astonishing array of 'firsts' the Museum holds and on the simple aesthetics of a 'sculpture court'. This was an appealing idea, given the licence to draw on the collections and existing galleries to show objects of the calibre of Stephenson's Rocket which demonstrated this priority over a large number of fields. However, the curatorial team in charge of developing the gallery believed that this strategy was too monolithic to be deployed uniformly over such a large space and an interpretive framework was essential. The resulting strategy was to order the key objects as a timeline, but to provide two parallel displays to left and right, marching in step with the central chronology. One side, 'the Technology of Everyday Life', served to contrast the exceptional and new with the everyday artefacts in every age, as encountered at home, at work or at play, and served, too, as a corrective to what could be read as a simple 'progressivist' account in the main area. On the other side, a series of historical bays, or studies, reflected on the epochs which had produced the exhibits in the main timeline under headings running from 'Enlightenment' through to 'Ambivalence', and these set out to form a well-researched 'cultural history of industrialisation', which would reflect, but not explicitly quote, recent scholarship in the history of science and technology.

It might be thought that the logic of this account of emerging gallery styles would have suggested that this historical framework in 'Making the Modern World' should be encountered more inescapably by the visitor, and that it would have clearly (rather than implicitly) ordered the main central area and have disciplined the whole space. That, however, was never feasible due, among other reasons, to the requirement that the gallery should be constructed to allow unrestricted visitor flow to the Wellcome Wing and Imax theatre while allowing clear

Illustration 8.5   The long view of 'Making the Modern World' (2000) was expressly composed to suggest a journey from the emerging industrial world of the eighteenth century to Modernity

sight lines which precluded large graphic devices or new structure. The point is that all exhibition projects are highly contingent on a range of institutional and external circumstances. Moreover, due to its history and architecture, the Science Museum has few 'white spaces' that can be populated by new, untrammelled design, and its galleries are interlinked and often serve as both throughways and exhibition spaces.

Conclusion: The craft of display

'Making the Modern World', with its high representation of real objects, shows that in recent years we have come to reaffirm that the Museum must 'trade', in part, on the presence of the real, not least because there is much general informational content available that can be provided easily and conveniently by other media accessed at home. Real artefacts, with real historical associations, are 'a unique selling point' and we have become bolder, again, in asserting that collections can be eloquent, particularly if marshalled by a narrative that embodies interesting propositions.

Many new techniques have been adopted for the craft of exhibition work over the years, each of which has been seen in itself as an essential component of communication – dioramas, video, interactive exhibits in close proximity to the real (and perhaps worryingly enigmatic) display object, computer terminals offering searchable information or graphical simulations that unpack and explain the exhibit. All are valuable, but all have to be deployed imaginatively, and they make high demands of us as producers. In an age with countless TV channels, including many with substantial content in science, technology or history – as well as the incredible resource of the Internet – the task for exhibition professionals to produce comparably 'finished' engaging material that can be accessed by a moving flow of visitors is highly demanding.

Notwithstanding these challenges, there is still strong demand for structured, bounded works of interpretation with a trusted authorial voice and which have interesting and culturally relevant points to make. There is also the issue of what some historians have recently started calling 'presence', which curators may sometimes refer to as 'the shock of the real'. The impact, say, of the Apollo 10 capsule or a melted rice bowl from Hiroshima depends not only on the narrative frame in which curators embed it, but also on its 'aura' and what we all, as consumers of our own culture, know, feel or believe about it.[28] Of course, this is not a state of affairs that is simply 'given', for the exhibition-maker is also an actor in this semiotic process. For example, Charles Babbage was an intriguing figure, little known beyond the circle of historians of early computing, but the exercise of replicating and building the originally uncompleted Difference Engine No 2 for his centenary in 1991, led by Doron Swade at the Science Museum, has now unequivocally inserted Babbage into contemporary culture and popular thought.[29]

Such changes in the broad knowledge and understanding of the audience mean that exhibition development must be essentially dynamic. Moreover, as design styles or specific techniques grow stale, or are shown not to have really worked, new solutions must constantly be invented and developed. Today we accept the commitment to select and order our exhibits in the light of what we intend as a compelling narrative that expresses a particular proposition, or 'conceit'. This parallels moves within art galleries to regard rehangs and new exhibitions as works of art in themselves, in a sense, or, at least, as artful constructions. Art or not, exhibition craft is necessarily a 'game of communication' and it aspires, in addition to the textual content, to communicate through 'the spectacular' something else – 'that which cannot be said with words'.[30]

Over the period considered here, the Museum has come to construct exhibitions as a particular, highly considered form of multimedia publication, using a synthesis of architecturally designed spaces, graphics, film, real artefacts and text, but we can never overlook, in our ambition to communicate and engage, that ultimately an exhibition is a publication that you read and experience standing up (or running through!), and this fact alone exposes hard truths about style and level of address, text length and the creative structure of all the media employed. Like all creative products, exhibitions require intense commitment and imagination to succeed. The rules of production are never completed and it will remain – hopefully – an experimental and evolving medium.

Notes

1. Nevil Shute, *Slide Rule; the Autobiography of an Engineer* (London: Heinemann, 1954). He wrote as Nevil Shute, though he used his full name, Nevil Shute Norway, for his parallel career in aircraft design and aviation. However, his novels and his engineering were linked, as, for instance, in his 1948 novel *No Highway*, which presaged, in an uncanny way, the metal fatigue failures of the Comet airliner.
2. *Catalogue of Machinery, Models Etc., in the Machinery and Inventions Division of the South Kensington Museum*, with descriptive and historical notes by E.A. Cowper (London: HMSO, 1890).
3. H.G. Lyons, 'A Memorandum on the Arrangement of Collections in the Science Museum to Serve as a Basis of Discussion', August 1922, SMD, Z 210/10/11.
4. 'Spatial Analysis of Existing Collections – as Distributed to Officers', June 1937, SMD, Z 100/1.
5. Ghislaine Lawrence, 'Museums and the Spectacular', in *Museums and Late Twentieth Century*. (Manchester: Manchester University Press, 1996), pp. 69–80.
6. Lawrence, 'Museums and the Spectacular'.
7. I mean 'scaled by ruling perspective' in the sense that individual models of differing scales may be used at different distances within the scene, or a particular model may even embody perspective in its own construction and so look correct only within its own diorama frame and at its designated position within it.
8. Jane Insley, 'Little Landscapes: Dioramas in Museum Displays', *Endeavour* 32 (March 2008), pp. 27–31.
9. Insley, 'Little Landscapes'.
10. Ludmilla Jordanova, 'Material Models as Visual Culture', in Soraya de Chadarevian and Nick Hopwood *Models: the Third Dimension of Science* (Stanford: Stanford University Press, 2004), p. 448. These essays provide much useful context on the development and wider use of models in museums, as well as in science and engineering generally.
11. Malcolm Baker, 'Representing Invention, Viewing Models', also in Chadarevian and Hopwood, *Models: The Third Dimension of Science*.

12. W.T. O'Dea, 'The Science Museum's Agricultural Gallery', *Museums Journal* 51 (1952), pp. 229–301, on p. 299.
13. O'Dea, 'The Science Museum's Agricultural Gallery'.
14. W.T. O'Dea, brochure for the new Agriculture Gallery, undated, c. 1950, in private hands but to be transferred to SMD. He noted: 'Numbers from the substantial proportion of younger visitors may be persuaded thereby to consider the attractions of agriculture as a career.... As many exhibits as possible will be shown working either constantly or by push-button operation'. Rather redundantly, it proclaimed that 'each exhibit will have a descriptive label, prepared in the Museum from information supplied'.
15. Science Museum Annual Report for 1961, p. 7 and Science Museum Annual Report for 1967, p. 1.
16. T.C.S. Morrison-Scott, Note to Higher Staff on labels, 25 June 1956, SMD, Z 183/2.
17. M.R. Preston, Note to Higher Staff on labels, 25 June 1956, SMD, Z 183/2.
18. F.A. Hartley, Minute to Director on space allocation, 5 December 1950, SMD SCM/8252/71.
19. *Sunday Times*, 3 February 1963 in SMD, Z 183/2.
20. Jane Insley, notes in 'Little Landscapes' that 'named artists received most of the commissions, and their signatures can still be seen in inconspicuous corners of the backgrounds. Anecdotal evidence from the files indicate how hard the curators campaigned to get the funds to order the dioramas for their shows, how they tussled with the artists for delivery dates to be met (with varying degrees of success), and the ongoing loyalty of the artists who frequently were held to low prices for work which was in most cases of excellent technical quality. However, once the show for which they were originally made was closed, the survival of the dioramas as objects rather than redundant display material became more problematic, as they were not only rather large and cumbersome, but also rather frail'.
21. D.H. Follett, Briefing for Minister's speech at the opening of Aeronautics, 25 June 1963, SMD, Z 183/2.
22. Ibid.
23. '...this fundamental difference affects every stage of museum arrangement'. Col. Sir Henry Lyons, 'Heirlooms of Science and Industry in the Science Museum', *Nature*, June 30 (1928), p. 1.
24. Minute of 30 May 1979, WKEG [Keith Geddes] to Dr Robinson 'STC proposals for centenary exhibition', SMD, SCM/2008/019, Standard Telephones and Cables – 1983 (on the occasion of their centenary).
25. Ibid.
26. Note: 'STC – Attached "topics" list', WKEG, 22 November 1979, SMD, SCM/2008/019.
27. Pepper's ghost is an illusion technique which uses a partially silvered mirror to give a view of the scene directly ahead, or of another scene set off to one side. By controlling the light balance either scene can be revealed, or they can be merged. It is often used in 'haunted houses' displays.
28. Eelco Runia, 'Presence', *History and Theory* 45 (February 2006), pp. 1–29.
29. See, for example, Francis Spufford and Jenny Uglow (eds), *Cultural Babbage: Technology, Time and Invention* (London: Faber, 1996).
30. Lawrence, 'Museums and the Spectacular', p. 69.

9

Beyond the Children's Gallery: The Influence of Children on the Development of the Science Museum

Anna Bunney

Children have constituted a large proportion of the visitors to the Science Museum for many years. In 1949 it was estimated that 25 to 30 per cent of all visitors were children, in 1951 40 per cent, and by 1999 the percentage of visitors under the age of sixteen had risen to 48 per cent.[1] The Museum usually considers these large child audiences in terms of the provision of special galleries. The Children's Gallery opened in 1931, closing in 1995, having been superseded in 1986 by interactive galleries such as LaunchPad.

Despite this large child audience and the general perception of the Museum as a 'children's playground', the Museum has rarely defined itself as an institution which exists for children.[2] Unlike in many institutional museum histories, children have been mentioned by historians of the Science Museum and other science and technology museums, but often in passing references or through implicit suggestion.[3] The opening of the Children's Gallery in 1931 has been frequently mentioned, and the popularity of the Science Museum is often associated with these facilities. William Stearn makes reference to children in his history of the Natural History Museum, London, when discussing the opening of their Children's Centre in 1948.[4] Although claims are made that working exhibits entertain and inspire curiosity in science museums such as the Deutsches Museum in Munich and the Museum of Technology and Industry in Vienna, whether their audiences include children is seldom stated explicitly.[5]

So why include children in the history of the Science Museum? One reason is that, as a large part of the Museum's audience and rationale for funding, they should rightly be considered in its history. Another reason is that, where historians of science and technology have studied children, it has raised interesting issues in terms of how science is viewed and communicated by society, gender and class.[6]

This chapter examines the special provision for children and how this is best understood in the context of the Museum reacting to children already visiting and in terms of segregating children from the rest of the Museum. It will explore the connections between children's exhibits and toys and the attempts to argue the educational values of these. Displays created for children were centred upon contextualisation in

194

the real world, especially through the use of the history of science and technology. As these displays were also seen to be of benefit to an adult audience, they can be viewed in terms of the public understanding of science. Consequently, it was through children that realism and history were adopted in Science Museum displays.

This chapter will consider children who visited the Science Museum before and after special provision was made for them in the form of the Children's Gallery. I have concentrated on those 'children who do not come in school parties and who outnumber those who do to a ratio of five to one.'[7] This statement is from 1955, but until recently school children have only represented a small proportion of child visitors. To a certain extent the recent rise in school visits to museums can be accounted for by the implementation of a national curriculum in British schools in the 1980s.[8] The Science Museum's view of itself as a national institution also has an impact on how direct school provision should be considered. For example, Director Francis Ogilvie was commissioned to write a report on links with museums and schools during the First World War, but as a national museum he thought the Science Museum should not provide facilities for schools or children.[9] However, I would argue that the underlying attitudes towards children were similar whether they were in school groups or not.

'Usual influx of young people'

Although the South Kensington Museum and, later, the Science Museum were not targeting children – their avowed aim was to educate adults – there is evidence to suggest that children were visiting the Museum from its inception. 'In the evening, the working man comes to the South Kensington Museum accompanied by his wife and children.'[10] While this statement from 1858 is the only reference to children in the official reports of the South Kensington Museum in the nineteenth century, the fact that children were visiting the Museum was being commented upon. For example, in 1857, the *Building News* argued that visitor figures might be inflated as children 'amuse themselves out of school hours by running in and out of the Museum, and putting the turnstile through more revolutions than are absolutely necessary to mark their visit.'[11]

Some of the child visitors seem to have been attracted to the wide variety of items in the Education Collections of the South Kensington Museum, such as globes and working models. In 1863, Alexander Beresford Hope, a wealthy connoisseur, noted 'the happy groups of children I see there, about and around me, prying with eager eye under the direction of governesses and tutors, into the varied and instructive treasures of the glass cases.'[12] The implication is that the children were taken to the Museum for educational reasons. However, many of the other comments about children suggest that the Museum provided them with a playground. For example, in 1883 William Stanley Jevons remarked that 'neighbouring wealthy residents are in the habit on a wet day, of packing their children off in a cab to the so-called Brompton Boilers, in order that they may have a good run through the galleries.'[13]

While Jevons was not keen on the 'neighbouring wealthy residents' leaving their children in the Museum, there are other indications that those children of a lower class were even less welcome.[14] The situation was summed up by an unverified story from 1888, 'that Sir P Owen endeavoured to enforce a standing rule excluding children unless accompanied by adults; that this roused feeling outside – letters in the "The Times" etc. – a Kensington Clergyman putting the rule as a differentiation between children of the well-to-do who had nursemaids and those of the poor who had not.'[15] This anecdote was given in a series of documents relating to a notice to keep out unruly visitors:

> Many of the objects are arranged so that they may be put in motion for more complete inspection. The facilities afforded for this purpose must not be abused.
>
> **BOYS**, in particular **ARE WARNED** that noisy behaviour, hustling and mere playing with objects intended for study will not be permitted. Instructions have been given that any visitor whose conduct is such as may interfere with the proper use of the Museum is to be required to leave the building.[16]

From the time of the Bell Report in 1911 through to the late 1920s the Science Museum was attracting rapidly rising numbers of unaccompanied child visitors, without any apparent wish to do so being stated. The Bell Report noted the decline of the role of the Museum in direct provision of school activities (such as demonstration of scientific apparatus and loan collections) and recommended extension of the loan of lantern slides to schools and better publicity for new methods of demonstration for schools.[17] In 1912 Ogilvie commented that the elementary 'school holidays this year have brought the usual influx of young people.'[18] But some of these child visitors were not to be encouraged. Lyons described how 'children who are ragged and not always clean are a serious distraction, but boys who attempt to use the Museum as a place of amusement are more annoying.'[19] Inspector George Bonnyman, the chief warder, reported to Ogilvie that 'children from the poorer neighbourhoods come to the Museum in groups of all ages and generally one or two take the lead and the others troop after them and appropriate all the "push-buttons etc." to the disadvantage of the individual child.' He proceeded to suggest that any adult accompanying that child will think that 'those rough children ought not to be admitted.'[20]

The Science Museum Annual Report for 1925 noted that:

> boys and girls of school age visit the collections in large numbers, but some of the children who come are too young to gain much knowledge from them, but those are gaining a useful familiarity with the collections, and many who now come to study seriously are recognised by members of the Museum's staff as having belonged formerly to this category which comes mainly for entertainment.[21]

The Annual Report for 1927 and 1928 declared that 'on two days at the end of October, 1928, which were holidays in the London schools, the attendances were 7,000 and 6,000 respectively, and this increase was due to the large numbers of school children visiting the collections in their holidays.'[22] The official claim, from 1929, that 'more children visit the Science Museum then any other Museum in London' was echoed in contemporary reports.[23]

The behaviour of children in the Science Museum appears to have compared badly with those visiting the Deutsches Museum. Various members of the Advisory Council made special note of this during their visit to Germany in 1929.[24] Throughout the 1950s and 1960s there are repeated references to children behaving badly. In 1958 the Children's Gallery was closed to unaccompanied children on Sunday afternoons, as the Museum was 'receiving many complaints about the turbulent conduct of youths in the Children's Gallery and mines on Sundays.'[25] As another example, in 1969 Stanley Janson, the Museum's Duty Officer, reported a verbal fight between boys and their fathers about the use of the periscope exhibit.[26]

Keeping children in the Children's Gallery

The Science Museum reacted to the large number of children by providing a special gallery.[27] The Children's Gallery opened in December 1931 and was a combination of working models showing scientific principles in action, such as time measurement and lifting apparatus, and dioramas showing the development of subjects such as transport and lighting. Other exhibits included a model of Battersea Power Station and an artificial rainbow.[28]

Illustration 9.1 General view of the Children's Gallery when it was opened in 1931

The gallery was an immediate success. A total of 171,263 people visited the Museum in January 1932, an increase of 80,652 over the previous year, which was claimed to be the result of the opening of the Children's Gallery. It was also very well received in the press. The popular science magazine, *Discovery*, enthused about 'Brighter Museums' and the *Observer* announced 'A real show for children, No Don't Touch'.[29]

Additional exhibits were transferred from the 'Photoelectric Cells and their Applications' temporary exhibition in 1932, including an automatic door and a machine to sort coloured marbles.[30] Two years later the gallery was doubled in size and the number of visitors had increased by nearly 100 per cent.[31]

The Children's Gallery was one of the first galleries to reopen after the Second World War. During the 1950s and 1960s the gallery continued to be popular and the exhibits were being well used. The 1956 Science Museum Annual Report stated that Formica had to be used on a new metalworking diorama, as it was the only material hard enough to survive the visitors, and in 1957 new 'light-coloured plastic film' surrounds for the dioramas had 'so far proved impervious to attack.'[32] During 1961 the counter on the 'Disappearing Golden Ball' exhibit was used a remarkable 4.5 million times.[33]

The Children's Gallery was partially relocated in the basement of the Centre Block in 1969, more or less in its original form, although the press notice announced

Illustration 9.2   A posed picture of some schoolboys by the time measurement exhibit in the Children's Gallery

the new red, pink and orange walls.[34] New exhibits included a periscope, through which visitors could view the floor above and a diorama of street lighting in the 1960s. The opening was covered widely by the press, almost all of which mentioned the diorama of the new 'Brent Cross flyover'.[35] The gallery fully reopened in 1971 and 'it continued to be one of the most crowded galleries in the Museum.'[36]

The significance of the Children's Gallery has usually been expressed in terms of special provision for children. In 1963 the gallery, with its exhibits with 'special appeal to children', was described as 'an Aladdin's cave of wonders and power over the genie of science. After a few visits the child's face will light up at the mention of the word science.'[37] Adjacent to some of the exhibits, such as time measurement, there were references to similar objects in the rest of the Museum. In 1931 the Science Museum's Annual Report proclaimed that 'the gallery is undoubtedly serving its purpose by arousing the interests of younger visitors, who afterwards go on to other parts of the Museum.'[38]

However, there is plenty of evidence to suggest that an equally, if not more, important function of the Children's Gallery was to keep children out of the rest of the Museum, to leave the more serious visitors in peace. Although these serious visitors were often defined, how they were intended to use the Museum is not made so clear. Director Henry Lyons suggested in 1922 that special provision for children would 'attract them from the Main galleries where they are at times rather a nuisance.'[39] Following the opening of the Children's Gallery this theme of segregation is continued. For example, the *School Science Review* applauds the gallery, which 'attracts a very large number of children from the main galleries, where the mass of exhibits and difficulty of discrimination tend to confuse their minds.'[40]

Underpinning these arguments for the segregation of children is the idea of the provision of separate galleries for different audiences. Lyons articulated this in an attempt to define the Museum's audience in ascending order of importance, as 'a) ordinary visitor, b) technical visitor, c) student, d) specialist', making no direct reference to children.[41] The Science Museum Annual Report of 1933 suggested separate galleries for different audiences in relation to the redisplay of physical phenomena, and this cry was repeated by the 1949 report of the Display Panel, which was chaired by Frederick Hartley, Keeper in charge of the Children's Gallery.[42] Herman Shaw, the Director in 1949, ascribed to the Children's Gallery a secondary aim of removing children and their 'natural exuberance' from the rest of the Museum and claimed that this segregation of facilities should be extended for other visitors.[43]

The gallery was described in a psychological study published in 1955 as a 'place that will attract them sufficiently to reduce the amount of disturbance they might otherwise cause to more serious students in the Galleries.'[44] Frank Sherwood Taylor, the Director, repeated the argument: 'A great many younger children, who, if they came to the Museum, might distract older visitors, are drawn off to the Children's Gallery and leave the main body of the Museum.'[45]

The location of the Children's Gallery within a basement area not originally intended for exhibition space could be taken to represent another expression of the intended segregation of children. The success of the Children's Gallery in excluding children from the rest of the Museum is debatable. The Science Museum Annual Reports frequently made reference to children enjoying exhibits. In 1957 it was noted that 'schoolboys in particular take much interest in the operating radio station.'[46]

Within the Science Museum, the Children's Gallery was usually considered to be an innovation. Director David Follett described the gallery as 'very much the Science Museum's own, born in conception and execution.'[47] However, in the context of the contemporary provision for children, as, for example, at the other South Kensington museums, nationally and internationally, and the debate about children in museums, the Science Museum's actions can be considered to be reactionary. The *Museums Journal* critically assessed a preview of the Children's Gallery in 1930, arguing that 'boys and girls should make models for themselves; that is the educational method on which the organizers of Children's Museums in the United States have seized with remarkable results.'[48]

Children also visited other museums in London. Records of 'Instructive Rambles for Boys and Girls' in 1893 and the *Guide-Companion for the Young* in 1903 suggested the presence of children in the Natural History Museum.[49] In 1893 the Association of Women Pioneer Lecturers arranged 'Instructive Rambles for Boys and Girls' at the British Museum, the Natural History Museum and the National Portrait Gallery.[50] In 1915 the V&A Museum organised some Christmas holiday activities.[51] In the late 1930s the Geffrye Museum developed participatory activities for unaccompanied child visitors, such as model-making.[52]

Working models

One reason why children were attracted to the Museum was that they 'appreciate the models of locomotives, aeroplanes, ships, etc. but are attracted by the numerous working models solely because they can be set in motion though they do not understand them.'[53] Ogilvie commented that children visited the Museum in large numbers, 'as the opportunity of turning handles and seeing something happen – of watching live fish – of playing with a magnet now afforded in the Science Museum.'[54]

Whilst it is unclear when the Museum started displaying moving models and machines, they are mentioned in Annual Reports of the Department of Science and Art from 1869. In 1891 compressed air facilities were installed to operate models, such as a Newcomen engine.[55] The number of working machines rapidly increased so that by 1919 there were 219 'models in motion' and by 1929 this number had increased to 300.[56] A guide–lecturer was appointed in 1924 and many of the audience appeared to be from school groups.[57] The demonstration of wireless began in about 1925, joined by an automatic telephone exchange in 1926.

Although it appears to have been universally believed that children 'do not understand' working models, the Museum responded by providing more working models in the Children's Gallery.[58] There were clear links with toys; for example, a 1931 article in the *Daily Telegraph* uses the word 'toy aeroplane' to describe the model of a wind tunnel.[59] The most direct reference to toys made by the Museum is through the appearance of advertisements for toyshops and models in the Children's Gallery Guide despite the Museum's anxiety to avoid any endorsement of commercial activities in its exhibitions and exhibition guides (see, for example, the following chapter in this volume on temporary exhibitions). These advertisements appeared from the 1930s to the 1950s and were for model-makers, toy and confectionery firms and books. A 1946 advertisement for the toyshop Hamleys makes the connection explicitly:

> After your visit to the Science Museum may we suggest that Hamleys of Regent Street is the place for your next exciting journey of discovery. There you will find many of the models, or kits to build them, of the things old and new which have aroused your enthusiasm for science and invention at the Children's Gallery.[60]

Furthermore, the link with toys usually associated with boys is also very clear. Although in an 1938 advertisement (see Illustration 9.3) 'Atalanta' kites are described as 'Excellent sport for Boys, Girls and Adults', the rest of the toys mentioned in the gallery guide are traditionally boys' toys, such as railway models and chemistry sets.[61] There appear to have been more boys than girls visiting the Children's Gallery. The 1955 psychology study of the gallery found, in a questionnaire of 100 children, that sixty-eight were boys and thirty-two girls, and that boys stayed longer and made more return visits than girls.[62] The Natural History Museum also appears to have attracted more boys than girls. Its Children's Centre had opened in 1948, and by 1950 70 per cent of the visitors were boys and 30 per cent girls.[63]

From the nineteenth century onwards the Museum had been keen to stress the 'educational value' of working models.[64] In the gallery guide, Hartley stated, 'Do not think of the exhibits as things to play with, but try to understand to learn from them.'[65] Lyons was keen to establish the educational credibility of the Children's Gallery by seeking advice from school inspectors, who from 1930 constituted part of the membership of the Advisory Council.[66] In 1949, Shaw defended the use of push-buttons, as a means by which the Museum could bring 'instruction and entertainment' and encourage children to read labels.[67] Sherwood Taylor continued the theme in 1955 by stressing that the special function of the Museum was 'to provide instruction through pleasure'. He admitted that 'the infant sorcerer can perform his magic by pressing a button and can be regaled by an uncomprehended sequence of lights and motion.' However, he proposed that these children would be inspired by science later in life and that some older children would work out the science behind the 'magic'. Sherwood Taylor vehemently disagreed with those who 'stigmatize this as "funfair" stuff.'[68] Follett also spoke about working models inspiring the child. He described the view that the Museum is a 'children's playground' as 'quite unjustified and is a very narrow view.' He considered that if a child is educationally engaged for five minutes that this makes the whole visit worthwhile.[69]

Illustration 9.3   Page of advertisements in the 1938 edition of the guide to the Children's gallery

Illustration 9.4 Schoolgirls using the pulley block interactive in the Children's Gallery, 1951

Tony Simcock has argued that one of the policies of Sherwood Taylor as Director 'was to try and make the Science Museum a really effective educational resource.'[70] He points out that Sherwood Taylor began his career as a chemistry teacher and, by the time he became Director, was an established popular science educator as well as a historian of science. Sherwood Taylor set up a special committee of the Advisory Council to examine how the existing educational provision for children could be improved.[71] The committee then commissioned two psychologists from the Institute of Education, Joyce Brooks and Philip Vernon, to carry out psychological studies into how children used the Museum, centred upon the Children's Gallery. In their report, Brooks and Vernon argued that children's facilities should be based on educational psychology but also that the Museum should 'appeal to their senses of wonder.'[72] Following this report the nature of exhibits for children did not change, but there were improvements in the provision for schools. A school liaison officer was appointed in 1956, followed by a promotional schools bulletin and the development of the film and school lectures programme especially for secondary schools. The schools lecture service expanded until the mid-1960s, when a decline in attendance was accounted for by the rise in science broadcasting on television.[73]

Display techniques

There is a strong connection between the display techniques used in the Children's Gallery and the use of history of science in education.[74] This 'Science for All' movement argued that history and context provided a way to make science more understandable and also to humanise science. W.H. Brock has argued that some proponents pursued a 'view that an individual child's mental developments paralleled that of civilisation.' In the history of civilisation nature and science were studied first out of wonder, then because the study was useful, and finally for 'intellectual gratification'. Science, therefore, should first be taught through utilising a child's sense of wonder, then by introducing the relevance of science to everyday life. It was often assumed that many children's, and often adults', understanding never developed beyond wonder and utility. As a result of this, a generalised science curriculum was suggested based on practical applications and the history of science.[75]

Raphael Samuel noted that visual representation, including scenes of historical and contemporary science and technology in action, were at the 'cutting edge' of the introduction of social history into elementary schools in the 1920s. He argued that the 'picturesque element' and the making of models to show 'material culture – in the form of houses, food, dress and means of locomotion' were techniques used in schools, as it was thought that children more easily understand the concrete rather than abstract concepts. Samuel links this to a long tradition of historical illustrations in literature and public places.[76]

This 'Science for All' movement had probably already influenced other science and technology museums and some displays in the Science Museum before the

Illustration 9.5   This photograph shows how the dioramas of 'Early Man' and 'Primitive Man' and their means of transport were displayed in a naive manner presenting a certain perception of reality.

Children's Gallery opened, such as dioramas in American natural history and children's museums and a 'realistic coal and salt mine' in Germany.[77]

The Science Museum experimented with displaying science in historical context on a large scale in the Children's Gallery. Hartley argued that 'the history of Science is closely bound up with the history of mankind like one of the strands of a heavy cable, and a knowledge of it leads to a better understanding of the person, and often points the way to the future.' For this reason the idea of historical development runs through most of the permanent exhibits in the Children's Gallery.[78]

An enduring theme of the Children's Gallery, which fits within the 'Science for All' tradition, was that the gallery was aimed at the non-specialist adult as well as the child. This view was being expressed as early as 1915, when Ogilvie argued that 'objects provided for the edification of adults may be very useful for the edification of the quite young visitors, especially when conducted by teachers or intellectually serious.'[79] From its inception the Children's Gallery has often been described as the 'Introductory Collections'. Hartley, writing about the origin of the gallery in 1951, repeatedly asserted the connection:

> It was felt that something should be done to help the layman and the child to see the development of science as a connected whole and with this end in view plans were prepared in 1930 for a children's gallery which should in effect constitute a general introduction of the Museum's collections and a broad statement of the part played by pure and applied science in the development of our material culture, for the enlightenment of the layman as well as the child.[80]

External reports of the gallery noted that adults also visited the Children's Gallery: 'it was not only children who came; the attention of a wider circle had been attracted, and the grown-ups are still coming, though in smaller numbers.'[81]

As a popular science educator, Sherwood Taylor was very closely linked with the 'Science for All' movement.[82] He pointed out that the 'infantile pleasure of button-pushing still persists, even up to ten or twelve (and indeed into adult life!).' He argued that 'a children's gallery must have a systematic plan, for in the Science Museum at any rate it is visited by older children as well as many younger ones and by many adults (who are children in the world of science).'[83] Certainly by 1956 *The Times* was claiming that 'adult visitors to the gallery normally exceed children. Many of these are accompanying children, but an appreciable proportion go on their own.'[84]

The success of using realism and history within the displays in the Children's Gallery and its appeal to both the young and the non-technical adult meant that it was applied to other areas of the Museum. This can be thus regarded as the main impetus for the introduction of contextual history into the Science Museum's displays. The 1931 Science Museum Annual Report stated that 'in each of the boat models for the Children's Gallery, made to show the evolution of the built boat from the log, one or more human figures were placed to give the scale and also to show the method of working the boat; the success achieved suggests that the addition of similar figures to some of the models in the main galleries would enable the public to appreciate more readily the size and purpose of the boats represented.'[85]

Reference to the Children's Gallery was also made in connection with a temporary exhibition on lighting in 1936. 'The technique is a development of that originally adopted in the Children's Gallery, aided by the experience of special exhibitions, and modified to suit the permanent exhibitions. The aim is to give a broader and more human aspect to the subject under treatment, and render it more attractive to the eye and easier to follow.'[86] The exhibition consisted of a mixture of dioramas, full-size replicas and displays of historical context.

The Science Museum Annual Report for 1949 described how lessons learnt from the Illuminations exhibition were to be applied to the 'new experimental Agricultural Gallery'.[87] This gallery was apparently such a new experience for the Science Museum that the annual report found 'it not easy to convey in writing the atmosphere created in this new display which has been devised to have the maximum appeal.'[88] Techniques of displaying science and technology in context were further developed in the Shipping and Aeronautics galleries that opened in 1963. The nature of the display techniques used in the Aeronautics gallery – a 'war against boredom is being waged in the Science Museum' – even merited an article in *The Sunday Times*, which is discussed in Chapter 8 in this volume.[89]

Science Museum publications increased their use of images and displays of science and technology in context. In 1963 a new type of book for the Science Museum was published, the first two in the series being *Ship Models* and *Timekeepers*, which each featured '20 fully colour plates'. These were a departure from the usual text-based publications aimed at specialists, and it is interesting to note that pictorial representation was considered to 'meet the requirements of the general visitor.'[90] *In The Science Museum*, [91] a guide to the Museum, was published in 1968, 'which through a readable text and stylised black and white and colour sketches provided a background social history to the museum's exhibitions.'[92] Stella Butler directly links the format of the guides with the success of the Children's Gallery in putting the human element into science and technology exhibitions.The authors of the guide suggested that the Science Museum was visited by more young people than any other museum in London, because, as well as engineering becoming more influential, 'it has many familiar applications close to everyday life. This popularity means that many of its visitors receive their first true impressions of historical perspective through the dioramas and such models as the man-powered mine lifting gear... or through such full-size exhibits as the tramcar.' They also expressed the 'Science for All' argument that 'even to those whose living is not earned in science or engineering the Museum offers a greater understanding of their world.'[93]

Conclusion

In this chapter I have argued that children need to be integrated into the history of the Science Museum, as it provides useful insights into both how the Museum was experienced by an audience segment and how this audience in turn influenced the development of the Museum. Without visitors museums would not exist, but in histories they are often not included. When people visit museums they often do so in ways that the museum itself did not expect.

Children came to the Science Museum because the way in which science and industry was demonstrated through working models and demonstrations appealed to them; they were perceived by children as being fun. The Science Museum had not intended children to be a major section of their visitors, but, as they came, the Museum reacted by providing more of what children wanted – the Children's Gallery. The Children's Gallery was so successful that its display techniques influenced how the Science Museum developed exhibitions and displays more generally, both in terms of interactivity and in terms of the contextual use of the history of science. Without this influence of the child visitor, the Science Museum might well have developed in very different ways.

More recently, the opening of LaunchPad and other interactive galleries, and the growth of performances such as Science Shows and drama, can both be viewed through the context of the Children's Gallery, in terms of the educational value of working models and the 'Science for All' tradition. The development of LaunchPad was greatly influenced by the 1969 opening of the Exploratorium in San Francisco. Influenced by the 'Science for All' movement, this hands-on centre created by Frank Oppenheimer, the younger brother of physicist Robert Oppenheimer, aimed to demystify science and technology. 'There is a need to develop public understanding of science ...' wrote Oppenheimer; 'the phenomena of basic science are not easily accessible by direct observation yet they are natural phenomena as intriguing and beautiful as a butterfly or flower.'[94]

LaunchPad opened in 1986, and the technology-based interactives had been developed by applying educational theory and evaluation techniques. Three years earlier the vision for the new gallery stated that it would 'provide a place where people of all ages can discover that exploring and experimenting in technology can be a satisfying and worthwhile experience.'[95] The application of educational ideas and research, such as the concept of multiple intelligences, has continued in the development of LaunchPad and other interactive galleries.[96]

The Science Museum roots of interactive performances can also be traced to elements that influenced the Children's Gallery, such as the development of guide lectures, working models and interactive demonstrations. By the late 1970s the interactive lectures of the Science Museum included elements to capture children's imaginations, such as live ducks and exploding dustbins.[97]

Notes

1. Herman Shaw, 'The Science Museum and Its Public', *Museums Journal* 49 (1949), pp. 105–13, on p. 105; Frank Sherwood Taylor, 'The Physical Sciences and the Museum', *Museums Journal* 51 (1951), pp. 169–76, on p. 169; Yvonne Harris and Ben Gammon, 'Science Museum Audience Profile. A summary of data from the MORI polls and admission data' (internal Science Museum report, 1997).
2. 'Children's playground' is how David Follett described some people's view of the Science Museum: David Follett, 'The Presentation of the Museum's Collections', 31 January 1961, SMD, Z 182/2 (No. 1).
3. Where children have appeared in other museum histories this tends to be in discussion of the children's museum movement in America or museum education services: for example, Eileen Hooper-Greenhill, *Museum and Gallery Education* (Leicester: Leicester

University Press, 1991); Marjorie Caygill, *The Story of the British Museum* (London: The British Museum Publications, 1981, p. 45; and David Wilson, *The British Museum* (London: The British Museum Publications, 2002), pp. 15, 88, only include children in their histories by use of this quote: ' "Babes in arms" were not allowed in the British Museum until 1878'.

4. William T. Stearn, *The Natural History Museum at South Kensington. A History of the Museum, 1753–1980* (London: Heinemann in association with the British Museum (Natural History), 1998), pp. 154–6.

5. Eugene Ferguson, 'Technical Museums and International Exhibitions', *Technology and Culture*, 6 (1965), pp. 30–4; Hellmut Janetschek, 'From the Imperial-Royal Collection of manufactured products to the Museum of Technology and Industry in Vienna', *History of Technology*, 17 (1995), pp. 191–213. An exception to this is the story of William Rosenwald, the 8-year-old son of the head of Sears in Chicago, who was instrumental in establishing the Museum of Science and Industry, having been entranced by a visit to the Deutsches Museum in 1911; Edward P. Alexander, *Museum Masters. Their Museums and Their Influence* (Nashville, Tennessee: American Association for State and Local History, 1983), p. 356. However, no children are mentioned in Otto Mayr et al., *The Deutsches Museum, Munich. German Museum of Masterworks of Science and Technology* (London: Scala Publications, 1990).

6. Examples of histories and science communication studies include James A. Secord, 'Newton in the Nursery: Tom Telescope and the Philosophy of Tops and Balls, 1761–1838', *History of Science* 23 (1985), pp. 127–51; Jackie Britton, 'Technology in Toyland: A Study of Miniature Technology', unpublished MSc dissertation, London Centre for the History of Science, Medicine and Technology, 1995; Aileen Fyfe, 'Science for Children', in Aileen Fyfe, ed. *Science for Children* (London: Thoemmes Press, 2003), pp. xi–xxii; and Alice R. Bell, 'The Childish Nature of Science: Exploring the Child/Science Relationship in Popular Non-Fiction', in Alice R. Bell, Sarah R. Davies and Felicity Mellor, eds *Science and Its Publics* (Newcastle: Cambridge Scholars Publishing, 2008), pp. 79–98.

7. Frank Sherwood Taylor, 'Children and Science in the Museum', *Museums Journal* 55 (1955), pp. 202–7, on p. 204.

8. Sophie Forgan, ' "A Nightmare of Incomprehensible Machines": Science and Technology Museums in the Nineteenth and Twentieth Centuries', in *Museums and Late Twentieth Century Culture* (Manchester: Manchester University Press, 1996), pp. 46–67, on p. 66. In contrast to the Science Museum, in many regional museums schoolchildren constituted a significant proportion of visitors; see, for example, Samuel Alberti, *Nature and Culture: Objects, Disciplines and the Manchester Museum* (Manchester: Manchester University Press, 2009).

9. A memorandum from Sir Francis Grant Ogilvie to Sir Cecil Smith, director of the V&A, 1915 , SMD, Nominal File 1776 Pt 1; Gaynor Kavanagh, *Museums and the First World War. A Social History* (Leicester: Leicester University Press, 1994).

10. Science Museum Annual Report for 1958, p. 127. This annual report also noted in a tone of incredulity that during evening opening 'not a single case of misconduct has occurred' and made reference to unaccompanied children being denied access to the Museum.

11. Anthony Burton, *Vision & Accident: The Story of the Victoria and Albert Museum* (London: V&A Publications, 1999), p. 50.

12. Burton, *Vision & Accident*, p. 50.

13. Forgan, 'A Nightmare of Incomprehensible Machines', p. 55.

14. Forgan, 'A Nightmare of Incomprehensible Machines', p. 55.

15. 'Memorandum from Francis Ogilvie, 1912', SMD, ED 79/1.

16. Admissions, 'Notice', SMD, ED 79/1.

17. *Report of the Departmental Committee on the Science Museum and the Geological Museum* Part 1 (London, 1911) and Part 11 (London, 1912), SMD, Z 170.

18. 'Science Museum Memorandum. Conduct of Visitors (especially Children and Boys) No. 18 F.G.O. 3 August 1912', SMD, ED 79/1.
19. 'Science Museum Memorandum No. 18', SMD, ED 79/1.
20. 'Memorandum from Inspector Bonnyman to the Director, 20 March 1915'. SMD, ED 79/1.
21. Science Museum Annual Report for 1925, p. 10.
22. Science Museum Annual Report for 1927 & 1928, p. 12. The two years were published in one volume.
23. Science Museum Annual Report for 1929, p. 4.
24. 'Set of Papers for the Meeting of the Science Museum Advisory Council on Friday June 13 1930', SMD, Z 193/1.
25. Frederick St A Hartley, 'Children's Gallery Memorandum, 16.12.58' 1958, SMD, Z 253/1.
26. 'Report of Museums Officer on Duty, Sunday 27 April 1969', SMD, Z 253/2.
27. Science Museum Annual Report for 1931, p. 4.
28. Science Museum Annual Report for 1932, p. 40; Anonymous, 'The Science Museum's Children's Gallery', *Museums Journal* 31 (1932), p. 555.
29. (M.H.L) 'Brighter Museums', *Discovery* 13 (1932), p. 9; *The Observer*, Sunday 22 July 1934.
30. Science Museum Annual Report for 1931, p. 40.
31. Science Museum Annual Report for 1934, p. 43.
32. Science Museum Annual Report for 1955, p. 31; Science Museum Annual Report for 1957, p. 24; Science Museum Annual Report for 1961, p. 17.
33. Science Museum Annual Report for 1961, p. 17.
34. Science Museum Annual Report for 1969, p. 11; in the Science Museum archives, a Press Notice, 'Children's Gallery at the Science Museum, London, 3 April 1969', SMD, Z 242/2.
35. 'Flyover in the Museum', *Willesden Mercury*, 11 April 1969; 'Science for Children', *Kensington News*, April 1969.
36. Science Museum Annual Report for 1971, p. 13.
37. Anonymous, 'A Science Museum's contribution to Science Education' (1963), SMD, Z 253.
38. Science Museum Annual Report for 1931, p. 14.
39. Henry Lyons, 'Letter to H.M. Richard, 4 January 1930', SMD,Z 253 Box 1.
40. Anonymous, 'Extension to the Children's Gallery at the Science Museum, South Kensington', *The School Science Review* 15 (1934), pp. 415–16, on p. 416.
41. Henry G. Lyons, 'A Memorandum on the Arrangement of Collections in the Science Museum to Serve as a Basis of Discussion' (1922), SMD, Z 253 Box 1.
42. Science Museum Annual Report for 1933, p. 9; Frederick St A. Hartley, 'Memorandum. Advisory Panel on Display' (1948), SMD, Z 157.
43. Herman Shaw, 'The Science Museum and its Public', p. 105.
44. Joyce A.M. Brooks and Philip. E. Vernon, 'A study of Children's Interests and Comprehension at the Science Museum', *British Journal of Psychology* (1955), pp. 175–82, on p. 175.
45. Sherwood Taylor, 'Children and Science in the Museum', p. 206.
46. Science Museum Annual Report for 1957, p. 19.
47. David Follett, *The Rise of the Science Museum under Henry Lyons* (London: Science Museum, 1978), p. 116.
48. Anonymous, 'Science Museum, Staff Reception', *Museums Journal* 31 (1931), p. 39; Gaynor Kavanagh, *Museums and the First World War*, p. 82.
49. Archives held in the Natural History Museum, BMNH(M) Printed Forms, 1889–1903; pp. 293–4, 296–8, 312, 314, 317, 336–7. Elizabeth Oke Gordon (daughter of geologist William Buckland) wrote a guide book for the Natural History Museum: Elizabeth Oke Gordon, *Natural History Museum, South Kensington. Guide-Companion for the Young* (London: British Museum (Natural History), 1903).

50. Flier for Lecture Association of Women Pioneer Lectures, 1893, held in Natural History Museum BMNH(M) Printed Forms, 1889–1903; pp. 293–4, 296–8, 312, 314, 317, 336–7.
51. Kavanagh, *Museums and the First World War*, p. 82.
52. Barbara Winstanley, *Children and Museums* (Oxford: Blackwell, 1969).
53. Science Museum Annual Report for 1929, p. 4.
54. The 'live fish' mentioned in this quote are presumably those in the Buckland Fish Collection, as referred to in the Science Museum Annual Report for 1868, p. 251. The Museum disposed of this collection – donated by Frank Buckland, son of William Buckland – in 1928, to make more room for industrial chemistry. 'Notes as to Admission of Children', (F.G.O, 1911), SMD, ED 79/1.
55. 'South Kensington Museum' in *Report of the Department of Science and Art for the Year 1869* (London: HMSO, 1870), pp. 352–3 and 'South Kensington Museum and Bethnal Green Branch Museum' in *Report of the Department of Science and Art for the Year 1891* (London, HMSO, 1892), p. xlii.
56. Science Museum Annual Report for 1919, p. 13; Science Museum Annual Report for 1929.
57. Follett, *The Rise of the Science Museum under Henry Lyons*, pp. 103–5.
58. Science Museum Annual Report for 1929, p. 4.
59. The article also included the following description of the gallery: 'Boys and Girls there were of every class, and they revelled in the exhibits. Which derived most benefit from the various demonstrations, it would be hard to say. But if one might judge from the exclamations of discovery and surprise, the "wildflowers of the streets" had the advantage in the matter of sheer enjoyment.' 'Scientists in the Making', *Telegraph*, 21 December 1931.
60. Frederick St A. Hartley, *The Children's Gallery* (London: HMSO, 1946).
61. Hartley, *The Children's Gallery*, 1938 edition.
62. Brooks and Vernon, 'A Study of Children's Interests and Comprehension at the Science Museum', p. 179.
63. In the Natural History Museum unofficial archives, Children's Centre Papers DF5006/11.
64. 'South Kensington Museum and Bethnal Green Branch Museum' in *Report of the Department of Science and Art for the Year 1891*, p. xlii.
65. Hartley, *The Children's Gallery*, 1935 edition.
66. H.G. Lyons, 'Letter to H.M. Richard, 4 January 1930', SMD, Z 253, Box 1.
67. Shaw, 'The Science Museum and its Public', p.110.
68. Sherwood Taylor, 'Children and Science in the Museum', pp. 202–7.
69. David Follett, 'The Presentation of the Museum's Collections' (1961); David Follett, 'The Purpose of the Science Museum' (1965).
70. Anthony V. Simcock, 'Alchemy and the World of Science: An Intellectual Biography of Frank Sherwood Taylor', *Ambix* 34 (1987), SMD, Z 194, pp. 121–39.
71. Science Museum Advisory Council. Committee on the Provision for the needs of Children in the Science Museum. Minutes of the first meeting, 1 October 1951 SMD,Z 194.
72. Brooks and Vernon, 'A Study of Children's Interests and Comprehension at the Science Museum'. Thad Parsons has pointed out that in SMD, Z 194 there are minutes of meetings which indicate that the value and relevance of this study were debated fiercely by staff.
73. Science Museum Annual Report for 1956, p. 3; Science Museum Annual Report for 1964, p. 4.
74. Anna-Katherina Mayer, 'Moralizing Science: The Uses of Science's past in National Education in the 1920s', *British Journal for the History of Science* 30 (1997), pp. 51–70.
75. W.H. Brock, 'Past, Present, and Future', in Michael Shortland and Andrew Warwick, eds *Teaching the History of Science* (Oxford: Clarendon Press, 1989), pp. 30–41.
76. Raphael Samuel, *Theatres of Memory* (London: Verso, 1994).

77. Donna Haraway, 'Taxidermy in the Garden of Eden, New York City 1908–1936', *Social Text* (1984–1985), pp. 20–64; Ghislaine Lawrence, 'Museums and the Spectacular', in *Museums and Late Twentieth Century Culture* (Manchester: Manchester University Press, 1996), pp. 69–80, on p. 69; Anonymous, 'Boston (Mass.) Children's Museum, Transportation Exhibition', *Museums Journal* 31 (1931), p. 121.
78. Hartley, *The Children's Gallery* (1935 edition), p. 1.
79. A memorandum from Sir Francis Grant Ogilvie to Sir Cecil Smith, director of the V&A, 1915 SMD, Nominal File 1776 Pt 1.
80. F. St A. Hartley, 'Science Museum, Children's Gallery'. Paper prepared by Mr Hartley and tabled at meeting of Children's Committee on 1 October 1951 SMD, Z 194.
81. Anonymous, *Museums Journal*, 31 (1932), p. 555.
82. Simcock, 'Alchemy and the World of Science: An Intellectual Biography of Frank Sherwood Taylor'.
83. Sherwood Taylor, 'Children and Science in the Museum', p. 205.
84. 'Butterflies in the Science Museum. Children's Response to "Amusement Arcade"', *The Times*, 31 August 1956, p. 11.
85. Science Museum Annual Report for 1931, p. 32.
86. Science Museum Annual Report for 1938, p. 145.
87. Science Museum Annual Report for 1949, p. 21.
88. Science Museum Annual Report for 1951, p. 15.
89. *The Sunday Times*, 3 Februrary 1963 in SMD, Z 183/2.
90. Science Museum Annual Report for 1963, p. 2.
91. John van Riemsdijk and Paul Sharp, *In the Science Museum* (London: HMSO, 1968), p. 4.
92. Stella V.F. Butler, *Science and Technology Museums*, Leicester museum studies series (Leicester: Leicester University Press, 1992), p. 35.
93. Riemsdijk and Sharp, *In the Science Museum*, p. 126.
94. Frank Oppenheimer, 'Rationale for a Science Museum', *Curator* 11 (1968), pp. 206–9.
95. Science Museum funding brochure, 1983.
96. Howard Gardner, *Frames of Mind* (London: Heinemann, 1983).
97. I would like to thank Rebecca Mileham and Anthony Richards for their input into this section. 'Hands-on at the Science Museum', Rebecca Mileham (2006); Hilde Hein, *The Exploratorium: The Museum as Laboratory* (Washington: Smithsonian Institution Press, 1990).

10

'An Effective Organ of Public Enlightenment': The Role of Temporary Exhibitions in the Science Museum

Peter J.T. Morris

Between 1919 and 1984, temporary exhibitions (called special exhibitions) were a major feature of the Museum's displays. They had been advocated by the Bell Report of 1911 as a 'means of keeping the Museum in direct touch with the movements of the day,'[1] and the first such exhibition, on aeronautics, took place in 1912.[2] As Mackintosh notes in his internally produced handbook, 'Special Exhibitions at the Science Museum':

> up to 1926 various exhibitions and celebrations of anniversaries were held, all on a fairly modest scale and in the galleries appropriate to the subject: considerable dislocation of the permanent collections was entailed, which sometimes evoked legitimate grumbling from the more studious and regular visitor.[3]

The number of temporary exhibitions greatly increased after the formal opening of the East Block in 1928, and Sir Henry Lyons 'introduced a new nature of exhibition with the object of interesting the public in the work which was being done by the research institutions of the country.'[4]

However, temporary exhibitions were also an easy way for the Science Museum to satisfy the demands of external bodies to have their activities displayed and thereby attract their support without allowing them to have influence over the permanent galleries. In 1932, perhaps encouraged by the success of 'Modern Glass Technology' arranged by the Society of Glass Technology in the previous year, Lyons had remarked in a memorandum to the Advisory Council entitled 'Museum Policy':

> It would, however, be possible, and probably useful, to move very actively in the direction of interesting industry in the Museum and its aims, ascertaining in what ways its use to Industry [sic] might be increased without turning [a temporary exhibition] into an ordinary trade exhibition. ... Large numbers of those engaged in various industries already visit the Museum, but it is much less well-known to those occupying high positions in the various branches of Industry, and it is with the object of interesting these latter and of working out

with them new ways in which this Museum can assist Industry that I antici-
pate a fruitful trend of future policy.[5]

The industry-wide (and thus commercially neutral) Research Associations were
pioneers in this field, with both the British Woollen and Worsted Research
Association and the British Non-Ferrous Metals Research Association holding temporary exhibitions in 1927. The plastics industry, under the aegis of the Society
of Chemical Industry (Plastics Group) and the British Plastic Moulding Trade
Association, was the first industrial grouping to create its own exhibition, 'Plastic
Materials and their Uses', in 1933,[6] and this was quickly followed in 1935 by the
first exhibition put on by a company, albeit a state-owned firm, 'The Empire's
Airway' by Imperial Airways (later BOAC and now BA). By meeting the expectations of these external bodies in this way, the Museum was able to obtain some
objects and support that might have otherwise been withheld, but the desired
stream of industrial funds never materialised. Nonetheless, it had become a deliberate policy by the late 1970s to use temporary exhibitions as a way of filling
spaces in the Museum without using Museum funds or drawing on the services
of Museum curators.[7]

In his handbook, Mackintosh mentioned that some exhibitions covered 'the
scientific aspect of some social betterment such as the reduction of noise or
smoke.'[8] Because of their authoritative but sober approach, these temporary exhibitions were often used by the establishment to prepare the public for change
of one form or another. It might be adopting smokeless fuel, buying a television, using computers, conversion to FM radio or natural gas, or, in an extreme
case, to accept a lack of coal to ensure the wheels of industry continued to turn.
One might have assumed the initiative for such exhibitions came from outside –
Mackintosh noted that 'an industrial or public utility exhibition usually results
from the suggestion of an outside body'[9] – but it appears many, if not all, of the
'socially useful' exhibitions were proposed by curators and offered to the relevant
external body.

After the East Hall opened, the number of temporary exhibitions per year
(see Table 10.1) was about 5.5 for Henry Lyons and Ernest Mackintosh, declining somewhat after the Second World War to around four for Herman Shaw,
Frank Sherwood Taylor and David Follett. The average in Terence Morrison-
Scott's time was very low, however, at 1.3. As noted in the first post-war
annual report, Mackintosh was keen on temporary exhibitions, and his average at 5.8 per year was higher than anyone's but Margaret Weston's. But, at 7.4
exhibitions a year, Weston's average is astonishing, and there was a remarkable peak in 1978–1980 with ten temporary exhibitions in 1978, twelve in
1979 and an all-time record of thirteen in 1980. This was partly the result of
extra space produced by the Infill Block, which was handed over on 30 June
1977, creating five new galleries, and partly due to the desire to draw visitors
to the Museum using externally developed temporary exhibitions while the
permanent staff concentrated on the development of new or renewed galleries, in particular the two new Wellcome medical galleries in the Infill Block

Table 10.1   Number of temporary exhibitions by period

Period	Director	Total number	Average per year
1912–1927	Pre-East Block	20	n/a
1928–1933	Lyons	31	5.2
1934–1938	Mackintosh	29	5.8
1946–1950	Shaw	20	4.0
1951–1956	Sherwood Taylor	24	4.0
1957–1959	Morrison-Scott	4	1.3
1960–1973	Follett	56	4.3
1974–1983	Weston	74	7.4

and the associated Wellcome collections formally acquired by the Museum on 21 June 1976.

Types of exhibitions

External institutions accounted for about a quarter of temporary exhibitions between 1928 and 1939, around 40 per cent up to 1972, and very nearly half of all temporary exhibitions (48 per cent) from 1973 to 1983. In 1979, it was remarked in the Science Museum Annual Report that:

> nearly all the special exhibitions in 1979 were sponsored by outside bodies, with assistance in varying degrees from Department 8 [Museum Services] in their mounting, and in some instances from one or the other of the curatorial Departments in their planning. This is a pattern which has been increasingly forced upon the Museum in recent years, because of the commitment of the Museum's own resources to the development of its permanent collections.[10]

Foremost among those collections was the newly acquired Wellcome Collections.

The most surprising result of breaking down the temporary exhibitions by type is the relatively low number of industrial exhibitions, by which I mean exhibitions put on by industrial companies or organisations themselves or at least heavily influenced by them – there were other exhibitions on industrial topics which are not 'industrial' by this definition. 'The Challenge of the Chip', organised by the Museum in collaboration with the Design Council and the Department of Industry, is a good example. Across the whole period between 1928 and 1983, they accounted for only 12 per cent of temporary exhibitions. After the often-mentioned spurt of such exhibitions in Mackintosh's era, they dwindle away, and even in Follett's period, which accounts for nearly a fifth of these exhibitions, they made up only 11 per cent of the total. By contrast, anniversaries – always

Table 10.2 Number of temporary exhibitions by type

	Total	External	Industrial	Anniversary	Books
Pre-1928	20	4	3	8	0
1928–1932	27	4	2	4	0
1933–1938	33	12	7	8	0
1946–1949	15	7	1	3	0
1950–1956	29	10	4	9	4
1957–1959	4	1	1	0	0
1960–1972	54	23	6	13	2
1973–1983	76	37	7	26	0
Total	259*	98	31	71	6

Note: *Includes the 'Aircraft in Peace and War' exhibition, which is not counted in Table 10.1

a popular reason for holding an exhibition – accounted for over a quarter of all temporary exhibitions. The academically inclined Sherwood Taylor introduced book-based exhibitions, but after his time most book exhibitions were small-scale exhibitions in the Library. There were only two major book exhibitions between 1957 and 1983.

Table 10.3 Types of institutions associated with temporary exhibitions

Company, industrial association or research association (inc. Kodak museums)	44
UK government and state bodies	43
Professional and learned bodies, conferences	41
Museums and art galleries (exc. Kodak museums)	18
Others, inc. charities and foreign governments	13

If one looks at the type of external organisations involved (Table 10.3) – allowing a weaker association than used for the 'external' and industrial categories in Table 10.2 – industry in all its forms (inflated by the Kodak museums and Kodak itself, which account for eight photographic exhibitions) accounts for the largest number, but is almost equalled by the British government (in all its historical manifestations, including the British Council, the Arts Council and state-owned industries). Organisations representing whole industries and industrial research associations do not account for many exhibitions, despite their early start, inter-war importance and commercial neutrality. A surprisingly large number were developed in collaboration with learned societies and professional associations. Exhibitions to run alongside important international scientific or technological conferences account for a notable proportion of temporary exhibitions, especially before 1939.

Table 10.4   Exhibitions covered in this chapter

Date	Topic	Associated institution(s)
1935	Noise Abatement	Anti-Noise League with the Ministry of Health, the Office of Works etc.
1937	Television	Radio Manufacturers' Association, BBC
1947	Chemical Progress	Chemical Society and DSIR
1947	Home and Factory Power	Ministry of Fuel and Power
1954	The Story of Oil	Shell
1971	Natural Gas	Gas Council
1976	Science and Technology of Islam	The World of Islam Festival Trust
1980	Challenge of the Chip	The Design Council and the Department of Industry

In this chapter, I will look at eight exhibitions between 1935 and 1982 (Table 10.4) that tried to deal in some way with a specific technological change or challenge to British society. I have avoided the term 'sponsor' as the associated institutions did not usually underwrite the exhibitions financially. These exhibitions involved government departments, professional institutions, public health bodies, state-owned industries, large corporations and cultural foundations.

'Noise Abatement', 1935

Putting on an exhibition on noise might seem surprising at any time, and particularly during the Depression.[11] One might have thought even the government had bigger social problems on its plate. However, industrial noise and industrial deafness were becoming to be recognised as an important environmental problem. Noise and smoke were seen as twin evils of modern industrial society. This was a problem which was technically soluble without any social upheaval and, furthermore, this was an issue the British government was capable of fixing, unlike many other problems in this period. The driving force behind this exhibition was the Anti-Noise League, a voluntary health organisation founded in 1933. This was dominated by patrician physicians, notably the royal physician Thomas Horder, and was also closely associated with the National Smoke Abatement Society. Not only was Horder an important member of the establishment, but he also wielded particular influence on the Board of Education. 'Noise Abatement' – along with its associated smoke abatement exhibition a year later – was a rare type of exhibition. The only remotely similar example was a short exhibition on 'Freedom from Hunger' in 1965. Public health, and even medicine, was not seen as part of the Science Museum's mission – the only other medical exhibition before 1978 was 'The History of Surgery' in 1947 – but noise and smoke were seen as a product of

Illustration 10.1   'Noise Abatement' exhibition, 1935

industrialisation which was within the Museum's remit. It was also unusual – at least in retrospect – in being proposed by an outside body rather than the Museum.

The Prime Minister, Ramsay Macdonald, opened the exhibition on 31 May 1935 (he had already decided to step down, and formally resigned as Prime Minister seven days later) and it ran during the month of June.[12] There were three sections of roughly equal size: research and development, transport and machinery, and buildings. Within each section, each exhibitor had its own stand, so it was more like a trade show rather than a unified thematic exhibition under the Museum's editorial control. This was typical for the period. The problem of 'displaying' noise was solved by having interactive sections where one could hear and experiment with noise.

The Science Museum's annual report noted that:

> The exhibition attracted a large number of visitors, particularly architects and members of the building trade, and organised visits were made by many interested societies and school parties... During its 30 days' duration over 44,000 persons visited the exhibition. This large attendance, together with some 200 notices and reports in the press was proof of the interest aroused; and there is

no doubt that it performed a useful and public service in placing before the community the various means now available for mitigating noise.[13]

While this exhibition was clearly successful, the Museum still saw its role as one of providing technological and scientific information rather than social enlightenment or even public health. The less successful smoke abatement exhibition a year later probably reinforced this view. Up to 1983, temporary exhibitions could be 'an organ of public enlightenment' on socially important issues, but only if the technology behind these issues was paramount.

'Television', 1937

Television broadcasting in Britain is often regarded as being a post-Second World War development, forgetting that the BBC officially began experimental broadcasts from its London transmitter 2LO in September 1929 using John Logie Baird's thirty-line electromechanical system.[14] In 1933 Baird Television began to develop a new 240-line electromechanical system at its Crystal Palace studios. Meanwhile, from 1932 onwards, EMI was developing alternative television systems, eventually producing an all-electronic system of 405 lines in its Hayes laboratories (close to the future Science Museum store; see Chapter 12 in this volume) after joining forces with the Marconi Company to form Marconi–EMI Television Ltd. The completed system was ready for use by the BBC in November 1936. A Television Advisory Committee chaired by Lord Selsdon, a former Postmaster General, had been set up in April 1934, and it decided to organise a competition for the contract to supply the BBC. Several companies entered – most of whom, such as Scophony, were also involved with this exhibition – but only Baird and Marconi–EMI had the resources to reach the final stages. When a regular television service from Alexandra Palace in north London – for two hours a day, six days a week – began on 2 November 1936, it was dual-standard, and there was a switch on the television sets to move between the competing 240 and 405-line systems. The final comparison was supposed to last for six months, but it was cut short after only three months, when the Selsdon Committee decided in February 1937 that the BBC should use the Marconi–EMI electronic system.[15] This period of television broadcasting came to an end when the BBC closed down its service two days before war broke out in 1939. A temporary exhibition on television was both an obvious attraction and yet at the same time problematic. Although the state-owned BBC was the only broadcaster, the television set manufacturers were vying for the public's custom. Similarly, the makers of television equipment were competing for the BBC's business. This created problems, given the Science Museum's ban on commercial advertising (and indeed the BBC's own ban), as the state could not be seen to favour one company over another.

One might have thought that the BBC would have proposed the exhibition as a way of promoting interest in its television channel, but it appears that it was the Science Museum that put the idea to the BBC in the summer of 1936, as the BBC gave its 'official blessing... to the proposal.'[16] The Museum then attempted to tie the BBC closely to the organisation of the exhibition by proposing the Chief Engineer of the BBC, Sir Noel Ashbridge, as the chairman of the organising committee, otherwise to be 'composed primarily of representatives of the chief television interests.' Mackintosh argued that the 'proposed Exhibition will have a very wide influence in developing public appreciation and interest in television and, in view of the vital interest which the Corporation now has in the matter, I feel that a public duty would be well discharged if you would personally accept the Chairmanship of the Committee.'[17] Ashbridge 'passed the parcel' back to the Museum, saying that if he took the chairmanship 'it might perhaps tend to make the Exhibition rather too closely linked with the B.B.C. and the Television Advisory Committee, of which I am a Member.'[18] However, his deputy T.C. Macnamara attended the committee regularly. Mackintosh himself took the chairmanship, but he never attended the meetings after the opening one, the chair being taken by Frederick Hartley, then Deputy Keeper of Industrial Machinery and Manufactures, or Macnamara. At this opening meeting, it was agreed that the exhibition would be run as a co-operative exhibition rather than a trade show and that 'publicity to individual exhibitors would be limited to an acknowledgement on the description label, in accordance with the general practice in the Science Museum.'[19] The development of the exhibition was then delegated to an Executive Committee, nominally chaired by Mackintosh, which comprised representatives of the Museum, the BBC and the major television manufacturers. Initially the Museum had hoped that the Radio Manufacturers' Association would help to finance the exhibition to the tune of £250. This was not forthcoming and the Museum had to meet all the installation costs, but the exhibitors had to pay for the tea they insisted must be laid on at the opening ceremony.

The first meeting of the Executive Committee, on 8 January 1937, strengthened the ban on advertising by abandoning individual labels in favour of a general label at the entrance informing visitors that 'receivers [i.e. television sets] had been supplied by the following firms ....'[20] The second meeting, a week later, was faced with one of the most important issues confronting the Exhibition, that the BBC only transmitted programmes between 3 p.m. and 4 p.m. during Museum opening hours. Fortunately, the firm of A.C. Cossor Ltd had already offered to transmit material (scanned from film) within the Museum, enabling visitors to see a television programme whenever they were in the exhibition.[21] The organisation of the exhibition threw up a number of interesting issues. Marconi–EMI was selling televisions under two different labels, HMV and Marconiphone, and asked if they could both be displayed as 'these organisations operate in commercial competition,' which was agreed.[22] Scophony Ltd wanted to display a 2-foot projection television screen (aimed at cinemas rather than domestic viewers) alongside the other commercial receivers. But the committee 'felt that the showing of a 2-foot picture alongside the Cathode Ray receivers would be misleading to the general public and might seriously affect the stocks now held by individual manufacturers.'[23] This row was resolved at the next weekly meeting, when it was agreed that the cathode ray receivers (which were commercially available) would be displayed on one side of the gallery as 'Receivers exhibited in this section are now available to the public' and the Scophony hybrid electronic and electro-mechanical 5-foot and 2-foot receivers would be displayed on the other side as 'Modern Mechanical Receivers'.[24] In the event, the 5-foot screen (77-inch screen in modern terms) was one of the key objects in the exhibition, attracting a lot of attention.[25]

The opening ceremony on 10 June was chaired by Ronald Norman, Chairman of the BBC Board of Governors – John Reith declined the invitation – and the exhibition was declared open by Lord Selsdon of the Television Advisory Committee.[26] *The Times* published an editorial the next day which emphasised the need to produce larger screens, on which more than two people could be seen clearly, and noted favourably the large Scophony screens on display. The exhibition was T-shaped, with a historical section in the long cross-bar of the T. The television receivers were placed in individual cubicles in the short arm of the T as it was decided that it would not be possible to see the picture easily under normal lighting conditions. The Scophony sets, as we have seen, were placed in a separate alcove on the other side of this arm.[27] The exhibition ran until 21 September and attracted more than a quarter of a million visitors. After it closed, the Museum took the opportunity to establish a television collection by acquiring some thirty-five items from the exhibition. In agreeing to these acquisitions,

Illustration 10.2   'Television' exhibition, 1937

Mackintosh remarked that '[television] is a newish subject & we must keep our perspective clear from the outset as to what principles are involved & how we can illustrate them.'[28] His comment shows that it was the value of these objects for showing the basic scientific principles behind television that mattered to the Museum in the 1930s, rather than their intrinsic historical interest or value, thereby creating the 'propriety' – to use Mackintosh's term – for keeping historical items. Otherwise it might have looked as if the Museum were lending historical significance to purely commercial articles.

The exhibition handbook was well produced, with a stylish image of the top of the Alexandra Palace transmitter on the green cover, but it is far more technical than most Science Museum exhibition handbooks and was probably incomprehensible to the non-technical visitor.[29] Even when the material could have been of interest to prospective viewers – for instance, a map showing the strength of the signal from Alexandra Palace – it was unhelpful to the ordinary viewer, showing the strength of the signal in mV/M rather than in terms of the quality of the picture obtainable at that strength (although, to be fair, the quality of the picture – now as much as then – is partly affected by the viewer's own aerial). This is a reflection of the exhibition's intended audience, namely people who would be working with television in one way or another – manufacturers, technicians, retailers and repairers – rather than the general public as potential viewers. Ten thousand copies were printed by His Majesty's Stationery Office and, in an unusual step, it was distributed to booksellers. Many copies were purchased by the television manufacturers for their own staff, and nearly the whole print run was sold, although the Museum still had unsold copies in a basement store in the 1990s.

Despite the earlier decision not to put commercial names on the individual labels, the final labels did bear the brand names when the exhibition opened. This was a result of an 'ambush' at the end of the last meeting of the committee, when Harry Barton-Chapple of Baird Television Ltd proposed that the 'manufacturers' names should be indicated over the various cathode ray receivers,' rather than the discreet acknowledgement on the label which was the Museum's standard practice. This proposal was carried unanimously by a committee dominated by the manufacturers.[30] This breach of protocol provoked a personal letter from Clifford C. Paterson, head of GEC's research laboratory, to Mackintosh on 5 July, which said:

> In all the other exhibitions in which we from here have lent a hand the exhibits have been anonymous and the exhibition has been truly one for public education. It appears to me that in the Television Exhibition you have got right away from this principle. The Exhibition appears to be now mainly a competition between manufacturers...[31]

Paterson threatened to withdraw GEC's exhibits, as the firm had designed these exhibits 'solely for their educational value' and it was alleged (by others in GEC) that they were 'not showing up well in comparison with other people's.' He concluded by saying 'I write this frankly, not with the object of embarrassing you, but in order to make a point of view clear which I feel is rather vital for the future welfare of Science Museum Exhibitions.' Mackintosh was clearly embarrassed, as he wrote back to Paterson to say:

> I was present at the first meeting of the Committee which organized the Television Exhibition, and mentioned the usual desirability of anonymity; I understood that at a subsequent meeting it had been accepted (though without enthusiasm in some quarters!), and gave no further thought to the matter. I am afraid I could not have been explicit enough on this point, either to my own staff or through them to the Committee, for the printed labels for the exhibits, which arrived from the Stationery Office only a day or two before the official Opening, all bore the names of the lenders or providers. This is, of course, quite contrary to our tenets in Special Exhibitions, and was a most unfortunate slip-up on our part, for which I must assume all responsibility. I didn't happen to notice the fact till lately, and then decided to leave things as they were rather than risk a complete disruption of the exhibition...I regret most sincerely that this should have happened, and must place myself in your hands.[32]

It was now the turn of Paterson to be discomfited, as he had just been appointed a member of the Advisory Council as the representative of the Institution of Electrical Engineers, and he was evidently taken aback at this grovelling response from someone he respected. He replied, 'I should have avoided causing you any embarrassment which I could possibly help, and after your letter I shall be more anxious to do so. ...Please do not allow my letter to you to worry you any longer.'[33]

Later in July Barton-Chapple complained that the films being transmitted in-house by the Cossor company were boring and asked why the films lent by the Baird company were not being used.[34] Hartley replied that the Museum had stopped showing A-rated films following a recent complaint.[35] The British Board of Film Censors had been set up in 1912 with the aim of standardising the classification of films, hitherto in the hands of numerous local licensing authorities, and it introduced two classifications: U (universal) and A (adult). Children could see both types of film, but this became a matter of controversy in the 1930s and there was a public campaign to stop children seeing A certificate films.[36] Clearly the Science Museum had been caught up in the crossfire between the two sides. This left just three films, the Jubilee of George V (which Hartley described as 'not very suitable' and 'rather worn'), Mickie Mouse [sic] and a collection of film trailers. Hartley said he would try to combine the Mickey Mouse film with some of the trailers to make a 20-minute film.

'Television' and its handbook were very successful, despite being very technical, but the Museum struggled to retain its commercial neutrality in the face of strong corporate rivalries. The purpose of the exhibition was to explain the principles behind television rather than its history (which was obviously of short duration at this time) or to promote its take-up by the general public. The way the television manufacturers had managed to slip corporate identification into the exhibition, despite strenuous attempts by the Museum to prevent this, must have made this exhibition a bruising experience for Mackintosh despite the undoubted kudos of being able to present a new consumer technology in the Museum at the moment of its release. The struggle between the Museum's ban on advertising and the determination of many firms to capitalise on an exhibition must have made exhibitions linked to a single firm or a small group of firms unattractive, despite Lyons's stated desire only a few years earlier to use temporary exhibitions as a bridge between the Museum and the captains of industry.

'Chemical Progress', 1947

This was a typical example of a common theme for temporary exhibition, the anniversary of a learned society, the Chemical Society. This was founded in 1841 and was one of the oldest societies in the chemical field. It was a learned society rather than a professional association – this latter role was filled by the Royal Institute of Chemistry, founded in 1877.[37] It had two very different functions. Its members outside the Home Counties were mainly interested in the cheap rate for its eminent journal, while its academic members in Oxford, Cambridge and London regularly attended scholarly meetings held at the society's London base. Yet this was also an exhibition in which a government department, the Department of Scientific and Industrial Research (DSIR), played a prominent part. Its involvement must be seen against the background of the Second World War, a conflict which had ended only two years earlier and one in which science – chemistry as much as physics – had paid a prominent role. Now that the war was over, the government was anxious to show that debt-laden and battered Britain could still 'make it', in the words of the famous exhibition across the road at the V&A in 1946. Britain's prowess in industrial research was seen as being a major factor in Britain punching above its weight in the post-war global economy. At the same time the spectre of the atomic bomb now hung over scientific research and there was a need to show that scientific and industrial research could be beneficial. The centenary celebrations of the Chemical Society provided the DSIR with an opportunity to celebrate scientific research which it duly seized.

The actual centenary of the Chemical Society in 1941 was overshadowed by the war, and an exhibition to mark the event only became possible afterwards. The celebrations were to take place immediately before the XIth International Congress of Pure and Applied Chemistry and the XIVth International Conference of the International Union of Chemistry in London, and an associated exhibition illustrating the history of chemistry was planned. The Centenary Celebrations Exhibition Subcommittee was formed in May 1946 and its chairman was Sir Robert Robertson, who had been the Government Chemist between 1921 and 1936 (not to be confused with his famous contemporary, the organic chemist Robert Robinson).[38] The first actions of the subcommittee, when it met on 7 May 1946, were to invite the Science Museum to join the committee (along with Sherwood Taylor) and to suggest that the exhibition should be held in or near Central Hall, Westminster.[39] As an amateur historian of chemistry, Sir Harold Hartley may have proposed the involvement of the Science Museum, although he did not become a member of the Advisory Council until 1951. When Herman Shaw was invited to join the exhibition subcommittee he immediately offered the, then empty, chemistry galleries on the third floor and proposed that Alexander Barclay, as Keeper of Chemistry, should represent the Museum on the subcommittee.[40]

Tom S. Moore (a leading physical chemist, who produced the official history of the society[41]) and Alexander Findlay (the author of a popular history of chemistry[42]) both declined the subcommittee's invitation to draw up a list of chemists with major chemical achievements. Sherwood Taylor, the curator of the Museum of the History of Science in Oxford, was then asked by the Joint Exhibition Committee (the exhibition and congress subcommittees had been merged) at its meeting on 19 June to develop the content of the exhibition around the theme of 'notable British chemists'.[43] Sherwood Taylor later published his treatment independently of the exhibition as *A Century of British Chemistry*.[44] Sir Wallace Akers of ICI and James Davidson Pratt, Director and Secretary of the Association of British Chemical Manufacturers and Chairman of the British Road Tar Association, were invited to join the committee. Akers immediately offered the support of ICI's exhibitions department and Davidson Pratt became the liaison with the DSIR. Subsequently Sir Edward Appleton of the DSIR appointed his colleague Oscar Brown as his representative on the committee.[45] This was perhaps a surprising appointment, as Brown was a radio engineer and the co-author of a textbook on radio communication who had served on the pre-war Selsdon advisory committee on television. In practice, the DSIR was usually represented by Christopher Jolliffe, who nearly became Director of the Science Museum in 1956.[46] At the meeting on 8 August it was agreed that the exhibition would fall into three parts: a historical exhibition covering 'the achievements of great chemists who have worked in this country since 1841,' a book exhibition (presumably of contemporary books, as is usual in international congresses, but this is not clear from the minutes), and a chemical exhibition organised by the DSIR.[47] Subject panels – staffed by eminent chemists including J.D. Bernal, Cyril Hinshelwood, Ian Heilbron and N.V. Sidgwick – were established to develop the individual sections of Sherwood Taylor's historical survey. The weakest section (judging from the exhibition handbook) was analytical chemistry, which had been given over to George M. Bennett, Robertson's successor as the Government Chemist, rather than having its own panel.

At its meeting on 17 October, the committee decided that 'the reports prepared by the Advisory Panels were of value in themselves and agreed to recommend that they would form a basis for an admirable guide to the exhibition.'[48] It is not clear how far this handbook represents the exhibition on the floor, but the objects on display were selected by the panels, with the help of Robertson, and a list of the Museum's own objects by Barclay.[49] It was a large exhibition, stretching across three galleries on the third floor of the East Block, the former home of the chemistry galleries, which had been cleared during the war. The historical section was T-shaped, with visitors expected to follow a predetermined route through the exhibition.[50] Each subsection typically had about ten to twenty objects, but organic chemistry had thirty-seven and the solid state twenty-eight. Surprisingly, given the importance of instruments in that field, analytical chemistry only had seven. The general approach was historical, but with an obvious (if not exclusive) emphasis on the period since the foundation of the Chemical Society in 1841. It was highly internalist and somewhat whiggish, aiming to show how chemistry reached its then current eminence. Some sections went very much from the work of one notable chemist to another, but most (for example Bernal's section on the solid state) were similar to the context-free historical introduction to a chemical paper.[51] The press release prepared in December emphasised the discovery of benzene by Faraday (which had been the subject of a temporary exhibition in 1925), the preparation of mauve by Perkin and the work of Ramsay on the rare gases.[52] In fact the Museum acquired several important items from Perkin's family as a result of this exhibition.

The DSIR, assisted by the Central Office of Information and several industrial research associations, put together a more visitor-friendly display on 'Chemistry in Everyday Life', which did not appear to have any objects at all. It aimed to show how chemistry 'affect[s] the day to day life of every citizen.'[53] It was developed by a committee headed by Davidson Pratt. This part of the exhibition had sections on agriculture and food, health, homes and building, fuels, oil, and transport and engineering. As it focussed on the practical applications of chemistry, the overlap with the academic sections was less than one might have expected.

Illustration 10.3 'Chemical Progress' exhibition, 1947

The handbook description of this section ends with a conclusion which put the emphasis on what we would now call scientific citizenship:

> ... we must admit that willy-nilly, our present way of life is possible only by science [sic]. Here have been shown only the beneficent applications of science, leading to increased health, food and comfort, thanks to the co-operative efforts of scientists, engineers and industry. But scientific knowledge is neutral; its uses can be for good or evil – for life or for death. It is only by knowledge and understanding that the right uses of scientific discoveries can be chosen, and it is up to all of us to make this choice.

This concluding remark in the exhibition handbook, clearly influenced by the recent development of the atomic bomb, sums up the philosophy of many of the temporary exhibitions. The visitor was given a summary of the issue, rendered authoritative by the participation of the relevant government organisations, professional institutions and industrial bodies, all of whom were linked into a national scientific network. It was then up to visitors to draw their own conclusions, seemingly to make their choice. But there was no real effort to develop a real sense of choice, of a participatory technological democracy. The jury might decide, but this jury was given clear directions on what to think by the temporary exhibition.

Initially Shaw had proposed that the exhibition should run for a month, but in November 1946 he wrote to Robertson and suggested that the exhibition should run for a longer period 'in view of the widespread interest which is being taken by the general public at the present time in all matters of scientific interest,' a proposal the Chemical Society was happy to accept. In the event, the exhibition ran for eleven weeks, from 14 July to the end of September.

'Home and Factory Power', 1947

February and March 1947 were exceptionally cold and snow-bound. Matters were made worst by a shortfall in coal production, which also had an impact on the supply of gas and electricity in this period, as they were both derived from coal. This coal shortage was partly a result of blocked transport routes making it impossible for coal supplies to get through, but it was largely the end result of years of underinvestment, undermanning and poor labour relations. Coal production in 1946 was only 80 per cent of what it had been in 1937. Even the formation of the National Coal Board on the first day of 1947 did little to solve this problem. A combination of labour disputes and poor productivity following the introduction of a five-day week ensured that coal production remained low. Thanks to the big freeze, demand exceeded supply for the first three months of 1947. Thereafter production remained low, reaching its lowest level that year in July. Much of the blame for this was directed – probably unfairly – at Lord Hyndley, the Chairman of the National Coal Board. The coal shortage had a significant impact on industry, and the weaving industry in the north of England practically closed down for lack of power during January 1947. Only an emergency rationing plan developed by the President of the Board of Trade, Stafford Cripps, kept the wheels of industry turning. As the National Coal Board had failed to raise production by the end of the summer, there was a real possibility that the crisis would recur the following winter.[54]

Given the need to persuade the public at large to accept cold homes so that British industry could continue to operate, one might have thought that the government asked the Museum to help the cause by putting on a temporary exhibition, a classic top-down exhibition. In fact, although it is not clear who originated the idea, the initiative seems to have come from the Museum, since William O'Dea wrote to Alfred M. Rake, an Assistant Secretary in the Ministry of Fuel and Power, on 27 August to complain that:

> it is now over 13 weeks since Hartley & I saw your people at Millbank to discuss the Director's offer of facilities at the Science Museum for a small exhibition to emphasise, to 250,000 visitors, the importance of load conservation this coming winter...and it was found that the idea of presenting dramatically to the public the dire results of individual irresponsibility was very much in accord with the plans of your Department to cope with this vital task.

As an enthusiast for working exhibits, O'Dea also warned that 'the time factor...has passed through the desperate to the impossible if the exhibition is to be other than a static display to which the public would pay its usual scant attention.' He went on to say that the Director [Shaw] 'who had been increasingly perturbed at this lack of progress, authorised me to withdraw our offer if there was no apparent hope of effective implementation.'[55] In a note to Hartley the following day, O'Dea exclaimed: 'Never in a fairly extensive experience of various Departments have I found greater ineptitude than this display by the Ministry of Fuel.'[56] O'Dea's letter had an electrifying effect on the ministry, and financial authority for the exhibition was obtained less than a week later.[57]

W.H. Willson, the Director of Public Relations at the Ministry of Fuel, had already proposed Ian Jeffcott, 'who used to do the Ideal Home Exhibitions before he joined the army,' as the designer.[58] There was a budget of £2,500 and the major bottleneck was getting two working exhibits ready in time for a mid-October opening.[59] Shaw agreed that the exhibits should be made by the Museum's workshops out of hours as the Museum's contribution to the exhibition 'to help the exhibition along, and after the extraordinary delays of the Min of Fuel & Power this would seem the only way of getting it ready on time.'[60]

The Museum put pressure on the ministry to get the highest possible official to open the exhibition in the East Hall on 21 October, and was very pleased that Lord Citrine, Chairman of the British Electricity Authority, agreed to do the honours. He was chosen in preference to two political figures who are now better known – Manny Shinwell, then Minister of Fuel and Power, and Hugh Gaitskell, the Parliamentary Secretary in the ministry – because their attendance could not be assured. The Museum was also keen to avoid asking 'Lord Hindley' [sic],[61]

Chairman of the National Coal Board, who 'has perhaps not a very good press at the moment,' probably a reference to his problems with the five-day week and the continuing low level of coal production.[62] The press notice for the opening remarked that:

> the object of the exhibition is to show, in the Science Museum manner, the relationship between power in the home and power for industry and to show how, with reasonable care in the home, there need be the minimum of austerity all round....Thus starting from the fact that 24 persons are employed for each horse-power of marginal electric load in industry, a model has been constructed in which a workshop employing 24 persons is driven by a one horse power motor. Puppet workmen are busily employed until the visitor presses a switch to bring on a one-bar electric fire which is the equivalent of at least one horse-power in terms of domestic load. Immediately the workshop stops and the workers sit down to wait until the fire is switched off again. An illuminated label tells what has been done and at which hours of the day this disastrous result is most likely to happen....in addition there is full information, vividly displayed, about the relative importance of various domestic appliances. If, for instance, the visitor learns to take hot baths at night instead of in the morning, it is far more effective than breaking the cleaning routine by economising on the noisy but fairly harmless vacuum cleaner....It is hoped that up to quarter of a million people will see this exhibition during the three or four months it will run, and that many of them will learn how to avoid the worst types of discomfort – which we experienced last winter – while at the same time giving industry a fair chance.[63]

In his response to a letter of apology from the Director (who had missed the opening because of an unexpected trip to the Midlands), Lord Citrine remarked that he 'thought that the exhibition was really good, and I hope it will have a very substantial effect on reducing consumption, as it deserved to have.'[64]

The exhibition was eventually dismantled in the week beginning 9 February 1948 and was moved to Bromley, Kent, and from there to 'various Provincial centres.'[65] To complete this travelling exhibition, the ministry asked if it could have the Museum's power station model. The relevant Keeper, Hartley, felt that 'this would be a more useful home than we ever expect to find for a model which we could not hope to exhibit again in the permanent collns.'[66] The Director felt unable to give the model to the ministry ahead of the Board of Survey that would formally dispose of the model, but a loan was arranged in February, which was converted into a gift after the Board of Survey took place in May 1949.[67]

If the government had ever seen the Science Museum as a propaganda arm of the state – an offshoot of the Central Office of Information – this was surely the occasion when it would have asked the Museum to prepare the public for a potential rerun of the energy crisis of 1947. Yet it appears that it was the Museum that proposed the exhibition and it was the Ministry of Fuel and Power that procrastinated until it was almost too late. The exhibition was well attended, partly because of the lack of competition in this period before television became a major rival, but also because the public perceived the issue as an important one and the Museum's treatment of it as authoritative. Because it was on a topic – energy – which has always been associated with the Museum, and it was put on by a government ministry, the public was willing to believe in the message of the exhibition. In the event, the crisis was averted, not by the Science Museum's exhibition nor by the efforts of the Dowager Marchioness of Reading's 'sales force', which toured the country persuading people to save energy, but by the warm autumn and mild winter of 1947 and early 1948.

'The Story of Oil', 1954

If the coal industry was struggling in the late 1940s and 1950s, petroleum was clearly the fossil fuel of the future. By the early 1950s the vast oilfields in the Near and Middle East, first opened just before the Second World War, were turning out crude oil in prodigious quantities and at a historically low cost. The fact that this oil came from a strongly Islamic part of Asia and thus made the Islamic states there very wealthy was to have major political and cultural implications (as we will see when we come to the 'Science and Technology of Islam' exhibition), but this development lay in the future. The immediate problem for the Western oil companies was the rise of nationalism in the Middle East, fuelled in part by resentment at Western control over their oil-producing industries. In Iran, this wave of nationalism culminated in the nationalisation of the oil industry, hitherto controlled by the Anglo-Iranian Oil Company, by the popularist prime minister Mohammed Mosaddeq in 1951. Mosaddeq was deposed in a coup arranged by the CIA two years later, but the oil industry remained in Iranian hands. The Anglo-Iranian Oil Company (renamed British Petroleum in December 1954) was owned by the British government to ensure a reliable supply of fuel oil for the Royal Navy. Its major rival in Britain was the Anglo-Dutch firm Royal Dutch Shell.

Public relations was a major concern for the petroleum firms in this period. On one level the industry needed to persuade the public that oil was the fuel of the future and that the oil companies could be trusted to provide it. As the oil industry was an oligopoly controlled by a small number of international companies (the so-called 'seven sisters'), there was no effective competition on price (there were no supermarket petrol stations in the 1950s) and little difference in the quality of the products, but fierce competition between brands. This meant that every firm avidly sought opportunities to place its particular brand in front of the public in much the same manner as cigarette or detergent firms.[68]

The oil companies were also strongly dependent on science and technology, from geology and seismology to find the oil, to drilling technology to get it out of the ground, to process engineering to convert crude oil into fuel oil, diesel oil, petrol and petrochemicals. The industry's demand for scientists and engineers in the 1950s was almost insatiable. As the supply was limited – only 24.4 per cent of British science graduates went into industry and commerce in 1952[69] – the oil companies needed to encourage young people to become geologists and chemical engineers with the aim of working in this booming industry which offered high wages.

One way of achieving these two aims – of putting their brand before the public and encouraging young people to work for them – was to develop touring exhibitions. This was the period – before television became commonplace – when major companies had their own exhibition departments, including ICI (which assisted with 'Chemical Progress'), Shell and BP. The 'Story of Oil', a travelling exhibition created by Shell's exhibition department, had already toured the provinces when Walter Winton, then Assistant Keeper of Chemistry, suggested to Shell in the summer of 1953 that it could be shown at the Museum.[70] His Keeper, Barclay, was broadly in favour of this idea (he had already seen the exhibition) – 'all that we shall be required to do is provide a 5 KW electric supply' – and he proposed hosting it in the spring of 1954.[71] However, the Director, Sherwood Taylor, worried that the exhibition would be about Shell rather than the oil industry.[72] As we have seen, Shell's rival, the Anglo-Iranian Oil Company, was owned by the British government. The Science Museum did not wish to publicly endorse any company and certainly did not wish to be seen to promote one which was a rival to a firm owned by the British government, the Museum's paymaster. Barclay assured him that Shell was aware of the problem and that 'they have even used "Anglo-Iranian" in places' in their exhibition. The only potential problem was 'a discreet reference in their handout...which I do not doubt could be altered if necessary.'[73]

Although the Museum had accepted the exhibition on 27 October 1953,[74] the problem of potentially annoying Anglo-Iranian did not go away. This problem resulted in a correspondence between Sherwood Taylor, the former curator of the Museum of the History of Science, and A. Everard Gunther, a petroleum geologist and the head of the exhibition department at Shell, who also happened to be the son of Robert T. Gunther, the founding curator of the Museum of the History of Science in Oxford. He had worked with Sherwood Taylor to sort out his father's complicated affairs vis-à-vis the museum following the elder Gunther's death in 1940.[75] Sherwood Taylor wrote to Gunther in confidence on 18 November, fretting about 'the risk of repercussions from, say, Anglo-Iranian or other sections of the industry and I wondering how you would consider this risk may be avoided, if the exhibition be shown here.'[76] In his reply Gunther pointed out that Anglo-Iranian was one of the owners of Shell-Mex and the exhibition was billed as being sponsored by Shell-Mex and B.P. Ltd. Furthermore, the firms had a gentlemen's agreement that they informed each other of their exhibition schedule. Nonetheless, he offered to remove 'Presented by the Shell Petroleum Company, London' from the version of the handout printed for the Science Museum. He concluded by saying 'we do not seek to advertise ourselves, but you must forgive us if the tanker we show has a "Shell" on the funnel, but they do in real life as well.'[77] There is little doubt that the close relationship between the two men forged in Oxford – which is not apparent from this correspondence – helped to prevent this disagreement getting out of hand.

Sherwood Taylor was happy with the proposed solution, but the problem arose again when a copy of the handout reached the Museum in February 1954, ahead of the exhibition opening in April. The reference to Shell had been removed, but the Shell logo was still on the back cover. It is clear that Winton felt that this was intentional, rather than an oversight on the part of Shell, as he wrote to Barclay that 'whatever the assurances given by Gunther, there is a very strong commercial advertising interest in this exhibition which is circumventing rather than observing the spirit of a prohibition on advertising.'[78] Donovan Chilton, a Deputy Keeper who also served as the Museum's part-time publicity officer, contacted Gunther, asking him to take the handouts back and produce a new version without the logo, chiefly because it set a precedent for all such leaflets in the future. According to Chilton, 'Mr Gunther did not sound very pleased but said that he would feel it his duty to acquiesce in whatever we decided.' Chilton went on to say, 'I think not only that we ought to be firm in this instance, but also that we should reassert the Museum's supremacy in relation to exhibitions held here; but that is, perhaps, not strictly my job as publicity officer.' Chilton suggested that as a quid pro quo for Shell dropping the logo and reprinting the booklet, the Museum should acknowledge Shell's assistance in presenting the exhibition, a statement Chilton declared would 'not only restrain Shell's publicity ardour, but also indicate the museum's status in the undertaking.'[79] Although there is no reply from Shell in the file, it is clear that both sides were satisfied with the compromise, as the new version of the handout replaced the logo with the proposed acknowledgement. Winton, who may have been embarrassed about his role in bringing this contentious exhibition to the Museum, wrote to Chilton that 'I think this is a most statesmanlike Solution. It conforms to the established custom of acknowledging any object on loan. While giving credit where credit is due it makes it easier now to smack down hard on any further attempts to make the Science Museum a rattle for Shell Advertising Dept.'[80] The Museum's anxiety to avoid any link with Shell is made abundantly clear by the press notice issued by the Museum for the opening on 8 April, presumably written by Chilton. There is no mention of Shell at all in the text, not even in the reference to Gunther giving the introductory address in the lecture theatre, thus leaving his connection with the exhibition or the oil industry obscure.[81]

Despite all the fuss it caused, the exhibition was clearly of short duration, as Barclay could write to Gunther on 11 May that the exhibition 'has come to an end, and the panels safely returned.'[82] Both the Science Museum and Shell appear to have been satisfied with the visitor numbers of 23,000 over four weeks.[83] Gunther found it particularly gratifying that 'enough students have already called here who went to see the exhibition specially because they wanted to know more about the industry of which they had heard something.'[84] It is not clear what he meant by 'enough students' but it appears that recruitment to the rapidly growing industry was one of Shell's major aims in creating the exhibition. Gunther concludes by saying 'We are glad to think that the Exhibition has been a good thing for the Museum also,' although Barclay's comment in his letter of thanks that the exhibition had been 'well worth while' seems polite rather than enthusiastic.

There are two striking aspects to this episode: the Museum's great anxiety to avoid any overt association with Shell Petroleum and the lack of any financial aspect to the whole business. The Museum's concern about the link with Shell can be ascribed partly to the importance of the Anglo-Iranian Oil Company (BP) to the Museum and, once again, partly to the Museum's complete ban on any commercial advertising on the Museum's outputs. Anglo-Iranian was a large powerful oil company, but, more importantly from the Museum's point of view, it was government-owned. By appearing to support Shell the Museum could be accused of undermining another part of the Crown. Furthermore, the situation regarding Anglo-Iranian was very sensitive following the nationalisation of the Iranian oil industry and the overthrow of Mosaddeq in August 1953, which is perhaps obliquely mentioned in Sherwood Taylor's letter to Gunther in November 1953 when he doubts the wisdom of holding an exhibition linked to one particular firm 'at this time'.[85] Furthermore, the Museum felt that Shell was trying to use the exhibition to publicise Shell, and feared it would be the thin end of the wedge regarding other firms' expectations of publicity. Winton in his memo complaining about the Shell logo on the handout said that 'an avalanche of SHELL publicity has been loosed on all the schools in London and the Home Counties.' Shell's marketing manager, the Hungarian aristocrat A.E. Apponyi, was blamed for this and Winton remarked that 'Mr Apponyi is a good advertising man and will try anything in the hope that we shall let him get away with it.'[86] The exhibition was attractive to the Museum, allowing it to publicise an important aspect of industrial chemistry while the much-delayed Industrial Chemistry Gallery was still under construction. There appears to have been no attempt to use the exhibition to attract funding from Shell, and indeed this episode shows that such funding would have been problematic for the Museum in this period.

'Natural Gas', 1971

The coal gas industry in the 1950s was inefficient and increasingly old-fashioned, the result of decades of underinvestment and fragmented production.[87] In the early 1960s, the production of gas from coal was largely replaced by the manufacture of gas from petroleum using processes developed by the gas industry and the chemical industry. Furthermore, the other products of the coal gas industry, coal tar and ammonium sulphate, were losing their significance as a result of changes in the chemical industry. Natural gas (methane) became available as an alternative to manufactured coal gas in the mid-1960s from several sources, notably Algeria, and crucially from the geographically nearby and politically stable North Sea. The Gas Council decided to switch to natural gas, mainly from the North Sea channelled through a new terminal at Bacton on the Norfolk coast, as an alternative to the renewal of the coal gas industry. However, if natural gas were to replace coal gas, gas-burning equipment in the home and in industrial plants would have to be converted or replaced. Following a successful trial conversion programme carried out at Canvey Island, Essex, in 1964, the powers needed for this changeover were granted to the Gas Council by the Gas Act of 1965. It was a massive operation, taking ten years at a cost of £600 million pounds. It began at Burton-on-Trent in the Midlands in May 1967 and was completed at Comely Bank in Edinburgh in September 1977. The closing ceremony was carried out by Sir Denis Rooke, the Chairman of British Gas and subsequently Chairman of the Science Museum Board of Trustees. To some extent, the exercise was wasteful – I recall my family were given a brand new gas ring and gas poker, which are still in my loft, having never been used – but it had to be carried out to make a safe changeover to natural gas possible. So it can be seen that the temporary exhibition actually fell in the middle of the conversion programme rather than at the beginning, even in its own region of North Thames Gas.

In the case of the 'Home and Factory Power' exhibition we have seen that the Museum took the lead in an exhibition to promote state policy, namely the reduction in the use of domestic electricity at certain times. The 'Natural Gas' exhibition is another example of a temporary exhibition in the Science Museum fulfilling a state need, namely public acceptance and understanding of the conversion from coal gas to natural gas (gas being a state-owned industry at the time). As with 'Home and Factory Power', the initiative came from the Museum rather than the industry. However, the impetus for the exhibition actually arose not from the imminence of the nationwide conversion to natural gas but from the Museum's inability to renovate the existing Gas Gallery until at least 1973. Greenaway, the Keeper of Chemistry (gas then came under chemistry), proposed this exhibition to the Gas Council as a stopgap. The existing Gas Gallery was started in 1937 but work had been halted by the outbreak of war in September 1939 and it was not completed until 1954. It was unusual in being developed under the direction of the British Gas Federation. It was thus seen as being very much the industry's 'own' gallery. Mackintosh remarked in February 1939 that 'although our industries have always been helpful and generous in a piecemeal way, none have hitherto made themselves responsible for the entire section devoted to their subject[88] at this national museum' and he hoped that it would 'be a stimulating example to other industries.'[89] When the industry was nationalised, the responsibility for the gallery passed to the state-controlled Gas Council and thus came within the more usual arena of co-operation between the Science Museum and other state bodies.

When Greenaway was asked by T.E.D. Mason, Secretary of the Gas Council, in October 1970 to consider 'further work on the Gas Gallery', he was conscious of the need to 'maintain good relations and also experiment with public reaction to Gas as a museum subject.' He proposed a temporary exhibition on conversion, 'which will not feature notably in the eventual permanent exhibition.'[90] The Director, David Follett, felt that the conversion to natural gas was too narrow a subject for the Science Museum and the subject should be 'natural gas' with 'the aspect of conversion necessarily a prominent element in it.'[91] In his letter to Mason announcing that the Museum would make space for the exhibition as early as 1971, there is an interesting echo of the 'The Story of Oil', when Greenaway remarks that 'You yourself and the other Gas Council people I already know are so familiar with the custom and practice of the Museum that it hardly seems necessary to say that the exhibition ought not to have any advertising or competitive features in it. However, I am sure there is no-one who would wish it otherwise.'[92]

In an internal report written for the Executive Committee of the Gas Council written just before Christmas 1970, a very positive spin is put on this temporary exhibition:

> The Gas Council has been anxious that full advantage should be taken of the display facilities in the Science Museum to show the significance of the development of the Natural Gas Industry and to demonstrate natural gas in action. The Museum has itself proposed to go further than the original intention merely to refurbish the Gallery. It would, in fact, plan a complete reconstruction of the Gallery in the light of recent developments. ... in the next financial year provision had been made for experiments and trials to be carried out with a view to operating live gas equipment on permanent exhibition.
>
> The Museum is, however, well aware of the current interest in the introduction of natural gas, and the associated conversion programme. It has therefore suggested that a special exhibition at the present time would ... (a) inform the public on a technological development of national importance; (b) giving the Museum and the Gas Council experience in handling large live gas demonstrations under Museum conditions.
>
> A very rough estimation of the cost involved is in the range of £15–£20,000. This is clearly a P.R. activity and will have to be met from that budget.
>
> It is recommended that the Gas Council accept the offer of the Science Museum ... and that they mount a live exhibit at a cost not exceeding £20,000.[93]

It was also decided that the exhibition would be prepared by the Public Relations and Exhibition Departments of the Gas Council. Greenaway was informed of the Council's agreement to the exhibition in March 1971, just four months before the exhibition opened.[94]

The exhibition fell into three parts: natural gas, domestic conversion and conversion in industry.[95] So, despite Follett's concerns about suitability of conversion as a subject, the exhibition was largely about the conversion process, although admittedly not just the 'narrow aspect of the conversion of equipment.' There were models of a sea drilling rig and the Bacton pipeline terminal. Natural gas was 'sold' to the public by live comparisons of coal gas and natural gas flames, for instance in cooker burners. Under 'new developments' there was a live demonstration of piezoelectric ignition, which is now commonplace. It ran between 13 July and 30 November 1971 and attracted 250,000 visitors, an unusually high number, which included school parties from the south of England and 'a large number of foreign visitors.'[96] This reflects the public interest in the conversion process both as something which affected every home and as a technological feat.

In a letter to the Gas Council just before the exhibition closed, Greenaway said that the 'live gas exhibits have proved of continued interest and entirely justify

Illustration 10.4   Interactive demonstrating piezoelectric ignition, Natural Gas exhibition, 1971

the effort put into devising and installing them. ... On the basis of the experience gained, it is now possible to envisage the form which should be given to a reconstituted permanent Gas Gallery.'[97]

Once again we have the Museum, rather than the relevant body itself, putting forward an exhibition to cover an area of public concern, although that body – in this case the Gas Council – welcomed the initiative and was happy to fund it. The large number of visitors to this exhibition also makes the general point that the public flocked to exhibitions which reflected its own interests and concerns, but only if the exhibition was perceived as impartial or public-spirited. Exhibitions that put forward a specific point of view out of keeping with the Museum's own ethos did not fare so well. This appears to have been the case with both the smoke abatement exhibition and the subject of the next section, the 'Science and Technology of Islam' exhibition.

'Science and Technology of Islam', 1976

This temporary exhibition is of interest given the current debates about how to represent the cultures of diverse communities within British society.[98] Creating such an exhibition might now be considered problematic in terms of how to present

Islamic science, in particular how to avoid the impression that it was simply a pre-cursor of more recent and hence supposedly 'superior' Western science, and also the issue of avoiding 'orientalism' in presenting any aspect of Islamic culture.[99] There is also a debate about whether to present Islamic science as Islamic science on its own or to show the contributions of Islamic culture to specific sciences or technologies within an exhibition about that science or technology. Given the sensitivities within the Museum about Sherwood Taylor's apologetic activities in the 1950s, one might also have thought that there would be unease within the Museum about mounting an exhibition which combined science and religion, at least in a formal sense. Nowadays, in addition to building bridges with Islamic communities in Britain, such an exhibition might be seen as a way of developing a relationship with the wealthy Gulf states with a view to future funding oppor-tunities. In fact, hardly any of these issues arose during the development of this exhibition. Although much of the relatively modest funding for the exhibition (£30,000) came from the government of the newly formed United Arab Emirates, the Museum showed no interest in any longer-term relationship. Even more strik-ingly, the exhibition was seen by its organisers as presenting the riches of Islamic culture from the Near and Middle East to a non-Islamic British audience.[100] There was no reference to Muslims living in Britain at all. Although surprising with hindsight, it is also understandable as there were estimated to be only 369,000 Muslims in Britain in 1971, compared with about 2 million today.[101] The exhibi-tion was developed in terms of mediaeval Islamic science and technology without any concern about how this viewpoint presented Islamic achievements. Indeed, the leading curator at the Science Museum, David Thomas, admitted during the exhibition that he 'had little knowledge of Islam.'[102]

The driving force behind the exhibition was the World of Islam Festival Trust and its Director Paul Keeler.[103] Coming immediately after the Yom Kippur War of 1973 and the First Oil Crisis there were suspicions – voiced, for example, by Professor Donald Watt of LSE[104] – that the Festival, held in the spring and sum-mer of 1976, was a political exercise by Islamic states to present Islam in a good light rather than a genuine cultural activity. It was in fact very much the personal idea of Keeler, a former art gallery owner, who has remained active in this field to the present day. Keeler 'met Mahmud Mirza, the Indian classical musician, in 1968 and became interested in Islamic culture through studying the context of his music.'[105] Subsequently he put on a smaller festival of Islamic culture at the Institute of Contemporary Art in November and December 1971, before the Oil Crisis had even arisen.[106] Encouraged by its success and troubled by the general ignorance of Islamic culture that it revealed, Keeler then envisaged a much more ambitious festival to cover all aspects of Islamic culture, centred on an exhibition at the Hayward Gallery but supported by programmes on the BBC and exhibitions at museums such as the Museum of Mankind (part of the British Museum) and the Horniman Museum. In 1973 Paul Keeler was joined by Alistair Duncan, and they turned to the matter of raising funds by inviting prominent British Arabists such as Sir Harold Beeley, a former ambassador to Egypt and Saudi Arabia, Lord Caradon and Sir Anthony Nutting (who had resigned from the cabinet over Suez in 1956) to join the board of trustees in the hope that they would approach their contacts in the Middle East for donations. There was a breakthrough at the end of 1973, when the first ambassador from the United Arab Emirates and business adviser to the royal family of Dubai, HE Sayed Mohamed Mahdi Al-Tajir, became Vice-Chairman of the trustees. He was very successful at raising funds to the tune of $4 million, mainly from the United Arab Emirates.

Keeler approached the Science Museum in the spring of 1973 with the suggestion that the Museum should put on a supporting exhibition as 'the achievement of the Muslims in the Sciences and Craft Technology is truly remarkable.'[107] From the outset the Museum saw three problems: supplying the staff needed to develop the exhibition at a time when the Museum was very busy with the development of permanent galleries and the National Railway Museum; finding a suitable gallery space (although the space from the Infill project would be available); and a complete lack of suitable objects in the Museum's collections and to a lesser extent elsewhere, leading to a reliance on manuscripts. On the positive side, there was the opportunity to use material from the Wellcome collections as soon as they were transferred to the Museum[108] and to exhibit objects and manuscripts that would not otherwise be seen in London. The Science Museum clearly also felt the pressure to do something if the British Museum took part in the festival, as it eventually did. Opinion in the Museum, in particular the views of Frank Greenaway and John Wartnaby, swung away from cautious approval in the spring of 1973 to outright opposition by the autumn of 1974. Wartnaby noted in September 1974 that 'There is very little indeed in the way of instruments and I would not recommend mounting the exhibition in this museum.'[109] Greenaway also appears to have resented an attempt by Thomas to 'bounce' him into mounting a display about alchemy.[110] The exhibition survived because of the modest enthusiasm of Thomas (who had just taken over as Keeper of Museum Services from Winton) and the Museum's exhibition officer Michael Preston, and Weston's inclination to (reluctantly) support Thomas as a younger curator over the opposition of the 'old guard' of Greenaway, Wartnaby and Kenneth Chew.[111] The project was also given a much-needed shot in the arm by the successful background work of the temporary research assistant Anthony Turner.

The content of the exhibition was largely put together by Francis Maddison, curator of the Museum of the History of Science at Oxford, and Seyyed Hossein Nasr, Professor of the History of Science and Philosophy at Tehran University and Director of the Imperial Iranian Academy of Philosophy, who had written a doctoral thesis on the science and civilisation of Islam at Harvard University. He later became one of the first scholars to flag up the environmental crisis from the Islamic point of view.[112] There was a contrast in the approach of the two scholars. As a British museum curator, Maddison focused on objects which illustrated the practical side of Islamic science and technology, whereas Nasr, as a leading academic in an Islamic country, developed a rather abstract and sweeping presentation of the relationship between Islamic civilisation, the Quran and the sciences.[113] Thomas remarked that 'I have met Nasr and he has not come up with the sources of 3-D objects which we need for a fully successful exhibition.'[114] According to Keeler, this tension was resolved in a creative manner in the exhibition.[115]

The major features of the final exhibition were the earliest dated scientific instrument in Europe, a Moorish astrolabe of AD 1026–1027 lent by the Royal Scottish Museum, a replica of the first power-pump (used for raising water) as described by al-Jazari in 1206, a reconstruction of an Islamic pharmacy using material from the Wellcome collections, and a fishing vessel from Dubai called a Shahoof, lent by the Exeter Maritime Museum.[116] Keeler was pleased with the exhibition, remarking some three decades later: 'I thought the exhibition was very well done and nicely designed. It gave a holistic picture of the sciences and technologies in the context of another civilization and was, I think, ahead of its time.'[117]

In October 1974 Thomas had predicted that 'it is not a subject which would attract large numbers of people'[118] and the visitor figures were considered disappointing, amounting to just 32,236 up to the end of August 1976.[119] Just four weeks after the exhibition had been opened by Farah the Empress of Iran on 7 April 1976, the World of Islam Festival Trust wrote to the Science Museum and said 'of the many thousands of visitors who come to the Museum each day, very few trouble to look round this exhibition.'[120] The Trust blamed poor signage and the high price of items in the exhibition sales area. The ticket booth for the

Illustration 10.5   The Shahoof from Dubai on display in 'Science and Technology of Islam', 1976

exhibition was moved and extra signposts and newspaper advertising were produced to increase the figures, without any success. Part of the reason was doubtless the very hot weather in the summer of 1976, although part of the exhibition was air-conditioned to protect the ancient manuscripts. Another major reason was that it was a paid entry exhibition, which was a rarity at the Museum in this period. However, the exhibition did not fit well with the rest of the Museum's exhibitions and probably did not greatly interest the Museum's usual audience of families. Visitor figures for exhibitions which are perceived (rightly or wrongly) as the product of external organisations have tended to be low. Although the exhibition was well received in the press – Professor Watt was in a tiny minority in criticising the Festival in print – the Islamic oil-producing states were hardly popular in Britain at the time as a result of the Oil Crisis.

The 'Science and Technology of Islam' demonstrates how the Science Museum could produce an exhibition on almost any topic, even one for which its own collections were exceptionally weak. But the ability to develop such an exhibition did not guarantee success in terms of visitor numbers. This exhibition also shows how temporary exhibitions rarely had any long-term impact. The 'Science and Technology of Islam' exhibition did not create any new sources of sponsorship for the Museum, nor did it lead to any strengthening of the Museum's collections in the field of Islamic science and technology. If anything, the low visitor numbers and the paucity of available objects, coupled with demonstrations against the exhibition by the Federation of Students Islamic Societies – because it displayed a depiction of the Prophet in a miniature painting – appear to have reinforced the general opinion within the Science Museum that the Museum should not move into this new area.

The situation nowadays is very different. If an exhibition on this theme were to be proposed today, it would be approached in a more positive manner, perceived as reaching out to new audiences and an opportunity to present science in a new, more inclusive, light. More is known about the presentation of Islamic science, not least because of the continuing work of Professor Nasr in this area,[121] and there is now more of a focus on getting the topic across to an audience effectively than an overriding concern with having a large number of objects. The success of '1001 Inventions: Discover the Muslim Heritage in our World' held at the Manchester Museum of Science and Industry in 2006 shows that it is possible to develop a rigorous exhibition about Islamic science which appeals to the public.[122]

'Challenge of the Chip', 1980

If the Science Museum has a role in explaining technological changes that have a large economic and social impact, and that are already of concern to the public, there was no more obvious topic in the late 1970s than the rise of computer chips (or, as the Museum preferred to call them, microprocessors).[123] The Museum's role was well put by a memorandum on 'Microprocessors: the role of the Science Museum in educating the public', produced by Frank Greenaway, the Keeper of Chemistry, and Jane Raimes, the Assistant Keeper of Mathematics and

Computing, in October 1978. In the opening statement, written by Greenaway, the alarm is raised:

> Within the past few months public utterances about microprocessors have drawn attention to a situation of which informed technical opinion has been aware for a considerable time, namely that the rapid development of minute electronic circuitry devices has affected, is affecting and will enormously affect technology, the economy and social conditions. Discussion, even in high quarters, often seems confused and misguided and it is desirable that the Science Museum play a part in putting the record straight in the minds of the general public and of those who consult the Museum for technical information.

Under the heading of 'Urgency', Greenaway goes on to say:

> In view of the decision of Her Majesty's Government to create a well funded organisation to foster the use of microprocessors it is desirable that the Science Museum act quickly so as (a) to do something useful before it is overtaken by events and (b) to show that the Science Museum is an effective organ of public enlightenment in socially important areas.[124]

Although Derek Robinson, the new Keeper of Museum Services, found the submission 'clear and quite compelling in its argument,' he was concerned about the scale and ambition of the exhibition and the lack of time to complete it. He warned:

> Even given almost unlimited resources of people and money, only 10½ months remain for a scheme not yet adopted, refined or examined for viability. Providing guidance to these assumed resources to produce an exhibition to our standards and patterns, in the terms described, still seems well-nigh impossible.[125]

By contrast, John Anderson of the Department of Industry indicated to Greenaway that his department 'would probably welcome a Science Museum/Design Council Microprocessors Exhibition.'[126] When the representatives of the Design Council met Greenaway in January 1979, there was £65,000 funding on the table (£40,000 from the Department of Industry and £25,000 from the Science Museum) but another £50,000 was needed. The aim was to have a six-month exhibition opening on 6 November 1979.[127] At a meeting involving all three institutions, it was emphasised that time was very short for such a large exhibition (5,500 square feet) and all text contributions had to be finalised by 7 July. The basic scheme at this stage was manufacture and types of chips, programming a micro[-computer], and applications from education and leisure to manufacturing and medicine. In a shift from Greenaway and Raimes's initial ideas for the exhibition, 'it was agreed not to spend time on planning a separate section on "social consequences" at present but try to make these implicit in each section.'[128] At this stage the exhibition still lacked a title, but on 26 March the snappy title of 'Challenge of the Chip: How will Microelectronics Affect Your Future' was chosen, although the Director, Margaret Weston – an electrical engineer – queried whether the exhibition was really as wide as that and whether perhaps the 'microelectronics' in the title should be changed to 'microprocessors'.[129] Perhaps in the light of this objection (which Weston did not press), the subtitle was rarely used and it was thereafter always called 'The Challenge of the Chip'.

The period between April and November 1979 was dominated by frantic efforts to raise funds for the exhibition, mainly by attracting new sponsors with the promise of a new subsection. Considerable assistance was given in this difficult period by the Design Council. William Henry Mayall, a engineer and industrial designer who had formerly worked at the Design Centre, was effectively seconded to (and largely based at) the Museum for several months, and he established a close working partnership with Raimes, which covered both the content of many of the constructed exhibits and the content and layout of the main publication related to the exhibition.[130] The arrival of the new Conservative government in May made little difference to the planning of the exhibition. The Prime Minister, Margaret Thatcher, would have been enthusiastic about the exhibition's basic thesis that the microchip was an opportunity rather than a threat, and there was even an unsuccessful attempt to get the Prime Minister to open the exhibition.[131]

The completion of the exhibition was delayed by the addition of the new sections and the late arrival of text from the Science Museum.[132] The original target date of mid-November was probably always unrealistic, given the scale of the exhibition and the involvement of many different organisations, including the Department of Industry, Design Council, Post Office, Mullard, Marconi and Monotype. By August the opening date had been moved to the first half of December[133] and the timetable for an opening by the Duke of Kent, as Chairman of the National Electronics Council, on 13 December was in place by 2 November.[134] However, it was postponed again on 22 November – a delay which was blamed by John Becklake (who had taken over the day-to-day management from Raimes, who left the Museum in November) – on the failure of the contractors building the gallery to meet completion dates.[135] Furthermore, one of the subcontractors had gone bankrupt. The gallery was eventually opened on 26 February 1980 by Neil MacFarlane MP, the Parliamentary Under-Secretary of State at the Department of Education and Science and Deputy Minister for the Arts, who stood in at short notice for Norman St John Stevas, the Chancellor of the Duchy of Lancaster and Minister for the Arts, 'who was unavoidably prevented from performing the ceremony.'[136] The planned involvement of the Duke of Kent and the inventor of the microchip, Jack Kilby, fell through because of the delay, but the Duke of Kent visited the exhibition on 23 April.[137]

Even before it opened, the possibility of extending the exhibition beyond six months was debated. This exhibition was so popular that it was extended from its original closing date of 31 December 1980 to 5 May 1981, and a condensed form of the exhibition remained in the Science Museum until 15 August 1982.[138] It

Illustration 10.6   Transport section of 'Challenge of the Chip' showing the potential use of microchips in the motor car, a bizarre concept to most visitors in 1980. This photograph also shows how the design of this exhibition marked a sharp break with previous practice

may have been one reason why 1980 was the peak year for visitors to the Science Museum, at 4,224,000; 514,000 higher than in 1979.[139] Clearly the topic was one that attracted people, who were intrigued by the possibilities of the microchip and yet at the same time concerned about the possible impact on their own jobs. This was a period when computers were still unfamiliar to most people and personal computers were in their infancy. The combination of popular concern and the Museum's ability to get to the heart of complex technological issues produced a powerful response from the public.

Conclusion

Temporary exhibitions were taken up by the Museum initially as a means of displaying technological advances and then as a method of collaborating with industry, but they were also offered to the establishment as a way to influence the public. This process was hidden to some extent by the differing forms this establishment influence took, including personal crusades such as Horder's Noise Abatement League, government ministries, state-owned industries such as the Gas Council, and specific groups of influential people such as the ex-Foreign Office supporters of the World of Islam Festival Trust. The striking aspect of most of these exhibitions is that the initiative came from the Museum, not the external institutions. Indeed, the exhibitions originating outside the

Museum, from 'Smoke Abatement' in 1936 to 'Science and Technology of Islam' in 1976, were generally less successful than the Museum's own ideas. The most successful temporary exhibition of all, 'Challenge of the Chip', was almost entirely the brainchild of the young curator Jane Raimes. Whether this success was a result of the Museum proposing what it knew it could bring off successfully, an outcome of the curators knowing their audience, or a consequence of visitors detecting the Museum's authentic voice and authority in the resulting exhibition, is hard to say and probably varied from exhibition to exhibition.

Just as Lyons hoped industrial exhibitions would enable the Museum to make contact with top-level business leaders, these exhibitions enabled the Museum to show top-level Civil Servants, government ministers and other establishment figures how it could enable the establishment to shape public opinion. In this way the Museum and its curators could show that the Science Museum was more than just a museum (and especially more than just a children's museum) and that it was a powerful means of communicating with the public about technological and other issues 'in socially important areas'. Thus, through these temporary exhibitions, the Science Museum not only enlightened the general public but also established its claim to be a major player in the shaping of British opinion, alongside radio and television (especially the BBC) and newspapers. Although the audience for the temporary exhibitions was inevitably smaller than the audience for television, especially in the days when only three channels were available and a programme could attract over ten million viewers, the Science Museum could argue – with some justification – that this was more than counterbalanced by the intensity of the experience and the depth of information that was available in a museum exhibition, which often included practical demonstrations, lectures and seminars.

Notes

1. Quoted in E.E.B. Mackintosh, 'Special Exhibitions at the Science Museum', internal typescript intended for use as a manual, dated 30 March 1939, SMD, Z 108/4, 1.
2. As cited by Mackintosh, 'Special Exhibitions', p. 1. A complete list of special exhibitions can be found in Appendix One.
3. Mackintosh, 'Special Exhibitions', p. 3.
4. Ibid.
5. H.G. Lyons, 'Museum Policy', a position document for the Advisory Council, dated 11 November 1932, SMD, Z 186.
6. According to Mackintosh, 'Special Exhibitions', p. 4 and Follett, *The Rise of the Science Museum under Sir Henry Lyons*, (London: Science Museum, 1978), p. 120, although arguments can be made for adhesives (1926), glass (1931) and refrigeration (1934), the last of which was the first to be under the control of a single industrial association, the British Association of Refrigeration. See the annual reports for the relevant years for brief overviews of specific special exhibitions and their development.
7. See the Joint Science Museum Annual Report for 1976 and 1977 (p. 5), the Science Museum Report for 1978 (pp. 9–10) and the Science Museum Annual Report for 1979 (pp. 8–9).

8. Mackintosh 'Special Exhibitions', p. 1.
9. Ibid [or Mackintosh, 'Special Exhibitions'] p. 2.
10. Science Museum Annual Report for 1979, pp. 9–10.
11. This section is based on T.M. Boon, 'Noise and Smoke: Displaying the Costs of the Machine Age at the Science Museum in the Mid-1930s', Paper Delivered at the 5th Artefacts Meeting in Munich, August 2000, used with the author's permission.
12. Science Museum Annual Report for 1935, p. 9.
13. Ibid., p. 11.
14. The best history of television manufacture is Keith Geddes (with Gordon Bussey), *The Set-Makers: A History of the Radio and Television Industry* (London: BREMA, 1991). Also see Russell W. Burns, *British Television: The Formative Years* (London: Peregrinus, in association with the Science Museum, 1986) and Burns, *Television: The International History of the Formative Years* (London: IEE, 1998).
15. I am greatly indebted to Iain Logie Baird, the curator of television at the National Media Museum, for his assistance with the early history of British television in a personal communication dated 26 February 2009, and Paul Marshall of the University of Manchester in a personal communication dated 23 July 2009.
16. Letter from Mackintosh to Sir Noel Ashbridge of the BBC dated 16 November 1936 (but actually sent on the 18th), SMD, ED 79/178.
17. Ibid.
18. Letter from N. Ashridge of the BBC to Mackintosh, dated 20 November 1936, SMD, ED 79/178.
19. Minutes of the meeting held on 25 November 1936, item 4, quotation on 3, SMD, ED 79/178.
20. Minutes of the meeting held on 8 January 1937, item 5, quotation on 3, SMD, ED 79/178.
21. Minutes of the meeting on 15 January 1937, item 2 (amendments to the minutes of the first meeting) and item 6, SMD, ED 79/178.
22. Minutes of the meeting on 15 January 1937, item 5, SMD, ED 79/178.
23. Minutes of the meeting on 15 January 1937, item 7, quotation on page 4 of the minutes, SMD, ED 79/178.
24. Minutes of the meeting on 22 January 1937, item 4, quotation on 2, SMD, ED 79/178.
25. See Science Museum Annual Report for 1937, p. 11, and 'Television Exhibition at the Science Museum', *Nature* 139 (1937), 1077.
26. Science Museum Annual Report for 1937, p. 10. For Reith declining the invitation, see the minutes of the meeting held on 13 April 1937, item 4, SMD, ED 79/178.
27. There is a map of the exhibition and a picture of the cubicles at http://www.thevalvepage.com/tvyears/1937/tvy1937text.htm (accessed 7 May 2009).
28. Note from Mackintosh to Hartley agreeing to the acquisitions, dated 14 September 1937, SMD, ED 79/178.
29. G.R.M. Garratt, ed., *Television: An Account of the Development and General Principles of the Television as illustrated by a Special Exhibition held at the Science Museum, June–September 1937* (London: HMSO, 1937).
30. Minutes of the meeting held on 16 June 1937, item 6, quotation on p. 2, SMD, ED 79/178.
31. Letter from C.C. Paterson of the Research Laboratories of the General Electric Company Ltd, Wembley, to Mackintosh, dated 5 July 1937, SMD, ED 79/178.
32. Letter from Mackintosh to Paterson, dated 13 July 1937, SMD, ED 79/178. This is a significant time lapse in a period when letters were usually answered the following day; clearly Mackintosh had put considerable effort into composing his response.
33. Letter from Paterson to Mackintosh, dated 14 July 1937, SMD, ED 79/178.
34. Letter from H.J. Barton-Chapple of Baird Television to Garratt, dated 22 July 1937, SMD, ED 79/178.

35. Letter from Hartley to Barton-Chapple, dated 23 July 1937 (Garratt was away), SMD, ED 79/178.
36. For this controversy see Sarah Smith, *Children, Cinema and Censorship: From Dracula to the Dead End Kids* (London: I.B. Tauris, 2005), pp. 45–76.
37. For the history of the Chemical Society, see T.S. Moore and J.C. Philip, *The Chemical Society, 1841–1941: A Historical Review* (London: Chemical Society, 1947); D.H. Whiffen and D.H. Hey, *The Royal Society of Chemistry: The First 150 Years* (London: Royal Society of Chemistry, 1991); and Robert F. Bud, 'The Discipline of Chemistry: The Origin and Early Years of the Chemical Society of London', unpublished PhD Thesis, University of Pennsylvania, 1980. Also see C.A. Russell, N.G. Coley and G.K. Roberts, *Chemists by Profession: The Origins and Rise of the Royal Institute of Chemistry* (Milton Keynes: Open University Press, 1977). For the history of the celebrations themselves, see *A Record of the Centenary Celebrations* (London: Chemical Society, 1948). And for the Science Museum's treatment of atomic science, see Sophie Forgan 'Atoms in Wonderland', *History and Technology* 19 (2003), pp. 177–96.
38. See his biography G. M. Bennett, 'Robertson, Sir Robert (1869–1949)', rev. K. D. Watson, *Oxford Dictionary of National Biography*, Oxford University Press, 2004; online edn, Jan 2008 [http://www.oxforddnb.com/view/article/35785, accessed 19 Jan 2010], and for his work at the Laboratory of the Government Chemist see P.W. Hammond and H. Egan, *Weighed in the Balance: A History of the Laboratory of the Government Chemist* (London: HMSO, 1992), chapter 15.
39. 'Report of the first meeting of the subcommittee held on Tuesday May 7th 1946', SMD, ED 79/154.
40. 'Note of a discussion at the Science Museum at 11 am Monday May 27th 1946' and a letter from Shaw to D.C. Martin, General Secretary of the Chemical Society, dated 30 May 1946, both SMD, ED 79/154.
41. Moore and Philip, *The Chemical Society, 1841–1941*.
42. Alexander Findlay, *A Hundred Years of Chemistry* (London: Gerald Duckworth, 1937). It was reissued in a revised version by Trevor I. Williams in 1948, presumably to capitalise on the interest generated by the exhibition. For the centenary celebrations, Findlay produced a compendium of biographical memoirs, *British Chemists* (London: Chemical Society, 1947), with W.H. Mills.
43. 'Minutes of the second meeting held on June 19th 1946', SMD, ED 79/154.
44. F. Sherwood Taylor, *A Century of British Chemistry* (London: Longman Green, 1947). For the papers relating to this book (and Sherwood Taylor's involvement with this exhibition as a whole), see folders 132 to 137, MSS Taylor, Archives of the Museum of the History of Science, Oxford. I am indebted to Tony Simcock for this reference.
45. 'Minutes of the third meeting held on August 8th 1946', SMD, ED 79/154.
46. For Jolliffe's attendance in place of Brown see the minutes for the fourth and fifth meetings in SMD, ED 79/154. For his near-appointment as Director (he was placed third on a shortlist of four) see 'Results of Competition' dated 15 May 1956, TNA: PRO, Comp. No. S.4575/56.
47. 'Minutes of the third meeting held on August 8th 1946', SMD, ED 79/154.
48. 'Minutes of the fourth meeting held on Thursday 17th October 1946', SMD, ED 79/154. The handbook was published as *Chemical Progress: Handbook of an Exhibition*, a slim volume with a light blue cover, by His Majesty's Stationery Office in 1947.
49. See the letter from A. Barclay to J.R. Ruck Keene of the Chemical Society, dated 7 August 1946, with an attached list of possible exhibits from the Science Museum, SMD, ED 79/154.
50. 'Suggested Layout of Exhibition and Summary of Exhibits', undated, SMD, ED 79/154.
51. This description is based on the handbook, but the exhibition appears to have been similar.
52. A press notice entitled '100 Years of British Chemistry' dated December 1946 with a cover letter signed by D.C. Martin, also dated December 1946, SMD, ED 79/154.

53. Quotation taken from 'D.S.I.R. Section of Chemical Society Centenary Celebration Exhibition', undated but around October 1946, SMD, ED 79/154. The content was taken from this document and the handbook.
54. For a detailed study of this crisis, see Alex J Robertson, *The Bleak Midwinter: 1947* (Manchester: Manchester University Press, 1987). For a more general history of the coal industry in this period, see chapters 4 and 5 of William Ashworth, *The History of the British Coal Industry*, volume 5, *1946–1982: The Nationalised Industry* (Oxford: Clarendon Press, 1986).
55. Letter from O'Dea to A.M. Rake of the Ministry of Fuel and Power, dated 27 August 1947, SMD, ED 79/164.
56. Note from O'Dea to Hartley appended to a timetable of his dealings with the ministry, dated 28 August 1947, SMD, ED 79/164.
57. Letter from W.H. Willson of the Ministry of Fuel and Power to Hartley, dated 8 September 1947, SMD, ED 79/164.
58. Note from O'Dea to Hartley, dated 28 July 1947 and confirmed by Willson in his letter of 8 September, SMD, ED 79/164.
59. Note from Hartley to Shaw, dated 10 September 1947, SMD, ED 79/164.
60. Shaw's response to Hartley appended to the original note, dated 11 September 1947.
61. His title was spelt Hyndley although his surname had been Hindley.
62. Note from O'Dea to Hartley, dated 3 October 1947, and Hartley's response of 4 October. SMD, ED 79/164. The quotation is from Hartley's response.
63. Science Museum press release, undated but clearly composed in early October 1947 as it was sent to Lord Citrine on 15 October, SMD, ED 79/164. The projected visitor figures were probably inflated for political reasons, but the actual total was a very respectable 218,717.
64. Letter from Lord Citrine to Shaw, dated 29 October 1947, SMD, ED 79/164.
65. Letter from Willson of the Ministry of Fuel and Power to O'Dea, dated 6 January 1948, SMD, ED 79/164.
66. Note by Hartley to the Director, appended to a note from O'Dea to Hartley, dated 13 December 1947, SMD, ED 79/164.
67. Response from Shaw to Hartley, dated 16 December 1947, and subsequent internal exchanges in the file, SMD, ED 79/164.
68. There is currently no scholarly history of the oil industry as a whole. For popular accounts see Daniel Yergin, *The Prize: The Epic Quest for Oil, Money and Power* (New York and London: Simon & Schuster, 1991) and Anthony Sampson, *The Seven Sisters: The Great Oil Companies and the World They Made*, 3rd edn (London: Coronet Books, 1993). For the history of Royal Dutch/Shell, see Jan Luiten van Zanden, Joost Jonker and Stephen Howarth, *A History of Royal Dutch Shell*, four volumes (Oxford: Oxford University Press, 2007). For Anglo-Iranian/BP, see James Bamberg, *The History of the British Petroleum Company*, volume 2, *The Anglo-Iranian Years, 1928–1954* (Cambridge: Cambridge University Press, 1994) and Bamberg, *British Petroleum and Global Oil, 1950–1975: The Challenge of Nationalism* (Cambridge: Cambridge University Press, 2000).
69. *Hansard*, HC Deb 11 June 1953 vol. 5 16 c440, Response by the Financial Secretary to the Treasury (John Boyd-Carpenter) to a question by Cledwyn Hughes MP, available online at http://hansard.millbanksystems.com/commons/1953/jun/11/science-gradu-ates-employment (accessed 13 May 2009).
70. Letter from A.E. Gunther to Barclay, dated 7 September 1953, SMD, ED 79/179.
71. Note to the Director written by Barclay, dated 27 October 1953, SMD, ED 79/179.
72. Sherwood Taylor's response to Barclay, dated 28 October 1953, SMD, ED 79/179.
73. Barclay's response to Sherwood Taylor's comment, dated 3 November 1953, SMD, ED 79/179.
74. Formal letter from Barclay to Gunther, dated 27 October 1953, SMD, ED 79/179.
75. A. Simcock, *Sphæra*, 8 (Autumn 1998); available online at http://www.mhs.ox.ac.uk/sphaera/ (accessed 29 April 2009).

76. Letter from Sherwood Taylor to Gunther, dated 18 November 1953, SMD, ED 79/179.
77. Letter from Gunther to Sherwood Taylor, dated 19 November 1953. Sherwood Taylor sent a brief response on 25 November, accepting Gunther's offer to remove the reference to Shell, SMD, ED 79/179.
78. Memo to Barclay and Chilton written by Winton, date-stamped 21 July 1954, SMD, ED 79/179.
79. Memo from Chilton to Barclay, dated 26 February 1954, quotations on pp. 1 and 2. Also see the letter from Sherwood Taylor to Gunther, dated 1 March 1954, confirming the new arrangement. All SMD, ED 79/179.
80. Note from Winton to Chilton, dated 2 March 1954, SMD, ED 79/179.
81. Press Notice, dated April 1954, SMD, ED 79/179.
82. Letter from Barclay to Gunther, dated 11 May 1954, SMD, ED 79/179.
83. Figure given in Barclay's letter of 11 May.
84. Letter from Gunther to Barclay, dated 14 May 1954, SMD, ED 79/179.
85. Letter dated 18 November 1953, cited above, SMD, ED 79/179.
86. Memo to Barclay and Chilton by Winton, cited above, SMD, ED 79/179.
87. For the history of the British gas industry in the 1950s and 1960s leading up to the changeover to natural gas, see Trevor I. Williams, *A History of the British Gas Industry* (Oxford: Oxford University Press, 1981) and Andrew Jenkins, 'Government Intervention in the British Gas Industry', *Business History* 46 (2004), pp. 57–78. Surprisingly, there has so far been no scholarly account of the changeover process in any depth, but see chapter 17 of Williams's volume for a good account of the technical process of conversion.
88. Note that he says their *subject*, not their industry; the Museum was not a trade exhibition hall.
89. Letter from Mackintosh to J.H. Markham, Chief Architect of the HM Office of Works, dated 25 February 1939, SMD, ED 79/53.
90. Note from Greenaway to Follett, dated 22 October 1970, Nominal File 4584.
91. Response by Follett to Greenaway, dated 12 November 1970, SMD, 4584.
92. Letter by Greenaway to Mason of the Gas Council, dated 17 November 1970, signed by L.R. Day, SMD, 4584.
93. 'Science Museum – Gas Gallery', a document for the meeting of the Executive Committee of The Gas Council prepared by the Public Relations Adviser and the Manager, Publicity and Marketing Projects, dated 22 December 1970, SMD, 4584.
94. Letter from Mason to Greenaway, dated 17 March 1971, SMD, 4584.
95. For the content of the exhibition, see 'Natural Gas Exhibition: Summary of Contents', dated 8 July 1971, SMD, 4584.
96. Letter from Greenaway to C.A.R. Jones, Deputy Secretary of the Gas Council, dated 10 November 1971, SMD, 4584.
97. Ibid.
98. For a wide-ranging discussion of these issues see Yasmin Khan, 'The Representation and Interpretation of Islamic Science in the Museum', a paper presented to the 'Representing Islam: Comparative Perspectives' conference at the University of Manchester, September 2008.
99. Edward W. Said's path-breaking but controversial book on orientalism was published by Routledge & Kegan Paul in 1978, just two years after this festival. For a more recent edition, see Edward W. Said, *Orientalism* (London: Penguin, 2003).
100. Personal communication from Paul Keeler, dated 5 May 2009.
101. *The Guardian*, 18 June 2002.
102. Letter from Dr Thomas to Ahmad Bahafzallah, President of the Federation of the Students Islamic Societies, dated 2 June 1976, SMD, Z 228.
103. For general accounts of the 1975 'World of Islam Festival' see Harold Beeley, 'The World of Islam Festival, London 1976', *Museum* 30(1) (1978), pp. 10–11, and John Sabini, 'The World of Islam', *Saudi Aramco World* 27/3 (May/June 1976), available

online at http://www.saudiaramcoworld.com/issue/197603/the.world.of.islam-its. festival.htm (accessed 20 January 2010). For the Science Museum's exhibition (with illustrations of the exhibits) see David B. Thomas, 'Science and Technology in Islam: An Exhibition at the Science Museum', *Museum* 30(1) (1978), pp. 18–21.

104. A copy of Professor Watt's letter to the *Daily Telegraph*, published on 25 November 1975, is attached to a letter from Paul Keeler to Dr Thomas, dated 18 November 1975, SMD, Z 228.

105. Personal communication from Paul Keeler, dated 5 May 2009.

106. For the prehistory and background to the 'World of Islam Festival', see the draft letter to the *Daily Telegraph* written by Keeler and sent to Dr Thomas with his letter of 18 November 1975, SMD, Z 228.

107. Letter to the Director from Paul Keeler, dated 23 March 1973, SMD, Z 228.

108. Note by Greenaway, dated 3 April 1973, appended to a note by Winton of the same date.

109. Note by Wartnaby, dated 10 September 1974, appended to a note by David Thomas about Nasr, dated 9 September, SMD, Z 228.

110. Contrast David Thomas's breezy remark that 'Dr Greenaway is confident that the alchemy section can done successfully' in his memo of 4 October 1974 with Greenaway's marginal note of 5 October on this document that 'I am not "confident" that an alchemy section based on original material would be easy to do ..." (underlining in original). By 31 October, Greenaway was 'not at all happy with this'; a marginal note on Thomas's memo to the Director on the pros and cons of continuing with the exhibition, dated 28 October, SMD, Z 228.

111. Weston replied to Thomas's memo on the exhibition of 28 October with the marginal note 'I am going to let Dr Thomas go ahead with this, though I do regard it as a very borderline case', SMD, Z 228.

112. Personal communication from Paul Keeler, dated 5 May 2009.

113. See the four-page 'Proposed Outline of the Exhibition of Islamic Science for the Festival of Islam, London 1976, prepared by S.H. Nasr', undated, SMD, Z 228.

114. Note by David Thomas to Wartnaby (originally Greenaway but scored out), dated 9 September 1974, SMD, Z 228.

115. Personal communication from Paul Keeler, dated 5 May 2009. Maddison and Turner helped to produce a exhibition catalogue at the time, but Maddison's efforts to produce a scholarly 'catalogue raisonne' were probably defeated by his own perfectionism; personal communications from Tony Simcock, dated 3 March 2009 and 4 September 2009. Leonard Harrow and Peter Lambourn Wilson, *Science and Technology in Islam: An Exhibition at the Science Museum, London ... based on information and research by F. R. Maddison and A. J. Turner.* ([London]: Crescent Moon Press, 1976).

116. For the content, see Thomas, 'Science and Technology in Islam'.

117. Personal communication from Paul Keeler, dated 5 May 2009.

118. Thomas's memorandum on the pros and cons of the exhibition, dated 28 October 1974, SMD, Z 228.

119. 'Final Attendance figures', a handwritten and undated list of visitor figures for the period 8 April to 29 August, 'final' added afterwards and double underlined, SMD, Z 228. Later, in 'Science and Technology in Islam', Thomas stated that the visitor numbers were 'around 40,000' (p. 21), but it is possible that the figure was quietly rounded up to avoid any further embarrassment.

120. Letter to the Director from Guy Pearce of the World of Islam Festival 1976, dated 3 May 1976, SMD, Z 228.

121. See, for example, Seyyid Hossein Nasr, 'Islam and Modern Science', a lecture delivered on the eve of the Middle East peace conference in Madrid, 30 October 1991.

122. This assessment is based on Khan, 'The representation and interpretation of Islamic Science in the museum'. I wish to thank Yasmin Khan and Paul Keeler for their assistance with this topic.

123. Perhaps surprisingly, there is no scholarly history of the microprocessor itself, but its development and impact can be traced through histories of the computer, for which see Paul E. Ceruzzi, *A History of Modern Computing* (Cambridge, Massachusetts: MIT Press, 1998) and Martin Campbell-Kelly and William Aspray, *Computer: A History of the Information Machine* (New York: Basic Books, 1996).

124. Proposal by Greenaway and Raimes, dated 16 October 1979, sent to Weston on 18 October. Although the memorandum was produced jointly, the style is that of Frank Greenway rather than Jane Raimes, so we can ascribe the authorship of this passage to him with some confidence, SMD, SCM/2002/00/09 & 11.

125. Memo from D. Robinson to Weston, dated 25 October 1978, SMD, SCM/2002/00/09 & 11.

126. Note of a telephone conversation with J.G. Anderson of the Department of Trade by Greenaway, dated 20 November 1978, SMD, SCM/2002/00/09 & 11.

127. Minutes of a meeting between the Science Museum and the Design Council on 17 January 1979, SMD, SCM/2002/00/09 & 11.

128. Minutes of the 'initial meeting' (of the Microprocessors Advisory Group?) on 2 February 1979, SMD, SCM/2002/00/09 & 11.

129. Note of a meeting of the Microprocessors Advisory Group held on 26 March 1979 and a subsequent exchange of notes between Raimes and Weston on 27 and 28 March, SMD, SCM/2002/00/09 & 11.

130. Personal communication from Mrs Jane Raimes, dated 4 May 2009.

131. See the letter from Weston to the Prime Minister, dated 27 July 1979, inviting her to open the exhibition around 26 November, SMD, SCM/2002/00/09 & 11. There is no reply from the Prime Minister's Office on the file, and she may have declined (or never responded) because Norman St John Stevas was probably keen to open it himself; for his enthusiasm for the exhibition, see the note from Mary Giles to Mrs [Sheena] Evans of the Department of Education and Science, dated 11 July 1979, SMD, SCM/2002/00/09 & 11.

132. 'Financial Situation as at 26th October 1979' by Gordon Bowyer and Partners, dated 26 October 1979, SMD, SCM/2002/00/09 & 11.

133. A letter from M. Weston to HRH The Duke of Kent, dated 24 August, states: 'the exhibition will open during the first half of December'. A letter from his Private Secretary, Richard Buckley, dated 2 October refers to 13 December, presumably a date which suited the Duke of Kent, SMD, SCM/2002/00/09 & 11.

134. See the draft programme sent to St James's Palace by Derek Robinson with a cover letter from Robinson to Commander Buckley, the Duke of Kent's Private Secretary, dated 2 November 1979, SMD, SCM/2002/00/09 & 11.

135. File note by Becklake, dated 23 November 1979, referring to a meeting on 22 November, SMD, SCM/2002/00/09 & 11.

136. Science Museum Annual Report for 1980, p. 5. Mr MacPharlane's [sic] name is written by hand into the (undated) schedule for the opening of the exhibition on 26 February 1980, SMD, SCM/2002/00/09 & 11.

137. For the invitation to Kilby and his non-appearance, see a letter from Becklake (signed by J. Griffiths) to R. Mann, dated 7 February 1980, SMD, SCM/2002/00/09 & 11. For the Duke of Kent's visit, see Science Museum Annual Report for 1980, p. 5.

138. Science Museum Annual Report for 1980, p. 5.

139. Ibid., p. 12.

11

Collecting for the Science Museum: Constructing the Collections, the Culture and the Institution

Robert Bud

Introduction

The collections of the Science Museum are the finest of their kind in the world. This claim is both arguably true and deeply problematic. The collections' size and wealth are indeed beyond doubt. Catalogued by over a quarter of a million inventory numbers, many of them referring to numerous (in a few cases – such as the original workshop of James Watt – thousands of) individual artefacts, they cover the history and present state of science, engineering, transport and medicine from the measuring instruments of ancient Egypt to the latest electronic instruments.

The 'kind' of the collections is much more problematic. Over the last century, so many other museums with similar scope have been established across Europe that this combination might seem a category innate to our culture, like 'natural history' or national 'history'. Yet before the Science Museum most museums had been intended to engender respect either for the past or for the Almighty's creation. The Science Museum, by contrast, has – ever since its origins in the 1880s – been a modernist enterprise. It has been dedicated to using the past to illuminate the present and the future of the ingenuity of our own culture. For the combination of historic and the contemporary that they intended, and for the integration of the scientific and the technological, the promoters of the early Science Museum had just one example to take, the Musée des Arts et Métiers in Paris, and, despite many similarities, unlike that museum its formal focus was 'science' rather than technique. Again, in the subsequent development, the successors of those pioneers had few peers to look to as they sought to negotiate and renegotiate a proper balance. Even the closest parallels, such as those drawn with Munich's Deutsches Museum, were inexact.

In his official report of 1864 the founder of one of the ancestors of today's Museum collection, the Patent Office official Bennet Woodcroft, explained his acquisition strategy. He proposed an archaeological approach to technology as a means of addressing the rich heritage of the industrial revolution of the previous century.[1] Almost 150 years later we can approach the collections themselves in an archaeological manner. The process reveals a history as inviting as an archaeological dig going back through many layers of modern culture. Hopes and dreams, disappointments and triumphs are as evident as lathes and astrolabes. The collections expose not just the presuppositions of the Science Museum as an institution but also those of the cultures which sustained the Museum, and express the values and aspirations of the funders, audiences and staff within the institution who were recreating it on the ground.

Each acquisition has its own story, often revealed in its associated technical file in the Documentation Centre. Here we can tell only the big picture characterising each layer evident in this peculiar 'dig'. Above all, we can tell how the collections reflect layer upon layer of discourse on modern times. How they do this will be explored here by reflecting on both the enduring pressures that have characterised the entire century and a half of the Museum's collections and the changes that can be detected in collecting styles over time. Moreover, through its impact on millions of visitors and through the material culture it chose to preserve, the Museum has helped to construe the very nature of 'science' for the British public over its century of formal existence. The history of its collections illuminates in particular the relationship between science and technology as laid out to the British public. It reveals a process in which at first science was seen to be characterised by its technological applications, but a century later technology had come to be the leading quality of the Museum, characterising even science and medicine.

Enduring issues

One story the collections tell, over and over again, is the enduring conflict over cultural space waged by London's great museums. The Science Museum, at first as a junior institution and then through custom, has had to exclude certain areas for long periods. Even by 1857 there were dynamic institutions that would claim similar professional, cultural and physical space. These were, for instance, devoted to natural history. The British Museum's natural history collections found a wonderful home in what is now known as the Natural History Museum in 1883. The teaching of geology was kept separate from that of other sciences in the Royal School of Mines in the late nineteenth century, and the exhibition of geology was kept equally distinct in a separate museum first in Jermyn Street, near Piccadilly Circus (where it accompanied the first building of the Geological Survey and Royal School of Mines), and then in South Kensington. Aesthetic design has, of course, been the responsibility of the art collections of the South Kensington Museum, later the Victoria & Albert Museum, while painting has been the province of the National Gallery. The relationship to ships has also been problematic because of the tensions with collections at Greenwich (at first the Naval Museum from 1873, and from 1934 the National Maritime Museum), and the military museums have also circumscribed the Museum's attitude to the material culture of war.

The implications of this complex ecology can be seen in the division of the nation's timekeeping collection. Historic clocks have been predominantly housed at the British Museum, maritime clocks are at the National Maritime Museum, and clocks interesting for their technological significance are at the Science Museum. That is the principle, but the outcome has been that the workings of the Wells Cathedral Clock of the late fourteenth century were acquired by the 'Science Museum' in 1884, whereas the Cassiobury turret clock mechanism of the early seventeenth century is to be found at the British Museum. An early wooden John Harrison clock is in the Science Museum, though his much more famous chronometer is to be found at the National Maritime Museum.

Just as profound a constraint as the competition for cultural space by other institutions has been the limitation of physical space for the collections. This has been the strongest constraint, forcing decisions about priorities more than the shortage of any other resource. As John Liffen shows in the following chapter in this volume, it was only in 1952 that outlying stores were to become an enduring necessity. Late in the 1970s the Museum acquired its airfield high up on the Marlborough Downs near Swindon. Taking over old hangars from the Royal Naval Air Yard (RNAY) and building its own modern storage, the Museum acquired vastly more space, which temporarily allowed a much greater acquisition programme before even that space was exhausted.

Beyond such storage space, the galleries themselves were a substantial resource and their needs have been a continuing – even dominating – prompt and occasion for acquisition. The opening of new chemistry galleries in the 1970s was the occasion for a burst of acquisition in this area. More than four times as many objects dating from the 1970s than from the 1960s are found in the current Experimental Chemistry collection. A very similar disproportionate rate of growth was to be found in the Industrial Chemistry collection of the time, although the curators concerned were different. Sometimes a temporary exhibition, such as the 1936 exhibition on low temperatures, led to substantial acquisition of which the Museum is still a beneficiary, retaining, for instance, the low temperature liquefier of Sir Franz Simon. By contrast, and too often, objects were borrowed, and kept purely for the duration of the exhibition. Thus, of the immensely popular 1933 plastics exhibition, to which a million people came, very little was kept – not even an example of the bowls made on site during the show by a great compression-moulding machine, which stamped them out for all who wanted one.

Within the collecting agendas, even for a particular era, there has also been a profound and enduring tension between pure science, applied science and technology. This has been expressed in a variety of ways over time. The Museum adopted the unofficial subtitle of the National Museum of Science and Industry in 1923 to indicate its commitment to what was considered important by industry as well as academe. Curators in the Museum, alive to public interest and media engagement, have also had an enduring commitment to acquiring important inventions of the past and present, whatever their scientific or industrial origins.

At the centre of the Science Museum's collecting ambitions have been the joint aims of representing the most important relics of the past, the most significant products of the present and the indicators of the future. These have reflected the professional judgements of the curators and of external advisers, particularly, in early years, the professors of the Normal School of Science (which became the Royal College of Science) and professional engineers.

The relationship between the attitudes to past, present and future have varied, but an enduring link has been a particular attitude to 'progress'. The Museum was formed in a period when theories of progress and evolution were intermingled and Darwin's theory of natural selection and the tree of life provided a context for all discussions of progress. For many decades this provided a model of change in devices both scientific and industrial more cogent than any available from historians. How this evolutionary model has affected the collections in detail we shall see in the more detailed studies of individual eras.

Strikingly, academic historians have only very rarely been direct contributors to these discussions. Social historians have hardly been involved in the construction of the Museum's narrative. On the other hand, the Museum has been central to the development of the disciplines of history of science and technology. Both the Newcomen Society and the British Society for the History of Science held their early meetings there. A chart mounted in the Museum's entrance hall, dominated by the great pioneering steam engines of Newcomen and Watt, expressed the historical view at the time of the opening of the new building in 1928.[2]

The chart presented eleven parallel columns to enable visitors to correlate important events. The first dealt with 'Important events in pure and applied science whose influence on the development of prime movers was direct and indirect'. The second was 'lives of engineers', and the third reached 'timescale'. The subsequent six columns dealt with development in engines and boilers. These were followed by an integrated chart showing the population of England and Wales, coal and pig iron production and miles of railway open in the United Kingdom. At the very end were 'Important events in English and World History', in turn divided between economic and political history. There was no attempt to integrate the political developments – the most recent of which were the growth of Sinn Fein (1915–1916), the enfranchisement of women (1918) and the Irish Free State (agreement) Act (1922) – with the technical innovations. The chart clearly showed a commitment to mapping improvement and its linkage with underlying economic change, but hardly with social relationships.

The Science Museum's collections, therefore, express enduring continuities since the 1870s. Over a period of almost a century and a half we see the constant awareness of a complex cultural ecology and the dominance of space considerations and of the demands of exhibitions. There has also been a philosophy of celebrating the importance of science through its practical applications as well as laboratory achievements, but a continuing distrust of social history.

There have, of course, been great changes. If one looks at the Museum's current collection and then divides it according to year of formal acquisition as a percentage of the current collection, a pattern is seen in which the Museum has experienced waves of collecting peaking in the 1880s, the 1920s and the 1980s. If one looks at the current collections (excluding the Wellcome Collections), 5 per

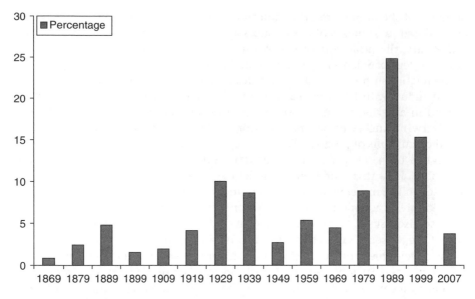

Graph 11.1   Percentage of Non-Wellcome items in the Science Museum collections in 2008, by decade of aquisition

cent still dates from the first of these decades, 10 per cent from the second, and 25 per cent from the third. So, out of the sixteen decades of the Museum's collecting, these three alone account for 40 per cent. These peak periods can therefore serve to point us to important enduring issues and to critical short-term turning points.

The formation of the collections

To understand the formation of the Museum's collections one needs to reflect on the 'reverse takeover' of two institutions, and of a brand created in 1883 that led to the new title 'Science Museum', a title used informally from that time and formally from 1909. The detailed process, described in more detail in the first chapter of this volume, was affected by the promoters of the 1876 Special Loan Exhibition of Scientific Apparatus. They were members of that scientific establishment known as the X-Club, the professors of the new Normal School of Science and the editor of the then-new science magazine *Nature*. This was a group dismissed by an opponent as 'a little "caucus" of philosophers desirous of persuading the Chancellor of the Exchequer to assist in what is termed "The promotion of science".'[3] Men such as T.H. Huxley had already called for the transfer of most of the Royal School of Mines from its accustomed perch in Jermyn Street to South Kensington. There it would be the basis of broad scientific education rather than specifically technical training in mining. This wonderful and unique efflorescence was one of the wonders of the nineteenth century. With a small staff and the technology of the mid-nineteenth century, *Nature*'s editor Norman Lockyer assembled 20,000 items from across the world and showed them to 300,000 people who visited over a period of three months. They were organised by five subcommittees of scientists, formed at the first meeting held in 1875. This drew upon some of the leading scientists in the land, and a vote of thanks to the Department of Science and Art was moved at the end by John Tyndall, then president of the British Association for the Advancement of Science. Following the order of the sections of the British Association, these subcommittees were:

a. Mechanics (including pure and applied mathematics)
b. Physics
c. Chemistry (including metallurgy)
d. Geology, mineralogy and geography
e. Biology

Although the majority of the large collection which had been assembled was returned, a significant number of items were retained. Some were donated and others purchased with a special fund from the Department of Science and Art. The most important items that could not be kept were reproduced using Elkington & Co.'s wonderful new process of reproduction by means of electrotyping. Thus, before it was returned to Prague, a copy was made of the quadrant of Tycho Brahe, which had once provided the data underpinning Kepler's picture of the Solar System whose planets rotated elliptically around the sun. Copied too were the two hemispheres of Guericke, who had demonstrated that two teams of horses could not separate what air had pressed together.

The exhibition closed at the end of 1876, but the intention was to make it permanent. Frustrated by delays, the promoters suggested they could purchase the loan collection items privately, but this was, in turn, put on hold while the government decided how to dispose of the collections of the existing South Kensington Museums to them. Ultimately, the professors were invited to raid the miscellaneous non-art collections of the South Kensington Museum and of the separate Patent Office Museum housed in the hideous Brompton Boilers on the east Side of Exhibition Road, taking what they wanted and rejecting the rest, borrowing, temporarily, the brand name and management structure of the 'South Kensington Museum'. The so-called science collections of the South Kensington Museum had been denounced by the Devonshire Commission of 1874 as being as poor as:

the play of Hamlet with the part of the Prince of Denmark omitted. There were antiquities; there were pictures and books; there were what the Americans call 'bugs', that is, natural history specimens; but there was nothing at all representing objects of interest to students of any observational or experimental science, or of utility to those industrial people who depended in any way upon the applications of science.[4]

While there were antiquities, pictures, books and 'bugs', there had so far been very little science.

The new advisers wanted very few of the existing collections of the South Kensington Museum, principally elements of the educational collection and, above all, the great collection of ship models. The rest, to follow the commercial metaphor, were 'asset-stripped'. Most of the non-art collections, the building and construction collection, the animal products, the fish and the food collections were dispersed.

The second victim of the takeover was the Patent Office Museum, then occupying space on the east side of Exhibition Road wanted for the art collections. It was the product of the enthusiasm and dynamism of their redoubtable champion and creator Woodcroft. Its collections, built up on a historic basis, included early steam beam engines such as the earliest surviving Boulton and Watt steam engine, the engine of the Comet, Europe's earliest commercial steam-powered ship, Puffing Billy (the earliest surviving locomotive) and its great successor Stephenson's Rocket. A delegation from the Society of Arts to the Lord Chancellor in January 1874, seeking better accommodation for the collections, described them as 'one of the most interesting in the world, in fact quite unique, containing as it did some of the first specimens of inventions which had revolutionised modern industry, modern travelling and the whole character of "civilisation" .'[5]

The solution to the problem of accommodation was a merger, and Woodcroft, who had retired in 1876 and passed away in 1879, was unable to defend his legacy against the raiders. Under the 1883 Patent Act responsibility for the collection was transferred to the Department of Science and Art, and the Treasury agreed that it would be amalgamated with other collections.

The 1880s

In 1882 the Department of Science and Art confirmed that the collections acquired already were to be supplemented by new acquisitions suggested by the professors of the Normal School of Science (as the victoriously transplanted Royal School of Mines was now termed). Thus the Professor of Physics, Frederick Guthrie, argued that in the presentation of physics 'The main object to be kept in view should be to illustrate by apparatus the various steps in physical discovery and its applications. I would thus have put together a complete collection of Electric Current Generators; from Volta's Crown of-Cups to some recent forms of Storage-Battery and Dynamo machine.' Having identified equivalent headings under heat and light, which he called the 'vertebrae' of the collection, he suggested:

> Of only less importance than the above would be exhibits showing Critical Discoveries or Methods: I mean such as have unquestionably given rise to new departures. Amongst these would of course be found: – Apparatus showing Boyle's law; the analysis of light by a prism; the turning of a magnetic needle by a current; Dialysis – These collections, which would not be extensive, would associate themselves naturally and in locality each with its proper vertebra. If these two main objects were kept prominently in view, I believe the other

exhibits would naturally co-ordinate themselves into two or three classes, namely:– Instruments used for Technical purposes: in the widest sense of the term. Such as, telegraphic and telephonic instruments; Musical instruments: Astronomical and Surveying instruments; Meteorological instruments. For many of these it is probable that other government departments such as the Admiralty, Ordnance, etc, might be asked to assist. Such Sciences as Astronomy, Meteorology, etc, would have common ground with Pure Physics in regard to many instruments such as Barometers, Thermometers, Spectroscopes, etc. There should be no difficulty in welding the two together.[6]

Then Guthrie recognised the importance of 'Objects of sentimental antiquarian interest. (Not Newton's prism, but Newton's pen.)' Finally he referred to 'So called 'Sets' of physical apparatus as exhibited by various firms of philosophical instrument makers. Such exhibits present considerable difficulty in acceptation, selection, and arrangement. In most of such sets there is a good deal of rubbish.' Similar approaches were taken by the professors of biology and of chemistry.

There was an important common factor behind most of the categories to which Guthrie had alluded: scientific apparatus or instruments, as we would call them today. This category was still fresh enough to have first been explored at length by James Clerk Maxwell in a lecture given to accompany the 1876 exhibition. This material and technological aspect of science fitted a museum much better than theories or books. The approach taken, as Guthrie had shown, was that these were the process of science. His analysis does not refer to the grand theoretical schemes or philosophical insights but to steps in discovery and applications.

At the same time a committee of engineers reviewed the Patent Office Museum collections. They had been asked to review them as a kernel of a science museum, and they used very similar criteria to those of the scientists:

> The principle of selection that we have followed has been to throw out such objects as have no historical interest, and are neither good examples of accepted practice or modern improvements, nor steps or links in invention.[7]

What is striking again is that the emphasis was on machines that could be interpreted as the cutting edge of applied science.

So by the late 1880s there was a collecting policy, and there had already been a generation of collecting whose various objectives had already resulted in a remarkable hoard. In terms of the gems of the collection as they would seem a century later, from the South Kensington Museum had come Humphry Davy's original safety lamps, the first to be used in any coal mine. These had been given to the Museum of Practical Geology and were transferred to the South Kensington Museum in 1857. On the other hand, the Rocket locomotive and the Comet steam engine had been acquired by Woodcroft for the Patent Office Museum in 1862. Herschel's prism and mirror and Thomson's mirror galvanometer were acquired for the 1876 Special Loan Exhibition. The loan of the Herschel apparatus was only converted to a gift eighty years later!

One item which the later list of 'gems' unfortunately overlooked had actually, by its size, contributed overwhelmingly to the acquisition total of the 1880s. In 1889 the Museum acquired the huge metallurgical sample collection of the late John Percy, for now it was the national repository for the great products of science. It was a telling victory, for Percy, himself a professor of metallurgical chemistry at the Royal School of Mines, had so objected to the move to South Kensington that he had resigned his post.

This diverse range of materials was held together by a belief, oft proclaimed, that the Museum's subject was 'pure and applied science'. So new and relatively undefined were these categories that it could fairly be said that not only did the categories make sense of the collections, the collections animated the categories. Certainly large parts of the public would first encounter them at the Science Museum.

In 1888 the collections were divided under two headings: 'Instruction and Research' and 'Industrial Applications'.[8] In an 1889 Parliamentary enquiry, great emphasis was put on the separation of the latter from a purely industrial museum. This was a museum whose collections illustrated the practice and application of science.[9] The engines and propellers were given significance not just because they were industrially important but on account of their place in the application of science. Equally, the spectrometers and rulers were significant not just because of their esoteric meaning to a few practitioners but also as the underpinnings of the wealth of the modern world. This reading was not universally accepted. The head of the Office of Works, Algernon Bertram Mitford, objected forcefully to the 1884 enquiry, 'Moreover, it is not by wandering through endless galleries of drain-pipes, specimens of terra-cotta, iron-work, and all the out-of-date and heterogeneous rubbish now in the Southern Galleries that a man will learn the builder's trade. To become a builder a man must work at the bench. To become an architect he must be apprenticed to a master of the craft.'[10] In other words, the practice of a craft required an apprenticeship, not science. To him the only value of a collection in South Kensington was as a reserve of artefacts for professors at the new Normal School to demonstrate. Milford was rebuffed, but his minority report rattled the scientists.

On the other side of the argument, Lockyer suggested that the collections that currently existed were but a tithe of what might be. He declared to an 1889 enquiry that ever since the 1876 exhibition 'the collections have been going on at half steam.'[11] After discussion of the potential space requirements it was suggested to him: 'with that you would be able to manage, from time to time weeding out, and replacing, so as to keep pace with the advance of science.' He answered 'Well, I do not know, but I think that has not been the history of other museums, quite.' He was further prompted: 'Museums will adapt themselves to space?' – a suggestion dismissed by Lockyer, tersely and prophetically, with the rhetorical interrogative 'National museums?'[12]

The acquisitions of the 1880s can be examined in terms of modern collections. Across the whole decade 6,728 inventory numbers were allocated. Of these about half (3,730) are accounted for by the metallurgical acquisition. After that we see a few hundred objects added to each of mining, electricity and astronomy. In the top quartile we see also a few dozen in the collections of time, water transport, motive power, heat, optics, experimental chemistry, art and textile machinery. This pattern was actually not hugely different from that in the 1870s, though in that defining decade mathematics (from the 1876 loan exhibition) had been the most active single collection.

The 1889 report into the Science Collections at South Kensington provides a very good overall account of the conditions of the collections at that time. The science collections closely followed the structure of subcommittees proposed for the 1876 loan exhibition and before it the British Association, with (perhaps as a concession to Huxley) Biology preceding Geology rather than succeeding it:

a. Mechanics and Mathematics
b. Physics
c. Chemistry, Metallurgy, and Principles of Agriculture
d. Biology
e. Geology, Mineralogy, and Mining
f. Navigation, 'Nautical Astronomy', and Physiography[13]

The very organisation of the collections in this way acted as a conservative force encouraging further acquisition according to the same categories. This structure, complemented by the engineering collections and minus geology and biology, would persist for a century in the organisation both of the collections and of staff.

The value of the collections was summarised by the committee in terms of the testimony of Huxley, then professor at the Normal School and secretary of the Royal Society as well as the well-known promoter of the evolutionary theory of Darwin. The committee reported that 'Professor Huxley, who, in one capacity or another, has had a very large share in the formation of these collections,' maintained that the link with the Normal School of Science was 'accessory and accidental,' and that much more important was firstly their accessibility to science teachers from around the country who needed to be familiar with modern instruments, models and specimens and secondly 'the preservation of apparatus and models which possess historical interest as marking stages either in the progress of discovery or in that of the application of scientific principles to art and industry.'[14]

So by the end of the 1880s the Museum had rich collections, a programme of acquisition and an ideology. Already deeply entrenched was an underlying belief in the metaphor of stages in the progress of scientific discovery leading up to the present and future glory of industry. There was also an underlying belief in the communicative power of objects. Lockyer told a story to the 1889 committee. He recounted a visit to the Manchester house of the great engineer Joseph Whitworth. The father of modern standards had taken the young astronomer aside and, pointing out a beautiful scale model of his future estate down to the wheelbarrows, emphasised: 'A great many people cannot read drawings, and they can read solids.'[15]

Illustration 11.1   The three foot theodolite made by Jesse Ramsden, 1792. It was used until 1862 and was capable of making observations of points more than a hundred miles distant. It was loaned by the Ordnance Survey for the 1876 loan exhibition and has been with the Science Museum ever since

The 1920s

Let us move forward thirty years. Even then, when the foundation of the Science Museum in 1909 made a fundamental turning point in the fortunes of the enterprise, the collection grew progressively rather than explosively. It was in the decade after the First World War that one saw another steep growth, under the new Director Henry Lyons, former protégé of Norman Lockyer. Lyons had under him a remarkable staff including the head of engineering, Henry Dickinson, and the head of science, David Baxandall. Nonetheless, because the Director personally signed every letter that left the Museum he was intimately involved in every acquisition.

The decade saw the acquisition of 14,340 items. One in eight of today's hundred greatest treasures were gathered in the ten years from 1920 to 1929. Each is now a great historic icon, but at the time some, such as Aston's mass spectrograph and quartz microbalance or the Logie Baird television, represented cutting-edge science and its perceived application in high technology. Some were slightly older but hugely respected pillars of new technology and science. These included the first aeroplane (the Wright Flyer), a Kodak 'No. 1' camera of 1888, Parsons's marine turbine and Crookes's radiometer. A few were icons of the industrial revolution; Trevithick's stationary engine is an example. To these we might have added the George III collection of scientific instruments from King's College London and the Watt workshop, on which Woodcroft had had his eye half a century earlier, but failed to acquire.

At one level the new growth reflected the provision of more space. The new building opened in 1928 could be filled. It expressed continuity too. Using language that would have been familiar half a century earlier, the 1920 catalogue of machine tools described the role of the Museum:

> The Science Museum, with its Collections and Library, aims at affording illustration and exposition in the fields of mathematical, physical, and chemical science, as well as their applications to astronomy, geophysics, engineering, and to the arts and industries generally. To that end the Museum includes objects which are of historical interest as marking important stages in development and others which are typical of the applications of science to current practice.[16]

Not surprisingly, among the most active collections of the 1920s were such old favourites as mathematics, experimental chemistry, astronomy, and water, road and rail transport. There were, however, new areas too, such as radio communications, telecommunications and aircraft. Such areas and others that were newly active, such as mechanical handling and machine tools, were not just further applications of science but entire new industries.

To what extent, one can ask, was this change a reflection of conscious policy? At one level, Lyons, the new Director, and the Chairman of the Advisory Council, Richard Glazebrook (Zaharoff Professor of Aviation at Imperial College, 1920–1923, and formerly founding director of the National Physical Laboratory), held to the traditional values laid down in the previous generation of intense planning and social cogitation. Lyons himself disclaimed any collecting strategy on the grounds that his budget was so limited he could not plan a programme of proactive collecting.[17] He could only respond to that which was offered. On the other hand, despite his new building, he was under great pressure of space and reported that the Museum was only coping by means of 'drastic sifting of material'.[18]

The Museum was also under new pressures and exposed to new kinds of opportunity. Great new companies, research laboratories and cartels had emerged early in the twentieth century in response to German and American competition: in chemistry ICI, in electricity supply GEC, in telecommunications the Post Office and in aircraft the Royal Aircraft Establishment. Although the distribution of both gas and electricity was fragmented, there were active industrial bodies fiercely competing one against the other and anxiously viewing the Science Museum as a venue for publicising their cause. The Parsons Marine Steam Turbine company presented the after-portion of 'Turbinia' (the first turbine-powered ship), with engines and propellers, free of charge to the Museum. They threw in Parsons's first experimental turbine engine, which was replaced in 1896 after only a year's service to demonstrate the history of their company.[19]

Inventors of the new technologies also saw the Museum as a way of proving their priority. Ambrose Fleming donated his pioneering valves as a means of establishing his claims against those of the American claimant Lee de Forest, whose 'vacuum tube' was not to be shown near the valve.[20] The Wright brothers were so appalled by the claims of the Smithsonian Institution's Secretary Samuel P. Langley that he had been the first to fly that they loaned their Flyer to the Science Museum until such time as the Smithsonian would admit their priority.

The Museum, of course, could not just passively rely upon the advantage others saw in their inventions coming to the Science Museum. The Wright Flyer was acquired after eight years of negotiation between Dickinson and Orville Wright. In 1920 Dickinson wrote to Wright asking for advice as to how to represent early aircraft in the Museum. Not getting a response, in 1923 Dickinson himself visited Wright in Dayton, Ohio. By 1928 the Flyer had arrived at the Science Museum.[21] After the Second World War the Flyer would be returned.

Many important objects came from other parts of the government. Poulsen's 'Telegraphone', an early magnetic tape recorder dating from 1903, was provided by the War Office in 1924.[22] Others were purchased. Baxandall acquired an original 1882 model Wimshurst machine for generating large electrostatic charges for £10.00 at auction in December 1926.[23] The following year the Museum paid £90 to the Cavendish Laboratory for Aston's first mass spectrograph, allowing the Cavendish to buy a magnet to replace the one acquired by the Science Museum.[24]

Artefacts first shown at the 1924 Empire Exhibition were a prime target for acquisition. An interesting angle on this was the Museum's loan to the exhibition of the tube (acquired in 1901) with which J.J. Thomson had discovered the electron. As Alan Morton has shown, this exhibition, although it seemed to replicate the 1876 Special Loan Exhibition in its presentation of science, foregrounded the knowledge

Illustration 11.2   John Ambrose Fleming's original thermionic oscillation valves, c.1903. These were acquired from Fleming in 1925 and represented the museum's commitment to acquiring the roots of then-contemporary high technology

gained from the instrument rather than instrument by genre.[25] It therefore fed on the Science Museum's narrative, but also in turn strengthened it. The development of handbooks on collections such as industrial chemistry by Alexander Barclay, the subject's curator, now told a historical story which led from science and crude experiment to the great industry of today.[26] But the chemical industry *itself* had no standing in the categories of the Museum's thinking, inherited from the nineteenth century.

So far-reaching were the implications of industrial research for science at the time that Glazebrook suggested in 1934 that the Museum be transferred from the Board of Education to the Department of Scientific and Industrial Research (DSIR). His fundamental argument was that increasingly the role of the Museum was to reflect an understanding of change rather than of eternal truths. The suggestion was rejected by DSIR, however, on the grounds that its grant would appear to have been increased while its core activities had not benefited.[27]

The problem of industry had already been raised in the 1929 Royal Commission on the National Museums and Galleries. In general the Science Museum itself was active in preparing responses to queries and only slightly covered by the report.[28] However, the tenor of the suggestions was all of a piece. More temporary exhibitions were required, particularly on contemporary themes. In response to a bid from the Chairman, Glazebrook, for a move to a system similar to the British Museum, the Committee recommended that there needed to be a more powerful and industrially representative advisory council with representation from major industries. Behind the admiring tones was stiff criticism:

> At present there is hardly a single group of the collections in which current practice can be properly studied. This need has been emphasised by many representatives of industry and by those responsible for advanced technical education. Since machinery, apparatus and models, illustrative of current practice, are readily contributed on loan by manufacturing firms, the cost involved in expanding this side of the Collections is inconsiderable.[29]

Changes were made in response. The informal title 'National Museum of Science and Industry' had already been adopted, and from 1929 seems to have been more widely used. The Advisory Council was enriched with industrial representatives. Temporary exhibitions on such industrial themes as plastics were introduced. The Royal Commission had made the suggestion that, for instance, the Museum should acquire models of ideal modern garages or factories. The model King's Fund Hospital built in 1932, later acquired by the Wellcome Collection (and subsequently loaned to the Science Museum), was an excellent example of what was contemplated.

The concern about representing industry was, however, more profound and harder to answer than merely through acquiring more model factories. The founder of the Patent Office Museum, Woodcroft, had taken as his historical model the ever-improving quality of individual kinds of machine as expressed in patents. In terms of the imagery of his time, it could be seen as a 'tree of life' of technical artefacts. The model was a powerful heuristic device for collectors. Lyons himself, in acquiring ancient Egyptian measuring devices in the 1890s when he was working for Lockyer, could be seen as investigating the earliest growth of modern science and technology.[30]

However, this endogenous model of scientific change was itself challenged by the integration between science and technology effected by the Museum after the death of Woodcroft. To a certain extent this could be addressed by reflecting on the progress of an integrated 'pure and applied science'. This was the approach taken by Dickinson in historical works such as his 1939 *A Short History of the Steam Engine*.[31] Although richly complex, this volume had little space for industry as such. Social and commercial dimensions, as opposed to the scientific bases, were ignored. As industry came to be a more important category from the 1930s, the objects acquired needed to represent broader social and cultural changes rather than just embodying key technical improvements. On the other hand, as Peter Morris points out the preceding chapters of this volume, as late as 1937, the Director, Ernest Mackintosh, refused to countenance the acquisition of television components as icons of television, but only as scientific devices.

The 1980s

Let us move forward now to the third period of rapid growth, the 1980s. The culture of this period was particularly significant for the shape of the whole collection; a full 25 per cent of all the current collections (excluding the Wellcome collections) were acquired in just that one decade. This era was characterised by ambitious collecting-oriented Directors and senior curators, who were keenly aware of the acquisitive traditions of other museums. By now the collections as a whole had grown to a size far greater than the number of objects on display. Yet collecting continued energetically, with a new emphasis on technology and medicine in their own right. Science per se would be just one component. There had completely ceased to be a hierarchy in which technological acquisitions were justified in terms of the application of science.

In addition to the 25 per cent organic growth, there was also the acquisition of the Wellcome Collections of the History of Medicine from the Wellcome Trust, formalised in June 1976 and implemented over the subsequent decade.[32] This loan was huge: the Wellcome Collections, consisting of about 100,000 objects, were as great in size as the entire Science Museum collections. They also added a major new subject to the Museum's orbit, the history of medicine. In addition, the Trust lent additional items found in its archives. So, for instance, in 1982 it provided the 'Shaped glass tube used by I. Curie and F. Joliot in an experiment on artificial radioactivity in 1934. Contained boron nitride that under alpha particle bombardment of the boron nuclide became nitrogen 13 nuclei.'[33] The newly produced nitrogen was itself radioactive. This tube, therefore, was a relic of the discovery of artificial radioactivity.

Growth was made possible by the large warehouse rented in Hayes, Middlesex, and in addition there was new space at the Museum's recently acquired airfield in Wroughton, near Swindon. Nonetheless, acquisition was fuelled by very active exhibition activity in the first half of the decade. Between 1977 and 1986, a space equivalent to half the entire exhibition area of the Science Museum was displayed or redisplayed. Space had been created for the Wellcome History of Medicine galleries on the new top floors, and, by filling a void below, space had been created for LaunchPad (1986) on the first floor, part of Telecommunications (1983) and Printing & Papermaking on the second floor, and Photography & Cinematography on the third. Below that, Industrial Chemistry and Nuclear Physics and Nuclear Power were redisplayed. Some of the many other areas redisplayed at that time included Gas, Plastics, Experimental Chemistry, and Space.

There were strong continuities from the past. Until 1984 there were seven curatorial departments: 1) Physics, 2) Chemistry, 3) Medicine, 4) Electrical Engineering and Telecommunications, 5) Transport, 6) Mechanical Engineering, 7) Earth and Space Sciences. Even the order of departments was familiar from the structure of a century earlier. Though, of course, the number of collections had grown as categories multiplied, the familiar departmental structure provided enduring groupings.

Illustration 11.3   Medicine chest made for Marchese Vincenzo Giustiniani, the last Genoese governor of Chios, c.1565. A part of the collection loaned by the Wellcome Trust in the 1970s

On the other hand, the space constraints were far less severe than in the past and annual acquisition rates reached 2,000 objects a year. At first Wroughton was used principally for the aircraft and transport collections. So, for instance, in 1982 the Museum could seek out and purchase a 1933 Boeing 247 aircraft, one of the first to include such innovations as an all-metal skin and retractable landing gear.[34] The same year a V2 rocket arrived (by lorry) from Cranfield Institute of Technology.[35] Agriculture was an early intended beneficiary, and a large collection of tractors was acquired in 1982.[36]

Not only aircraft and tractors could be stored in the large, if slightly damp, hangars, however. In 1983 the fiftieth anniversary of the discovery of polythene by ICI led to conversations about an appropriate temporary celebratory exhibition. ICI donated the original pressure bench in which ethylene was compressed to 2,000 atmospheres, leading to the discovery of a waxy solid in the reactor the following day.[37] For the 1977 redisplay of industrial chemistry, the Museum had already acquired the 9-litre reaction vessel and Michels bench with which the first ton of polythene had been made. The space now available meant that even the largest pieces of heavy technology, hitherto unavailable to museums, could be accommodated. In 1985 one of the reactors from the first synthetic ammonia plant in Ludwigshafen, Germany, was donated by the chemical company BASF. The reactor is 13 metres (40 feet) long and weighs 60 tonnes.[38]

In this period the government's purchase grant – a ring-fenced amount which could only be used for acquisitions – grew dramatically. In 1931 it had been just £1,000. By 1960 this had grown to £8,000, and by the mid-1970s to £18,000.[39] However, in the early 1980s it had reached £100,000. This jump had occurred because of the availability of the Robert Honeyman collection of antique scientific books and manuscripts. Parliament voted a one-off special grant for the purchase of such important volumes for the Library as Harvey's *De Motu Cordis*, a vote which was then annually incorporated, by intent or by oversight, in the Museum's grant thereafter. Moreover, bids for further support could now be made to the government's National Heritage Memorial Fund. This meant that, by traditional Science Museum standards, significant amounts of money could be paid for new acquisitions. Thus in 1983 an important 'Byzantine portable universal altitude sundial with geared calendrical device' could be purchased for £25,000.[40] A sum of £52,000 was paid for a collection of ninety-four photographs by the pioneering nineteenth-century photographer Julia Margaret Cameron dedicated to Sir John Herschel.[41]

The period was also an era of rapid change in the Museum and its collecting, culminating in the Museum's first formal collecting policy of 1987. The Directors (Margaret Weston and Neil Cossons) were no longer signing every letter as their predecessors had done (though copies of letters-out were still circulated in bulky files for them to skim) and only those acquisitions drawing on the purchase grant or significant space went to them.

On 6 October 1987 a policy was put to the Trustees (newly appointed under the 1983 National Heritage Act) by the Director and the Keepers under the leadership of Brian Bracegirdle, first holder of the newly created post of Head of Collections. Potential acquisitions were mapped to a two-dimensional model of significance.[42]

One axis specified the scope of the Museum's collections under five heads:

a. Association with particular important events, people, and/or institutions, historical or current
b. The practice of science or technology
c. The process of change in science or technology
d. An aspect of science or technology in the public eye
e. Significant aspects of non-Western science or technology

It was striking how, by contrast with the evolutionary model of half a century earlier, association with historic and current events came first. Moreover, the emphasis was not on the embodiment of science or of innovation but on their illustration. The policy specified that 'Desirable acquisitions have evocative, illustrative or explanatory functions.' Nor was the focus even nominally on science alone: science and technology, and by extension medicine, were being given equal weight.

The second axis specified the significant qualities a potential acquisition falling within this scope would have to have. These were:

a. The potential for the enhancement of an existing collection by enabling contrasts or comparisons to be made
b. The likelihood of being able to acquire a similar object on another occasion
c. The state of completeness of accompanying documentation – signatures, dates, origin, provenance, use
d. The short– or medium-term potential for exhibition and/or publication
e. The aesthetic appeal
f. The costs of acquisition – transport, handling etc:
g. Condition – potential cost of conservation/restoration
h. The costs of storage – security, size, weight, potentially dangerous components, labile materials etc.

These criteria were not intended to be radically new, but rather represented a formalisation of by-then existing practice. The policy then described how each of the collecting departments (now reduced from the previous seven to four: physical sciences, medical sciences, engineering and transport) would implement the policy for their areas to build on past strengths and repair recognised weaknesses.

The collections had individual histories rooted in their diverse origins. The physical sciences collections had owed their foundation above all to the 1876 Special Loan Exhibition and to ongoing work of the professors of the Normal School, who had endeavoured to keep the practical science collections up to date. Meanwhile, the Engineering and Transport collections had owed their origins to the 'naval models and engineering' collection of the South Kensington Museum, and above all to the Patent Office Museum. Transport had also benefited very considerably from the RAF during the 1920s. Rail, however, had not been significantly acquired by the Museum (beyond the great pioneering locomotives), and renewed opportunities in the late 1960s engendered a new museum, the National Railway Museum, which opened in 1975. The collecting of road vehicles again had to be negotiated within a rich institutional environment.

By contrast, the medical collection was rooted in the Wellcome Collection, which had grown relatively little from the period since the death of Henry Wellcome in 1936. Unlike the collections built up in South Kensington, Sir Henry Wellcome's had been consciously historical. Nonetheless, since there had been little collecting elsewhere in the technology of modern medicine, the collection's new curators felt a special responsibility to bring the collection 'up to date'. So a focus was placed on the acquisition of artefacts that illuminated the recent history of medicine, and their activity resembled the concerns of colleagues across the Museum.

Despite all the collecting since their origin, therefore, one could still see the diverse roots in their current state and policies. The collections differed, for instance, in their attitude to comprehensive collecting. The policy for physical sciences stated explicitly that it had two separate ambitions, of rather different kinds. One was to continue building those collections 'valued for their comprehensiveness'. This entailed both acquiring in new fields and also 'filling gaps' in old ones. The other approach was to treat objects as valuable windows into other worlds, and to seek 'objects to illustrate the conditions, process, nature and consequences of the

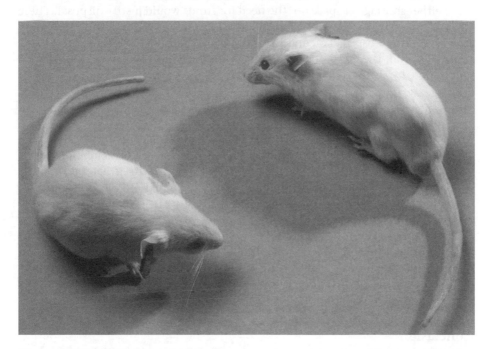

Illustration 11.4 Two freeze dried male transgenic mice (numbers 1134 and 1136) that are direct descendents of the first mammals to be granted a US patent, #4,736,866 dated 12 April 1988. These mice acquired in 1988 are examples of the 'modern' collecting on which there was substantial focus in the late 1980s

application of science and its relation to the practice of science.' Neither engineering, transport nor medicine aspired to the connoisseurial role, but each focused on the need to build up the post-Second World War collections and high technology. The transport curators were aware of the weakness in documenting the 'social usage of vehicles' and their special need to avoid overlap with other collections.

In addition to the issue of acquisition, the policy looked to disposal. During the pre-war years, as Lyons had pointed out, disposal was ruthless. Out of the 14,340 objects acquired in the 1920s, 4,637 have since been disposed of. This number, it must be admitted, includes loaned objects returned to their owners. Nonetheless, even this highlights how the rules for disposal were strict but self-imposed. A 'Board of Survey' of senior curators reviewed proposals for deacquisition, and those earmarked for dispatch were offered first to other museums before destruction was contemplated. Selling was never an option.

The self-management approach was ended in 1984 when the Museum, alongside its peers, was made into a non-departmental public body, governed by the conditions laid down in the National Heritage Act of 1983.[43] The act stated explicitly that destruction would be permitted only of objects that were useless by reason of damage, physical deterioration or infestation. Sale, gift or exchange were also rigorously limited either to other national museums, to duplicates, or to objects which could be lost 'without detriment to students or other members of the public.' Neither shortage of space nor the need for funds would justify disposal. These strict criteria have meant that, of the 50,505 acquisitions made in the 1980s, only 1,540 have been disposed of, and that number includes loans returned to owner. It must be said, therefore, that in recent years the collections have been shaped much more by their acquisitions than by disposals.

Science at the Science Museum

In 2007 the Smithsonian Curator Paul Forman published his provocative thesis that, whereas in the past technology had been seen as the embodiment of science, by the end of the twentieth century, technology itself had become the primary reference. Science had come to be widely valued only as one source of innovation and skill. Forman proposed that the shift in primacy had occurred around 1980. He associated this with a broader cultural shift, which he identified as the move from the era of modernity to postmodernity. Whether this was an appropriate interpretation of the intellectual trends at the Science Museum might be doubted, but the switch in primacy between science and technology could indeed be observed in South Kensington. After a century it seemed that the privileged place of science had been lost.

Conclusion

The 1987 collecting policy was periodically updated, but its basic approach shaped subsequent editions. Perhaps the continuities in the concerns about the shape of the collections are more remarkable than the turning points, at least during the twentieth century. The style of collecting and the anxieties about getting the balance 'right' between representation of the past and present have persisted. Moreover, the shift from embodiment of innovation to illustration of change had, by the early twenty-first century, been reinforced. In part this was driven by the changing nature of technology and the growing importance of electronics.

Less and less was close visual examination of an object a clue to how it worked differently from other devices. More and more, acquisition would be justified by its meaning within a story. 'Progress' was one such story, but others could be cultural-historical accounts, reflections on changing environmental concerns, or indeed industrial change across the world. Interesting stories about modernity and its past more and more became the framework for interesting objects and the collections as a whole. The relationships with popular culture have not been one-way. Such has been its influence on the preservation of material culture and on millions of people that, through its collections and exhibitions, the Science Museum has served to define, and progressively to redefine, the category of science which it has modestly claimed to track.

Notes

1. 'Report of the Patent Office Library', *PP* 1864, XII, p. 26, quoted in Christine Macleod, *Heroes of Invention: Technology, Liberalism and British Identity, 1750–1914* (Cambridge: Cambridge University Press, 2007), p. 261.
2. A copy of this chart is appended to H.W. Dickinson, *A Short History of the Steam Engine* (Cambridge: Cambridge University Press, 1939).
3. 'Y' [a pseudonym], 'Royal Commission on Scientific Instruction and the Royal School of Mines', *The Times*, 24 August 1871.
4. 'Science collections at South Kensington. Report of the committee appointed by the Treasury to enquire into the science collections at South Kensington; with appendix, index, and minutes of evidence', *PP*. 1886 (246), evidence of Lockyer, q. 853, p. 55.
5. Deputation of Society of Arts to Lord Chancellor, 17 January 1874, quoted in Bennet Woodcroft to Board of Management of the Commission appointed by her Majesty for the Promotion of the Exhibition of the Works of All Nations (hereafter 1851 Commission), Enclosure L, Minutes of the 105th Meeting of the Commissioners of the 1851 Commission, p. 13, Archives of the 1851 Commission, Imperial College.
6. Guthrie's report is reprinted in 'National Science Collections. Copy of report of the Interdepartmental Committee on the National Science Collections', *PP* 1889 [C.5831], p. 19.
7. Report of the Committee appointed on the 21st December 1883, to advise the Lords of the Committee of Council on Education as to the existing Patent Museum considered in connection with the Mechanical Section of the South Kensington Museum; especially as to what objects in their opinion should be retained by the Science and Art Department for the various sections of the Science Museum; and to offer suggestions as to the scope and development of the Mechanical Section of the Science Museum treated from the Scientific and Educational point of view. March 1884, TNA: PRO, Works 17/20/5.
8. 'Science Collections at South Kensington' (1889), p. 3.
9. 'Science Collections at South Kensington' (1889), Donnelly testimony p. 24, q. 181–8.
10. 'Report by A.B. Mitford', reprinted in 'National Science Collections' (1886, p. 37).
11. 'Science Collections at South Kensington' (1889), q. 854, p. 56.
12. Ibid, q. 866, p. 56.

13. Ibid, p. 3.
14. Ibid, p. 4.
15. Ibid, q. 899, p. 58.
16. Board of Education, *Illustrated Catalogue of the Collections in the Science Museum South Kensington with descriptive and historical notes: Machine Tools* (London: HMSO, 1920), p. 1.
17. Lyons to Beresford, 10 April 1929, 'Royal Commission on National Museums and Galleries, Organisation of the Science Collections at South Kensington', document 175, TNA: PRO, AF 755/11 part 5.
18. Royal Commission on National Museums and Galleries, 'Report of South Kensington sub-committee', p. 9, TNA: PRO, Works 17/288.
19. SMD, File T1927–479.
20. SMD, File T1925–814.
21. SMD, File T1928–186.
22. SMD, File T1924–188.
23. SMD, File T1926–1029.
24. SMD, File T1927–1085.
25. Alan Morton, 'The Electron Made Public: the Exhibition of Pure Science in the British Empire Exhibition, 1924–5', in Bernard Finn, ed. *Exposing Electronics* (London: The Science Museum, 2000), pp. 25–44.
26. Alexander Barclay, *Handbook of the Collections Illustrating Industrial Chemistry* (London: HMSO, 1920).
27. See TNA: PRO, DSIR 17/129. The suggestion was made by R. Glazebrook, minute dated 24 November 1934. The response was given in a meeting on 8 February 1935, minuted by F.E. Smith.
28. 'Interim Report of the Royal Commission on National Museums and Galleries (Royal Commission)' *PP* 1928–29 [Cmd. 3192]: VIII.699.
29. 'Final Report of the Royal Commission on National Museums and Galleries; Part II. Conclusions and recommendations relating to individual institutions (Royal Commission)' *PP* 1929–30 [Cmd. 3463], p. 47.
30. See, for example, the acquisition of 1913–573, a replica of ancient Egyptian measuring devices (Merkhat and Bey) used for timekeeping, from the Royal Museum in Berlin, donated by Lyons.
31. Dickinson, *A Short History of the Steam Engine*.
32. On the Wellcome collections, see Ghislaine Skinner, 'Sir Henry Wellcome's Museum for the Science of History', *Medical History* 30 (1986), pp. 383–418.
33. SMD, File T1982–548.
34. See SMD, File T1982–1172.
35. See SMD, File T1982–1264.
36. SMD, SCM/1982/295/306 and SCM/1982/975/1033.
37. SMD, SCM/1983/408.
38. SMD, File T198–1289.
39. V.K. Chew, 'The Purchase Grant of the Science Museum', SMD, Z 204.
40. SMD, File T1983–1393.
41. SMD, File T1984–5017.
42. SMD, Collections Division, Science Museum, 'Collecting Policy', October 1987, Science Museum Records Management Group, pp A1-A2.
43. The relevant chapter 47 of the National Heritage Act of 1983 is available on the web on http://www.opsi.gov.uk/acts/acts1983/pdf/ukpga_19830047_en.pdf (accessed 14 February 2009).

12

Behind the Scenes: Housing the Collections

John Liffen

The number of objects in the Science Museum's collections in 2009 amounts to some 220,000 or so individual objects. If the number of constituent parts were to be included (for example, the hundreds of tools contained in a tool chest), that figure would be considerably higher; but 220,000, representing the number of separately accountable objects on the Museum Inventory, is a convenient estimate. Of that number, only about 7 per cent are on display in the Museum's galleries at South Kensington; the remainder, about 204,000, are held in storage, either in a converted office building in west London or at a former airfield near Swindon, in Wiltshire. Although there is undoubtedly some 'dead wood' in these 'reserve collections', as they are sometimes called, on the whole they represent a judicious selection of significant items representing the material evidence of centuries of innovation in science, technology, industry and medicine. Yet, given any reasonable amount of resource, it is now impossible to display all this material in a normal museum setting. It is not what either the Museum's staff or its visitors would necessarily wish, but it has become an inescapable fact of life. It is all the more noteworthy to find that as late as 1936 nearly all the objects in the Science Museum's collections were, in theory at any rate, either freely on show in the galleries or available for inspection with little formality in basement study rooms or stores somewhere on the Museum's South Kensington site.

So what has happened to change this apparently utopian situation? How does the way the Museum added to or disposed of its collections in its formative years differ from the situation today? What have been the changes in societal attitudes to museums which may have influenced the Science Museum's collecting policy over the last hundred years? Indeed, what is the relationship between display and preservation? The story of the growth of the Science Museum's collections is one of acquisition always running somewhat ahead of available display space; and of a shifting understanding of the purposes for which the Science Museum exists, by both its managers and its users.

273

Storage before the stores

The issues surrounding the growth of the collections have been there since the Science Museum's two principal antecedents, the South Kensington Museum and the Patent Office Museum, first opened their doors in 1857. From the beginning the South Kensington Museum's art-based collections were developed more actively than those for science, and pressure on space was soon felt. In response, new groups of galleries began to be built on the land immediately adjacent to the Bromton Boilers. In 1862, building on the success of the Great Exhibition of 1851, another international exhibition was held at South Kensington. A vast brick building was erected on what is now the site of the Natural History Museum, with annexes extending a considerable way northwards along Exhibition Road and Prince Albert's Road (now Queen's Gate). Unlike 1851's Crystal Palace, this building was ugly and unloved, and in 1863 it was demolished with explosives. One narrow range of buildings on the north side, the 1862 exhibition's refreshment rooms, was designed to provide an attractive backdrop for the Royal Horticultural Society's gardens, which were being established in the main part of the South Kensington estate. They were retained, and in 1863 were leased by the government. At first called the South Arcade, the buildings were later usually known as the Southern Galleries. They offered convenient additional space at low cost for the South Kensington Museum. In 1864 the Animal Products section and the heavy building materials of the construction collection were transferred there. The space released allowed the formation of a Naval Models and Marine Engines collection. This too moved across Exhibition Road into the Southern Galleries in the following year. In 1866 the educational collection followed, as did the food collection in 1867. The Science and Art Department's annual report referred for the first time to 'The Scientific Collections'. These now included 'Machinery etc.', which in turn was established in the Southern Galleries in 1870. It is clear that a policy was developing of physically separating art and science to opposite sides of Exhibition Road.

The Patent Office Museum, under separate management, was free to develop in whichever way Bennet Woodcroft and his successive curators wished to take it. In their cramped and awkward space in the southern portion of the Brompton Boilers, room was somehow found to display, for example, the 1788 Boulton and Watt beam engine from Matthew Boulton's Soho Manufactory, the historic locomotives Puffing Billy (c. 1814) and Rocket (1829), and a group of Arkwright's textile machines. There was no room for physical expansion, and little appears to have been disposed of, so the display space became extremely cramped. Even those who might have been expected to be supportive were unimpressed, such as the editor of the weekly journal *Engineering* in 1869:

> In a shabby shed-like building at South Kensington – a building which is not only miserably deficient in size, but is badly lighted, and does not possess the slightest pretensions to being fireproof – there lie at the present time, crowded together, and piled one on another, without the least attempt at classification,

many of the most valuable relics which our nation possesses of its early engineers and inventors.[1]

This sounds more like the 'open storage' of today – indeed, perhaps not quite as good as that. By that time the Patent Office Museum was the only remaining occupant of the shortened iron building, most of it having been dismantled in 1867 for re-erection at Bethnal Green in east London (these buildings now house the V&A Museum of Childhood). For the remainder of its separate existence there was no change in the museum's approach to collecting (to call it a 'policy' might perhaps be going too far), though towards the end the display was tidied up somewhat.

In 1871 the Royal Commission for the Exhibition of 1851 organised the first of a new series of international exhibitions at South Kensington. Less comprehensive than those of 1851 and 1862, they were housed in new buildings, which flanked the east and west sides of the Horticultural Society's gardens at the southern end, and which were called the Eastern Galleries and the Western Galleries. These exhibitions came to an end after the one in 1874. Two years later the Western Galleries, and parts of the Southern Galleries, were used to house the Special Loan Collection of Scientific Instruments. At the close of the exhibition many items were left by the lenders for the South Kensington Museum. These formed the core of a greatly enlarged Scientific Division, which was permanently installed in the Western Galleries. The collections of ship models and marine engineering, and of models of machinery, were also added to at this time, and were placed in the Southern Galleries. The enlargement of all these collections drew the wider public's attention to the fact that the elements of a future Science Museum were emerging.

It will be clear that, even without a coherent forward plan, additional display space was continually needed. This need was met, or at least alleviated, by taking over existing exhibition buildings originally provided for other purposes. It was also met by releasing material considered no longer appropriate for display. For example, in 1881 a collection of Munitions of War was transferred in its entirety to Sheffield Public Museum and to Birmingham Free Library and Museum. In 1883, portions of the Building Materials and Models of Construction collection were removed from display, these 'comprising such objects as had become old and obsolete in practical application and which further illustrated ornament and decoration rather than construction.'[2] Two years later, much of what remained was 'suppressed', 'because the temporary outside wooden passage in which the collection was housed was demolished.'[3] I assume that 'suppressed' meant that the collection was retained, but was packed up in boxes in a storeroom.

The absorption of the Patent Office Museum brought further strain. It was thought, probably with good reason, that there was a good deal that could be discarded. The Science and Art Department appointed a committee under the chairmanship of John Slagg MP in December 1883 to carry out a review. The committee reported in March 1884. About 600 objects were to be retained, seventy-two borrowed items were to be returned to owner and a further 420 to be placed in store pending disposal. Despite this, considerable reorganisation of the displays in the Southern and Western galleries was needed before the selected objects could be brought across the road from the 'Brompton Boilers'. The Department's report for 1885 stated 'that these examples will be amalgamated with the Science Collections of the Department, and will in future be known as "the Science Museum".' The move finally took place in the early months of 1886.

Between the 1880s and the opening of the V&A Museum's new range of buildings in 1909 the purpose and development of the science collections were under continual review.[4] During this time it is probable that certain items from the collections were stored on site and inaccessible to visitors, but references to confirm this impression are very sparse. For all practical purposes, the term 'collection' can be considered to describe the items selected for display. If there was a printed catalogue available, it described the display as if that were the entire collection.

In 1909, then, the Science Museum comprised an inconspicuous main entrance on Exhibition Road leading to a narrow and confined shed-like gallery. This emerged into the more spacious but lightly built Southern Galleries. On their first floor, a corridor to the right led through to the impressive new reading room of the Science Library, opened the previous year. Carrying on to the west, more galleries, quite narrow, eventually turned north to an entrance in Imperial Institute Road. On the opposite side of the road were the Western Galleries, another long, narrow building running due north. In all these galleries the showcases were packed as close together as practically possible in regimented lines, with only the occasional large machine or antique locomotive to relieve the view. The total display area was about 98,000 sq. ft (9,114 sq. m).[5] Portions of the collections might be stored in temporary buildings around the site, but in the main what was held was on show.

The Bell Committee envisaged that all this would be swept away, to be replaced by a range of modern buildings running all the way from Exhibition Road to Queen's Gate. Coverage of science and engineering would be comprehensive, and nowhere in the Committee's Report is the idea that collections might be held for reference or simply in preservation, but not displayed.

The establishment of an independently administered Science Museum brought with it a review of procedures. Among these was the institution of a Museum Inventory to make the auditing and tracking of Museum objects more straightforward. Until this time acquisitions were recorded in large volumes called 'Divisional Registers', one for each collecting division.[6] There was a separate number series for each, using different initial letters. Those in charge of collections used special copies of the printed catalogues, interleaved with blank pages, to add new references or make location notes.

Francis Ogilvie, Director from 1911, devised a new inventory system, which was adopted in 1913. This was for a single numbering sequence, regardless of Division or collection, created by taking the year of acquisition hyphenated with the next serial number available, starting again from 1 each year, for example 1913–573.[7] There was a central Inventory Register in a new series of volumes, supplemented by duplicate sets of index cards which could be arranged in acquisition order and by subject classification within each collection. These sets of cards were issued to the officers in charge of each group of collections. Although new index cards are no longer issued, today's computer database Inventory retains the same basic numbering system. Then as now, the Museum's procedures were tightly drafted to ensure that collections were kept secure and in good order.[8]

In 1913 the buildings closest to Exhibition Road were cleared and then demolished to make way for construction of the East Block, the first of the range of new buildings recommended by the Bell Committee. This led to considerable consolidation in the remaining galleries, and a number of collections were stored in their entirety until the new building was ready. One early decision was made which has had a fundamental influence on the layout and grouping of the displayed collections ever since, and this was to construct special pits in the ground floor East Hall to accommodate six large stationary steam engines.

The construction of the East Block was slowed considerably by the First World War, and, although the Museum was never fully closed, further areas were appropriated for government use, such that when full opening was resumed on 1 January 1918 there was even less exhibition space available than in 1914. In 1923 the Western Galleries had to be vacated so that they could be taken over by the new Imperial War Museum. This brought the available exhibition space down to only 63,000 sq. ft (5,859 sq m).[9] As much as 25 per cent of the collections were stored, albeit on site, and on the galleries the exhibits were crammed into the showcases, which were themselves placed close together, making for cluttered and uncomfortable conditions for the visitors.[10]

It was at this time that the prospect of a proportion of exhibits remaining undisplayed well into the future first began to be articulated. Colonel Henry Lyons took over as Director in May 1920, and on him rested the responsibility of laying down the practicalities of the occupation of the soon-to-be-completed East Block. In August 1922 he circulated to his senior staff 'A memorandum on the arrangement of collections in the Science Museum to serve as a basis for discussion'. Topics were arranged under the following headings: Aim and scope; History of Invention; Exhibition of objects; and Labelling. Although the whole is of absorbing interest, it is the section on 'Exhibition of objects' which is of relevance here. Lyons stated:

The crowding of objects in the existing exhibition galleries must not be reproduced in the new buildings where only so much will be placed as can be shown there properly. Only rough estimates are possible but it is certain that the cases now in use need about 40 per cent more space in order that they may be placed properly, and to relieve the crowding of objects in the cases an even larger proportion of additional space will be required. From 100 to 150 per cent additional space could therefore be utilised for the collections now on exhibition, and probably 150 to 200 per cent of additional space should be reckoned for collections that are now stored. This will exceed what is available in the Eastern Block of the new buildings together with the Southern galleries, so that only a proportion of the collections which the Museum contains can be adequately shown as yet.

Lyons then went on to raise the question of reserve collections:

A visitor cannot examine collections with any benefit or satisfaction to himself for more than about an hour and a half or two hours, so that the question of what he may be expected to look at with comfort and appreciation must be considered. Related collections should be kept together so that the tendency to wander through a series of galleries instead of visiting one or two main collections only, may be discouraged. On the other hand for even a few of the existing collections to be visited in comfort it will be necessary for only a selection of objects to be shown in some of them. This raises the further point, where are the remaining objects to be kept? If they are stored in a basement remote from the gallery they will rarely be asked for, and if they are to be of any real use they must be brought up from time to time to replace others, which will involve very considerable labour. Reserve collections have been much recommended but the practical method of dealing with them in a technical museum has not been worked out. If they are to be kept in neighbouring but more crowded galleries for ready reference some difficulties are avoided but others will arise as space for new objects may then be difficult to provide.

In a very few words Lyons encapsulated what has become a central dilemma for all museum directors everywhere: what to do about collections that become larger than the space available for displaying them.

By 'reserve collections' Lyons appears to be describing rooms in which the objects are kept in a logical or coherent order, with labelling, but considerably crowded together in the showcases, themselves being placed very close together. Such rooms would be accessible to visitors with a special interest, who would not worry unduly about the cramped conditions. By contrast, objects in deep storage would be shelved unlabelled in random order or packed up in boxes or packing cases, making a visit impossible unless accompanied by a member of the specialist staff. The Science Museum's guide booklet, available from early 1923, noted that:

Owing to lack of space the following Collections have been temporarily withdrawn from exhibition, but can be inspected on application to the Officer in Charge: – Kinematics, Constructional Engineering, Textile Machinery (except a few examples in Room 1), Agricultural Implements, Paper-making and Printing Machinery, Writing and Copying, Lighting and Illumination, Gas Manufacture, Buckland Fish Collection.

This suggests that no unpacking was needed, so the conditions were closer to 'reserve collection' than deep storage. The situation would be regarded as temporary, as material flowed out to the new display spaces of the East Block. Indeed, this particular copy of the guide has several printed amendments pasted in to reflect successive changes in location. What these storerooms actually looked like has proved difficult to find out. It appears that at the time such places were considered too mundane to be worth photographing. At any rate, no photographs have been discovered.

The Museum's annual report for 1923 propounded a new collecting policy, which was perhaps the result of Lyons's 1922 discussion paper:

> As the National Museum of Science and Industry the aim of the Science Museum is firstly to form in each group a series of carefully selected objects representing the principal stages of development in the branch of science or technology which the group represents; and secondly to show a collection illustrating the current practice of the groups. These groups, each of which represent some special industry or branch of science, number about fifty or sixty, so, if the collections are not to become unmanageably large, strict limits must be laid down within which they are to be developed.[11]

How this was to be done was by establishing a 'historical series' which would remain little altered. The examples of current practice were expected to be provided on loan by manufacturers, and withdrawn and replaced in due course by fresh items. The way in which the historical series was to be developed under the scheme was not articulated. The underlying principle appears to be that the collections should solely comprise the material on exhibition. Storage, reserve collections even, were not envisaged once the Museum had been completed to the full extent laid down by the Bell Report. Certainly the new range of printed catalogues of separate collections, which were issued from the beginning of the 1920s onwards, seem to describe only what was visible in the respective galleries, and none of the associated items held elsewhere. These were passed out of the relevant curator's care and placed in what was called 'Custody' in the 'Director's Store'. If later required for display, an object would be reissued to the curator's charge.

A further option for easing crowded conditions was, of course, to institute a more rigorous disposal policy. It has already been noted that certain collections were heavily weeded or eliminated during the nineteenth century by scrutinising committees. In 1913, as part of the new accounting procedures introduced that year, Boards of Survey were established. A Board comprised three Museum Officers (senior members of the curatorial staff), who periodically examined objects no longer required for exhibition. It then made recommendations as to their retention or disposal, and, if the latter, whether by transfer to another institution, sale or destruction. Except for items to be returned to lender, objects were national property and the Board's recommendations required the agreement of the Accountant-General's office and the approval of the President of the Board of Education before action could be taken. The process ensured that objects could not be lightly discarded, and in principle the same system, still known as Board of Survey, continues to operate today, but without the need for permission for disposal from the government.[12]

The East Block was progressively prepared and opened from 1923 onwards, culminating in a Royal opening by HM King George V on 20 March 1928. Over the next few years, curators reviewed the material that had not been chosen for display to see whether it could be passed out of the Museum's collections altogether. By way of example, of the total of 291 items described in the 1890 Electrical Communications printed catalogue, no fewer than 159 were disposed of between 1926 and 1933. Many surplus items originating from the Post Office were returned to them, but other recipients included the Royal Scottish Museum[13] and the new regional museums at Hull and Newcastle upon Tyne.

Although annual disposals by Board of Survey continued throughout the 1930s, new acquisitions to the permanent collections continued to be made. Most were squeezed in to display on gallery, or placed in the various storerooms and huts scattered about the Museum's South Kensington site. However, one notable group of acquisitions in 1936 may well have marked the point at which external warehousing became unavoidable. During his brief reign King Edward VIII gave a number of horse-drawn carriages, surplus to Royal requirements, to the Museum. This was clearly an offer that could not be refused, but it must have created considerable difficulties in accommodating them.

Hopes that an early start would be made on constructing the next stage of the Museum's building plan, the Centre Block, were not fulfilled.[14] As an expedient, therefore, additional storage space was essential. Outlying warehouse storage amounting to about 20,000 sq. ft (1,860 sq. m) was provided in suburban London at Alperton, Hayes and possibly one or more other sites, but precise details of their location have proved elusive.[15] The Museum's annual report for 1937 stated:

> The gradual removal of exhibits temporarily to store, in order to make room for newer acquisitions, has now filled the Museum stores, and it has been necessary to use accommodation outside the Museum buildings. Acceptance of acquisitions is more and more carefully scrutinized, but the filling of important historical gaps must continue when opportunities occur, and progress must be selectively shown: otherwise the Museum will cease to fulfil its national purpose.[16]

A typed list showing the positions of vehicles in the Road Transport section in January 1939 has survived. Of thirty-nine cars in the collection, nineteen were at Alperton; of twenty-four motor cycles, nine; and of forty carriages, fifteen. Even so, in 1939 overall only about 5 per cent of objects in the Museum's collections were not on display.[17] Nevertheless, from this time on, a proportion of the Science Museum's collections (increasing as the years have gone by) has always been kept at locations away from South Kensington.

After the Munich crisis of September 1938 there was a strong expectation that war was likely before too long. As bombing of London could be expected, plans were made to evacuate the most important or rare objects to country houses well away from cities.[18] The Museum was closed to the public on the outbreak of war in September 1939. It later partially reopened, but closed again for the duration in September 1940. The upper galleries of the East Block were cleared and as many objects as possible were moved to the ground floor or basement. The Southern Galleries were emptied out completely, being lightly built and a considerable fire risk. The pre-war outlying stores were apparently no longer available, and about half the objects kept in them had to be squeezed into the Museum basement. Later in the war it was decided to send away out of London as many of the remaining objects as possible. Ultimately about two-thirds of the collections were stored elsewhere.

Herman Shaw, Keeper of the Department of Physics and Geophysics, acted as Deputy Director during the war. He gave much consideration to the form the galleries should take upon reopening and drafted a long memorandum to promote discussion. In a section devoted to gallery displays, he wrote:

> In spite of numerous Boards of Survey there is little doubt that a further severe weeding out is required in nearly all sections of the Museum, and this if carried out drastically would greatly improve the attractiveness and appeal of the display. Every effort should be made to scrutinize all sections of the Collections before the Museum is re-opened, to ensure that only those objects are replaced on view which are of outstanding importance or interest and which can be displayed in an attractive and dignified manner. By relegating to the 'Reserve Collections' some 75 per cent of the exhibits previously on view, it would be possible in nearly every section to display the remaining exhibits simply, logically and in a really attractive manner for the benefit of visitors who have no special knowledge of the subject.[19]

With regard to 'Reserve Collections', Shaw continued:

> The adoption of the above proposal would necessitate the withdrawal from exhibition of a large portion of the collections, but these objects could be assembled appropriately as 'Reserve Collections' to supplement the objects remaining on exhibition, and placed in galleries specially devoted to 'Reserve and Study' purposes, an arrangement which has proved both convenient and successful at the Geological Museum.
>
> The contents of these 'Reserve Collections' should be just as well labelled as those of the main collections, and although the objects may be placed closer together these collections should be carefully developed and intelligibly arranged, so as to be readily examined by students and other accredited visitors, who should be afforded ready access to them.

Though the Museum reopened to the public in February 1946, the galleries could only be reopened gradually as staff returned from war duties. In addition, the Southern Galleries were no longer fit for reoccupation, so the total display space was well below what had already been inadequate in 1939. In 1949 the Southern Galleries were demolished in preparation for building the basement and ground floor of the long-awaited Centre Block. The reason for this partial start was to provide accommodation for the Festival of Britain's 'Exhibition of Science' in 1951. However, after the Festival was over, financial restrictions meant that no further construction work was possible for the time being.

The first outstations

In the early post-war discussions about allocation of space, the Western Galleries, which were once again available for the Museum to use, were earmarked for reserve collections. Before this could be taken forward the question arose as to where the Aeronautics collection was to be displayed. In March 1946 it had been placed in Gallery 6, the ground floor gallery to the west of the main East Block. In 1949 this gallery was reallocated to the 1951 Exhibition of Science. Aeronautics was regarded as 'a subject of such current interest, reputation and popular appeal that it could not be immobilised in store even for a limited period of years.'[20] Displaying the collection away from South Kensington was briefly considered, but nowhere suitable was found. Instead the collection was given a temporary home in the Western Galleries, where it opened in mid-1950. It remained in this 'temporary' location for a decade. This precluded the use of the Western Galleries for reserve collections, and the idea remained an abstract concept in the discussions about allocation of gallery space that continued throughout the 1950s.

By 1952 the proportion of the collections not on display amounted to about 50 per cent.[21] The last few objects had only just been retrieved from dispersal storage, but various other outlying storage facilities had had to be found to house all this material until (it was hoped) it could be redisplayed. None were purpose-built, and all were more or less unsuitable for housing museum objects. From 1947 to 1950 Stockwell deep shelter, in south London, was used. In 1948 a wartime underground factory at Warren Row, near Reading, was allocated by the Ministry of Works. Though an awkward configuration, from a conservation viewpoint the environment was reasonable. Warren Row was retained until 1960. After the Museum moved out the site was converted to become a Regional Seat of Government (RSG6), for use in times of national emergency.[22] In 1949 an aircraft hangar at Pyrford, West Byfleet, Surrey was occupied. Another requisitioned factory building, the Britannica Works at Waddon, near Croydon, was allocated by the Ministry in 1950. This had to be given up in 1954 and the contents moved to another factory building, Allandale Works at Lower Sydenham, in south London. Again, in 1958 the collections stored at Byfleet were transferred to a former Government Communications Radio Station at Knockholt Pound, near Sevenoaks, Kent.

These frequent moves exposed the objects to risk of damage through mishandling, or even loss, and much staff time had to be spent in unpacking, checking and auditing. As early as 1952 the Museum recognised that outlying stores were going to be a fact of life for an indefinite time, and in that year the Advisory

Council recommended:

1. Provision of a permanent store, on the fringe of Greater London, with an area of about 100,000 square feet (9,320 sq. m).
2. Arrangement of objects therein so as to be visible without manipulations.
3. Disposal of objects, unlikely to be shown, by gift to other institutions, by sale or by destruction.[23]

In 1957 government approval was at last given for work on the Centre Block to resume. Construction of the building took place between 1959 and 1961, and it was progressively occupied between 1963 and 1973. One of the first new galleries to open was Aeronautics, allowing the Museum to relinquish the Western Galleries altogether. They were soon demolished, as the space was required for the expansion of Imperial College and the construction of a new joint library for the Museum and the College.

It is perhaps appropriate to say something here about how the collections were physically handled and generally cared for at the time the Centre Block galleries were being opened. The largest and heaviest items, such as aeroplanes or full-size railway locomotives, were moved by specialist contractors. Aside from that, as

Illustration 12.1 One of the storerooms at Warren Row, a former Second World War underground factory near Reading, Berkshire, seen in May 1953

much as possible was undertaken in-house. The Museum employed a large force of 'Manual Attendants' (familiarly called 'Man Atts') to clean gallery floors and show-cases. Whenever necessary, groups of them would be formed into 'gangs' to move, under curatorial supervision, objects and showcases as required. They received only minimal on-the-job training and used few specialised tools to help them, relying mostly on combined muscle power. Nevertheless, the Man Atts were a large rum-bustious group of people who were well liked by other staff and whose overenthusi-asm in tackling moves could usually be restrained with tact and diplomacy.

Cleaning the exhibits was carried out by 'Object Cleaners' (often transferred from the Manual Attendant grade), who likewise had no formal training and whose standards were of the spit-and-polish variety. Conservation as a concept was unknown in this period. Objects were 'restored to exhibition condition' by highly skilled fitters and carpenters in the Museum's workshops, this usually involving dismantling, thorough cleaning and repainting, and replacement of broken parts. Unfortunately, there was little awareness that historical evidence could be destroyed in the process. This is no reflection on their work, simply an acknowledgement that preservation and display standards are now very different from what was required and expected half a century ago.

To some extent the occupation of the Centre Block galleries relieved the pressure on the outlying stores, but during this time changes in collecting and display

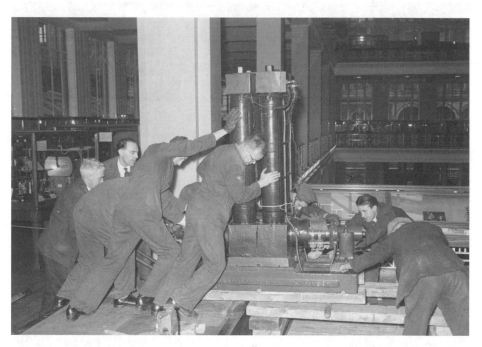

Illustration 12.2   A 'gang' of Manual Attendants moves an 1886 Edison dynamo, Inventory Number 1931-29, in November 1956. From left: John Flannery, 'Nobby' Walpole, (unknown), Dick Haynes, (unknown), Jim Lomax, (unknown)

policies meant that storage was to become a significant element in the Science Museum's future activities. During the 1950s the Museum had continued to juggle the needs of technical education with an increasing element of historical exposition. Frank Sherwood Taylor, who was appointed Director in 1950, was a historian of science and considered that the historical collections were a necessary part of the Museum's purpose and function. In an assessment of the Bell Report, written in 1952, he observed that:

> we do not seek so much to preserve isolated objects of interest, to form as it were a number of memorials of isolated events, as to present continuous histories of the sciences by means of connected series of objects, leading from the primitive beginnings of an invention to its present-day condition of development. This has become our primary historical function: to preserve and display famous objects is secondary.[24]

In retrospect, Sherwood Taylor's concern for the historical aspect reflected a new development in technical and industrial preservation: the contribution of amateurs working in their spare time. As Britain underwent a process of renewal during the 1950s and 1960s, older, more 'traditional' technologies that held a sympathetic appeal for onlookers were thought to be in danger of disappearing altogether. Preservation societies proliferated, seeking not only to keep individual artefacts such as stationary steam engines, trams, buses and railway locomotives but to acquire sites where these could be displayed as museum collections. Entire railway branch lines were purchased and operated; volunteers exhausted themselves with the physical restoration of derelict canals. The term 'industrial archaeology' was coined in 1956 and within a few years became a university discipline.

The policies of David Follett, Director from 1960 to 1973, whether consciously or not, mirrored the changing views of wider society and the emerging preservation movement. In March 1960, shortly before being appointed, he wrote a memorandum, 'A brief statement of aims and function of Museum'. In this he stated: 'The enduring aspect which must underlie all the work of the Museum is its historical function, in preserving for posterity machinery and scientific apparatus which marks significant stages in the development of engineering and science.'[25] In 1965 Follett stated the three main purposes of the Science Museum as:

1. to preserve for posterity objects of significance in the history of the development of the physical sciences and technologies based upon them, as part of the record of man's endeavours to understand and turn to his material advantage the forces of nature;
2. by using the collections to trace the development of science and technology from earliest times up to the present day, to present the world of science and technology to the public at large, and to offer to scientists and technologists the opportunity to acquaint themselves with the history of science and technology;

3. to advance the study of the history of technology, through the studies and publications of its own staff, and by making the collections and their knowledge of them available to students at all levels.[26]

It is significant that Follett put preservation for posterity ahead of the use of the collections for broader educational purposes. Although it was not set up as a museum of history, this was now the Science Museum's main purpose.[27] Paragraph (2) does not use the word 'museum' at all, but rather the oblique phrases 'trace the development' and 'present the world of science and technology' to suggest the display function. The implication for museum storage was significant. Acquisition was not now primarily about creating and maintaining coherent expositions in the Museum's galleries; it was an end in itself, a source from which various good things (including museum displays) would emerge.

This was at a time when Follett and his staff could perhaps feel fairly relaxed about the prospect of collecting for posterity. The Museum was undergoing a major enlargement and the emergence of museums dealing with specific industries was probably felt to reduce the burden on the Science Museum, hitherto virtually the country's sole repository for representing the history of industry and engineering.

The suitability of the Museum's outlying storage was brought into sharp relief three years later. The store at Sydenham was flooded to a depth of nearly 3 feet (900 mm) on 15 September 1968 for a period of 24 hours. Very heavy rain had swollen the River Pool alongside the store, which then flooded as a result of a partial blockage of a culvert further downstream at Lewisham. An immediate remedial exercise was undertaken by the Museum workshops and curatorial staff. Ninety per cent of workshop staff worked at Sydenham for several weeks after the flood. Tanks were purchased for washing, dewatering and derusting fluids for metal objects. Sodden straw packing had to be removed and replaced. There were severe access problems for the crated objects: the stacking of crates meant that the top four or five had to be removed to attend to the wet bottom crate, but the store was so full that there was little or no space for laying out the crate and contents.

The flood at Sydenham, together with concerns about the suitability and long-term tenure at Knockholt, brought the desire for higher-quality storage, which the Advisory Council had pressed as far back as 1952, to the top of the agenda. As the Science Museum was a department of the Civil Service, provision of storage was the responsibility of the Ministry of Public Buildings and Works. Follet therefore wrote to the Ministry in December 1968. After describing the long-term implications of the Sydenham flood, he went on, 'I feel obliged, therefore, to ask that your Ministry provide the Science Museum with alternative storage space not subject to flooding – or indeed to any other hazard which would militate against what is, after all, our prime purpose – the preservation of material for posterity.'[28] The Ministry, as was to be expected, prevaricated, suggesting that a flood relief scheme would provide the necessary protection without the expense of a move. In April 1969 Follett repeated his plea to the Ministry, politely refuting their counter-arguments and pressing home the reasons for requiring new accommodation. Perhaps rather surprisingly, the Ministry's representative, L. Potts, threw in the towel without any more fuss:

> You sent me a further letter on 23rd April about having one combined store to replace those at Sydenham and Knockholt. We are disarmed by the sweet reasonableness of your arguments and accept, after consultation with the Department of Education and Science, the need for alternative accommodation. If you will let me know when you would like to start to make a move, I will ask our Estate Surveyors to see what they can arrange to provide in time for the necessary adaptation work to be completed. We trust you will keep your requirements to an absolute minimum.[29]

Even Follett was probably surprised by this swift and complete capitulation. After a few months a suitable building was offered and, after inspection by the Museum's Keepers, accepted. This was a newly constructed warehouse at Hayes, Middlesex, adjacent to an extensive area occupied by Electric and Musical Industries Ltd (EMI). Despite the wish for security of tenure, only a 21-year lease could be obtained. However, it offered 75,000 sq. ft (6,975 sq. m) of heated warehouse space, 3,000 sq. ft (279 sq. m) of office space, good head clearance, rails for an overhead crane in two of the three bays, and relatively easy access by road and public transport. In order to make efficient use of the space available it was necessary to use the whole volume of the store rather than just the floor area. High-density racking was provided in one bay by installing Boltless Construction's Cubestore mobile racking system, where the objects were stored in unit loads on pallets, access being by reach truck. All this represented a long-term commitment to storage and was a radical improvement on what had been available before. Until this time, objects had been stored either free-standing on the floor, wrapped in brown paper or polythene on wooden racks, or packed in large wooden crates piled into stacks. In the crates routine inspection was all but impossible, and the amount of

Illustration 12.3   In August 1969 the Science Museum's Director and Keepers made a preliminary inspection of the new warehouse at Hayes, Middlesex, intended to become Hayes Annexe. From left: John Wartnaby (Astronomy and Geophysics), Keith Gilbert (Mechanical and Civil Engineering), Gerald Garratt (Water Transport and Mining), Sir David Follett (Director), Donovan Chilton (Electrical Engineering and Communications), Frank Greenaway (Chemistry), John Garratt (Deputy Museum Superintendent), Francis Ward (Physics), Margaret Weston (Museum Services)

handling required to unpack or unwrap objects subjected them to unnecessary risks. The adoption of industrial handling techniques enabled a standard range of fibreboard boxes to be used, objects being prevented from movement by polystyrene chips or 'worms'. Up to twelve boxes could be stacked on to one pallet to give real economy of storage space. It was only later that it was realised that polystyrene was an unsuitable packing material for museum objects (for example, it welded itself quite unpleasantly to rubber-insulated electric cables). The ensuing decade was a steep learning curve for Museum staff as, belatedly, proper conservation techniques were gradually introduced. It is now recognised that, provided the building's environmental conditions are satisfactory, it is better for objects to be kept unboxed wherever possible.

The move from Knockholt was completed in December 1971 and from Sydenham in July 1972. The name 'Hayes Annexe' was adopted, rather than 'Hayes Store', because ambitious plans were discussed for laying out parts of the store in semi-display and to allow occasional opening for the general public. It was intended to re-erect the Bull Engine, one of the pumping engines used until 1961 to keep the railway tunnel under the River Severn dry, and operate it in steam.[30] None of these hopes was fulfilled. The store began to fill far more rapidly than anyone expected. The 1970s were a period of great change in Britain, when 'smokestack' industries were contracting and manufacturing generally was losing out to foreign competition. Curators were under few restraints about what and how much to collect; all they needed to do was to run each prospective acquisition past their departmental Keeper, who usually gave the necessary permission to acquire the object in question. There was much that was thought to need preserving, and apparently unlimited space at Hayes. The result was that new stocks of pallets and boxes were continually needed, and within two or three years, incredibly, space began to run out. Extra rows of fixed pallet racking were erected wherever possible and new mezzanine floors were built in parts of two bays. By 1977 there was only 15 per cent spare capacity.[31]

Blythe House and Wroughton

Part of the need for extra space was prompted by a shift into a new area of applied science. In 1976, following several years of negotiation, the Museum entered a commitment to take on loan the Wellcome Collection of the History of Medicine, amounting to about 114,000 items. Though two substantial galleries on medical history were planned to occupy a new infill block being built within the East Block at South Kensington, the bulk of the collection would have to remain in store. The Wellcome Medical Collections were housed at a building in Enfield, north London, which would need to be cleared within four or five years. During 1977 another warehouse under construction close to the existing Hayes Annexe was considered, but it proved difficult to get funds allocated.

At the beginning of 1978 the situation was critical, but the solution came from an unexpected direction. Out of the blue the Property Services Agency (PSA) of the Department of the Environment (the successor to the Ministry of Public Buildings

Illustration 12.4 The Cubestore mobile racking system in use at the Hayes Annexe in January 1973

and Works) advised the Museum that a large office building in Blythe Road, Hammersmith, adjacent to the Olympia exhibition halls and only about two miles from South Kensington, was shortly to become vacant. It had been built in 1901 to the design of Henry Tanner to house the administrative staff of the Post Office Savings Bank, and until the mid-1970s had been occupied by several thousand office workers. The PSA was offering space in the building to a number of national museums and galleries in the London area. Eventually the offer was taken up by three museums: the Science Museum, the V&A Museum, and the British Museum. Each had roughly a third of the building, allocated to give each museum a sepa-rate secure space, with security and other services held in common. In the Science Museum's case, this meant a total of 128,414 sq. ft (11,943 sq. m).[32] As with Hayes, the Science Museum's earliest plans for Blythe House envisaged exhibition areas, but practical and financial restraints soon caused these ideas to evaporate. The conversion and refurbishment of the building necessarily took several years; in the meantime, a specially recruited project team began work at Enfield to catalogue and pack the Wellcome Collection. Soon, vast numbers of blue fibre boxes started to arrive at Hayes and seemed to occupy every available spare space. By the mid-1980s Blythe House was ready for occupation, and material from Hayes, beginning with the Wellcome collection, began to be moved in.

At the same time that the Blythe House opportunity emerged, plans for another great expansion were close to fruition. Now that the collections were being devel-oped to ensure preservation for posterity rather than primarily for exhibition pur-poses at South Kensington, size was no longer regarded as a constraint. Curators put forward schemes for the acquisition of certain types of object whose size or weight had hitherto precluded collecting, such as commercial road vehicles, civil aviation, space science and agricultural machinery. There was also a require-ment to assume responsibility for historic road vehicles transferred to the Science Museum under the terms of the 1968 Transport Act. All this would need a great deal of additional storage space, much more than a warehouse in an urban area could provide. Other major museums were successfully developing airfield sites, such as the Imperial War Museum at Duxford, and the Science Museum decided to follow suit. With the support of the Department of Education and Science and the Department of the Environment, a search was made for a site which would offer a visitor catchment away from other museum airfields. Eventually a site at the Royal Naval Aircraft Yard at Wroughton, Wiltshire, was found. The airfield itself had closed in 1978 but the Royal Navy for a while retained a helicopter maintenance facility in the south-east corner.

The Science Museum formally took over Wroughton Airfield and six hangars on 1 May 1980. Though it was only four miles south of Swindon and close to the M4 motorway, access by public transport was poor. At the time, however, this was not thought a material issue, and visitors to the planned annual open days and other attractions were expected to travel by car. With apparently almost infinite space available (the hangars, each 600 by 300 ft, looked huge when empty), acquisi-tion went forward apace. Airliners such as a Trident, a Comet IV and a Lockheed Constellation; buses, lorries, traction engines and such specialised vehicles as the

Tucker Sno-Cat, used for Antarctic exploration; guided missiles such as Blue Steel; all these and many more rapidly filled the limitless space. A long-term project for a national farming museum was helped along by the purchase of sixty-six historic agricultural tractors from farmer and enthusiastic collector Frank Smith. Within fifteen years the original accommodation was full. Additional hangars around the site were taken over, but increasingly their deteriorating condition has given rise to anxiety, as they were not originally built with such a long-term use in mind.

Director Neil Cossons initiated a review of the outlying stores to see whether financial savings could be made. It proved impossible to negotiate a new long lease for Hayes Annexe, though short extensions to the existing lease were allowed. It was decided to close Hayes and transfer the objects to either Blythe House or Wroughton. In the main, palletised objects went to a new store building at Wroughton constructed to good conservation and energy-saving standards. This building incorporated high-density moving racking in one part, with fixed racking for larger or individually heavy items in the other. Smaller items housed in cupboards or on shelves were moved to Blythe House. When these moves were completed, Hayes Store was closed on 31 March 1994. Concurrently, the new emphasis on object conservation (and security) meant that virtually all remaining on-site stores at South Kensington were given up. Basement storerooms or

Illustration 12.5   By the 1980s the pace of collecting was overwhelming the resources available to accommodate the material. A proportion was still kept within the Museum itself and this is the overcrowded basement store of the Department of Engineering in February 1984. Ian Blomeley, Museum Assistant, is in the background

cupboards under showcases were no longer regarded as acceptable environments for keeping objects. Such items were removed mostly to Blythe House, not without some initial opposition from those members of staff who felt that access was being compromised for both themselves and visitors. Conditions at Blythe House are good, with a conservation laboratory and the Museum's photographic studio.

The Science Museum's outlying stores currently comprise Blythe House and the Science Museum at Wroughton. Changing methods of exposition mean that the old-style collection-based gallery displays have been largely superseded by broader, theme-based exhibitions. Inevitably this has meant that a great many items which formerly provided much technological and historical detail no longer fit the requirements, so they have had to be removed to store. Acquisitions have continued, though at a more measured pace as more detailed scrutiny is given to each proposal. Consequently pressure on storage space is greater now than it has ever been. Such space costs a great deal of money, and much thought is currently being given as to what role these preserved collections will fill in the future, to justify the investment being made. In the longer term the aspiration is to build sufficient suitable accommodation at Wroughton to hold all the reserve collections, both those already there and the collections currently held at Blythe House. With a coherent layout, access to these collections for visitors would be significantly enhanced, and perhaps move some little way towards how it was in 1909 when, in theory, everything was on view.

Notes

1. *Engineering*, 22 October 1869, p. 275.
2. *Report of the Science and Art Department for 1883* (London: HMSO, 1884), p. 249.
3. *Report of the Science and Art Department for 1885* (London: HMSO, 1886), p. 286.
4. These debates are discussed in Chapter 1 of this volume.
5. Science Museum Annual Report for 1927 and 1928, p. 8.
6. In 1915 these were: Machinery and Inventions; Naval and Marine Engineering; Scientific Apparatus; Buckland Fish Collection.
7. 1913–573: Copy of a merkhet and bay, ancient Egyptian instruments used for astronomical timekeeping (the original is in the Royal Museum, Berlin). See Illustration 1.7 in Chapter 1.
8. *The Science Museum: Regulations* was a printed volume of forty-four close-packed pages first issued in March 1914.
9. Science Museum Annual Report for 1930, p. 4.
10. Science Museum Annual Report for 1921 and 1922 , p. 5.
11. Science Museum Annual Report for 1923, p. 5.
12. SMD, SCM/269/4, Board of Survey, general policy.
13. Now named the National Museum of Scotland.
14. See Chapter 3 of this volume for a discussion of the reasons behind the delay to the construction of the Centre Block.
15. Col. E.E.B. Mackintosh, *War History of the Science Museum and Library 1939–1945*, p. 79. SMD, Z 101.
16. Science Museum Annual Report for 1937, p. 14.
17. Science Museum Annual Report for 1952, Appendix 1, 'The Policy of the Science Museum, 1952', p. 32.

18. For a fuller discussion of the pre-war planning and evacuation of the collections, see Chapter 3 of this volume.
19. H. Shaw, *A Memorandum on the Future of the Science Museum* (18 June 1943), SMD, Z 183/1.
20. Memorandum, W.T. O'Dea to Director, 11 December 1950, in SMD, SCM/8252/7/1.
21. Science Museum Annual Report for 1952, Appendix 1, 'The Policy of the Science Museum, 1952', p. 32.
22. RSG6 Warren Row achieved national notoriety in 1963 when its location was discovered and publicised by an anti-nuclear activist group. By 2009 it had become a storage facility once again.
23. Science Museum Annual Report for 1952, Appendix 1, 'The Policy of the Science Museum, 1952', p. 32.
24. F. Sherwood Taylor, 'The Bell Report as Seen in 1952', 1952, SMD, Z 183/2.
25. D. Follett, SMD, Z 183/3 No 4.
26. D. Follett, 'The Purpose of the Science Museum', July 1965, SMD, Z 183/2.
27. This argument is developed in detail in Xerxes Mazda, *The Changing Role of History in the Policy and Collections of the Science Museum, 1857–1973*. Science Museum Papers in the History of Technology No. 3 (London: Science Museum, 1996).
28. Letter, Follet to L. Potts, 3 December 1968, SMD, SCM/8469k/1/1.
29. Letter, Potts to Follett, 28 May 1969, SMD, SCM/8469K/1/1.
30. Science Museum Annual Report for 1970, p. 2.
31. SMD, SCM/2008/206/1, Extension on lease, Hayes Store.
32. SMD, SCM/2000/402, Administration Storage Space for Objects – National Savings, Blythe Road.

13
The International Context and the Context of Internationalism

Tom Scheinfeldt

George Sarton, one of the principal founders of the academic discipline of the history of science, liked to make a distinction between facile 'cosmopolitanism' and serious 'internationalism'. Cosmopolitanism, he argued, makes no distinction between different countries and cultures, refrains from making difficult value judgments, and makes connections between people and places where no historic basis for them exists. Internationalism, on the other hand, is more difficult. It begins with a keen nationalism on the part of its participants, and proceeds through the coming together of representatives of various nationalisms, carefully considering the pros and cons of different cultures, finding connections with real cultural and historical roots, and building on those connections to create something greater than the sum of its parts. Reintroducing his journal of the history of science, *Isis*, after a wartime hiatus in 1919, Sarton wrote:

> The internationalism of *Isis* is not an artificial and subsidiary feature, but on the contrary a perfectly natural one. It is less an aim than a result of the very nature of our studies. The development of science is not the work of any single nation or race, but the fruit of the constant collaboration of men of many types and of many faiths. To avoid any misunderstanding however, it is well to state emphatically that the internationalism of *Isis* is very different from, and indeed hostile to that childish cosmopolitan spirit, which would ignore and despise racial and national peculiarities. That great internationalist Kant remarked a long time ago that international relations can thrive only when the national units are sound and strong and vice versa. The spirit of internationalism and the spirit of nationalism are not antagonistic but complementary. On the contrary, the cosmopolitan spirit is destructive of what is best in both the national and the international ideals.[1]

By Sarton's definition, science museums have always been profoundly international, as opposed to cosmopolitan, institutions. They have been simultaneously patriotic promoters of national achievements in science and industry and members of an international fraternity of similar institutions. They have been agents of national advancement and at the same time conscious participants in a 'science museum movement'. Britain's Science Museum is a very good case in point.

The Science Museum's mix of nationalism and internationalism dates back to its prehistory in the Great Exhibition of 1851. The Great Exhibition of the Industry of all Nations was the first of its kind. The brainchild of Prince Albert, from its earliest days it was designed at once to be a spectacular display of British industrial and technological prowess, and as an occasion for international communication. At one of the earliest planning meetings for the Exhibition, held at Buckingham Palace on 30 June 1849, Prince Albert said:

> It was a question whether this exhibition should be exclusively limited to British industry. It was considered that, whilst it appears an error to fix any limitation to the productions of machinery, science, and taste, which are of no country, but belong, as a whole, to the civilised world, particular advantage to British industry might be derived from placing it in fair competition with that of other nations.[2]

At the Exhibition's opening, Queen Victoria reiterated this sentiment:

> I cordially concur with you in the prayer, that by God's blessing this undertaking may conduce to the welfare of my people and to the common interests of the human race, by encouraging the arts of peace and industry, strengthening the bonds of union among the nations of the earth, and promoting a friendly and honourable rivalry in the useful exercise of those faculties which have been conferred by a benevolent Providence for the good and happiness of mankind.[3]

It is important to note the emphasis on competition amidst these heartfelt expressions of universal sympathy. Competition at the Great Exhibition was manifest and tangible. The exhibition awarded nearly 3,000 'jury medals' and more than 150 'council medals' for best entries in more than thirty categories for which all exhibitors were required to compete. Juries were composed of equal numbers of British and foreign subjects. Approximately 17,000 exhibitors participated in the exhibition, about one-half of them British.[4] Judging strictly by medal counts, this system worked fairly, with Great Britain receiving approximately half of the council medals and France not far behind. Council medals also went to Germany, the United States, Austria, Russia, Belgium, Switzerland, Tuscany, Rome, Holland, Spain, and Turkey.[5] Distribution of prizes occupied the greater portion of the exhibition's closing programme. Thanking the judges for their service, Prince Albert closed the Great Exhibition with a prayer, 'that this interchange of knowledge, resulting from the meeting of enlightened people in friendly rivalry, may be dispersed far and wide over distant lands; and thus, by showing our mutual dependence upon each other, be a happy means of promoting unity among nations, and peace and good-will among the various races of mankind.'[6]

This was precisely the kind of competitive internationalism described a half century later by George Sarton, and which would be institutionalised in the twentieth-century Science Museum. Throughout its first hundred years, the Science Museum has viewed itself as a member of an international community of like-minded institutions. Sometimes competitors, sometimes collaborators, and very much alive today, this international science museum community counts among its key members the Deutsches Museum in Munich and the Smithsonian's National Museum of American History in Washington, DC, along with major museums in Paris, Vienna, Ottawa, Sydney, Chicago, and cities around the world.

The Deutsches Museum

Though founded nearly fifty years after the first museums at South Kensington, the Deutsches Museum von Meisterwerken der Naturwissenschaft und Technik in Munich has had a tremendous impact on the development of the Science Museum. Unlike the Science Museum, whose earliest history was driven by several different – often competing – institutions and individuals, the early history of the Deutsches Museum was very much dominated by the larger-than-life personality of one man, the museum's founder, Oskar von Miller. von Miller was born in 1855, the youngest son of Ferdinand von Miller, a prominent Munich foundry owner. Early on the elder von Miller decided that his son would be an engineer, and Oskar was enrolled at the Munich Technische Hochschule. In 1878 he graduated at the head of his class and entered a successful career building roads, bridges and tunnels in the German Civil Service.

In 1881, von Miller travelled to Paris for the International Electrical Exposition, where – along with the rest of Europe – he was first introduced to Edison's incandescent light bulb. von Miller's interest was immediately and powerfully sparked by the possibilities of the new technology. Equally powerful, however, was his impression of the exhibition itself and the ability of the medium to move the public mind. These two intellectual strands – electrical engineering and the exhibition medium – would dominate the rest of von Miller's career. Indeed, as soon as he returned from Paris, he began planning a Munich International Electrical Exhibition, which opened to great success in 1882.

Soon after his Munich exhibition closed, Von Miller took on the directorship of the German Edison Company in Berlin. Then, in 1889, he struck out on his own as an independent consulting engineer in Munich, working primarily in the field of hydroelectricity. von Miller would spend much of the next forty years organising and directing many of the largest public works projects in Germany, including Bavaria's primary electrical grid, the famous Bayernwerk.

von Miller's ongoing interest and involvement in electricity was surpassed only by his ongoing interest in exhibitions. After the success of the Munich exhibition of 1882, he was selected to organise the larger and more broad-based Frankfurt Electrotechnical Exhibition of 1891. This experience solidified his belief in the exhibition medium and, along with visits to the science museums of Paris and South Kensington, impressed upon him the need for a permanent

German location for the public exhibition of science and technology. In 1903 von Miller and a group of fellow engineers and industrialists chartered the Deutsches Museum. von Miller and his group quickly secured the financial and political support of the German business community and Imperial nobility, including the direct involvement of Crown Prince Ludwig of Bavaria and the enthusiastic blessing of Kaiser Wilhelm II. By 1906 the museum had opened temporary galleries and broken ground on an enormous new building on the Kohleninsel in the middle of the River Isar. Due to the intervention of war the museum's grand opening, originally planned for 1916, was delayed by more than a decade, but on 7 May 1925 – von Miller's seventieth birthday – the Deutsches Museum finally opened as one of the world's largest and most modern museums.[7]

For most of the interwar period, the Deutsches Museum was roundly acknowledged as the world leader among science museums. It was certainly the most generously funded science museum, receiving money not only from the City of Munich and the Bavarian and German governments, but also directly from industrial concerns and visitor entrance fees. With 360,000 square feet of exhibition space, it was also probably the largest science museum in the world. By comparison, in 1930 the Science Museum was estimated to have less than 200,000 square feet of usable gallery space.[8] Not content with its vast exhibition halls, in

Illustration 13.1 A float in the procession for the grand opening of the Deutsches Museum, 7 May 1925, Oskar von Miller's 70th birthday. Coutesy of the Deutsches Museum

the following years the Deutsches Museum continued its building programme, adding separate buildings for its library and meeting rooms by 1935.

Beginning in 1933, however, the Deutsches Museum's international leadership began to wane. Two unrelated, yet for the Deutsches Museum equally important, events marked the beginning of nearly twenty years of turmoil. First was the rise of National Socialism in Germany. From 1933, under Nazi governance, the Deutsches Museum found it increasingly difficult to operate independently of political concerns. The strong role of government and industry (which itself was increasingly subject to Nazi interference) in its governance and funding made the problem even more acute in the Deutsches Museum than in some other German institutions. While the main exhibition halls remained largely intact, the library's galleries and the museum's meeting rooms became a favourite location for Nazi Party propaganda and events. Second was the retirement in 1933 and unexpected death in 1934 of Oskar von Miller himself. Sadly the Deutsches Museum lost its charismatic, well-connected and steadfast leader just when it needed those leadership qualities most.

Like much of Germany, the Deutsches Museum emerged from the Second World War a battered institution. Physically, the building was severely damaged by multiple air raids and was not completely restored to its pre-war state until the 1960s. Intellectually and institutionally, it was faced with an equally daunting challenge: remaking itself as a post-war institution and regaining its place among the first rank of national science museums. As Connie Moon Sehat has shown, immediately following the war popular images of German science and engineering that 'mingled with scientific racism and technological genocide' presented obvious difficulties for the 'German Museum of Masterpieces of Science and Technology'. In a Cold War that divided Germany into East and West, moreover, the Munich museum was faced with the challenge of turning that image completely around, refashioning itself – and indeed West German science and engineering – as overt symbols of the superiority of Western democratic ideals, as vehicles for liberal education, and as aids to 'responsible cultural engagement' with science and technology.[9]

The Deutsches Museum met both its physical and intellectual challenges with rapid and remarkable success. By the mid-1960s the Deutsches Museum occupied reconstructed gallery space equal to that which it had enjoyed before the war. During the 1960s, 1970s and 1980s, the museum increased its focus on original research, housing history of science departments from two universities and expanding its publication programme to include the journal *Kultur & Technik*. It also remade its educational mission, with a positive new emphasis on the supranational nature of science and technology and a forward-looking new focus on children's (as opposed to workers') education. At the same time, throughout the Cold War, the Deutsches Museum struggled to accommodate difficult issues such as atomic energy, the space and arms races, and environmental pollution and explain them to a new generation of Germans and growing numbers of foreign tourists. In more recent decades, the Deutsches Museum has faced a set of different, but in some ways no less daunting, challenges, including rebuilding large portions of its galleries following a large fire in 1983 and reimagining its mission and message for a newly reunified Germany. Today, as part of its continuing reinvention, the Deutsches Museum operates in three locations: the original site in Munich (which continues to expand its exhibits, most notably with the opening of a vast new transport hall in 2003), a facility for air and space technologies at Schleissheim airfield (which opened in 1992), and a centre for late twentieth-century and current science and technology in Bonn (opened in 1995).

Since the Deutsches Museum's grand opening in 1925, the Science Museum has been highly influenced by developments in Munich. Within a few years of its opening, even Science Museum officers recognised the Deutsches Museum as 'the outstanding example' of science museum technique.[10] This recognition had a tonic effect on the Science Museum. Calls for expanded construction at the Science Museum under the Centre Block took on new urgency after the opening of von Miller's enormous facility on the Kohleninsel. Tellingly, after the opening of the Deutsches Museum, these calls also began to include accommodation for a planetarium, one of the Deutsches Museum's most spectacular and highly acclaimed innovations.[11] Less dramatic, though no less important, were changes to existing galleries brought on by comparisons with the Deutsches Museum. The Science Museum's special exhibitions programme and its new Children's Gallery are good examples.

We have seen how in the 1930s the Science Museum's programme of industry-specific, industry-led special exhibitions became progressively nationalistic in its aims and rhetoric as it responded to pressures from domestic industrial concerns and the build-up to the Second World War. Yet even this nationalist endeavour was consciously formulated in a context of international competition, and indeed was first instituted in response to perceived competitive pressures from the Deutsches Museum, which, through von Miller and his associates, enjoyed much closer ties to industry.

In 1927 the Home Office instituted a Royal Commission on National Museums and Galleries to investigate the overall situation, performance and provision of the various national museums, which were then governed by a confusing assortment of boards and departments. In considering the Science Museum, the Royal Commission dedicated a substantial portion of its investigations to the state of foreign science museums, in particular the Deutsches Museum, asking nearly every witness at some point about his knowledge of and relationships with that museum. The Commissioners were especially interested in the Deutsches Museum's close relationship with industry, repeatedly asking witnesses about what could be done to improve the Science Museum's relationship with British industrial concerns.[12]

At the Commission's behest Director Henry Lyons travelled to Munich in 1929 to investigate these matters first-hand, as did Advisory Council member Richard Glazebrook and curator William Plummer. Plummer pointed out not only that 'manufacturers are very generous in presenting objects to the Museum,' but also that ordinary citizens 'are contributing considerably to the cost of the Museum extensions.'[13] Reporting to the Royal Commission and the Advisory Council, Glazebrook recalled a conversation with von Miller: 'As Dr. von Miller said to me: "The rich gave; not a kilogramme of steel, wood, stone, or other material was paid for; the workmen received their ordinary wages – they could not live otherwise – but even they were ready to give their services freely on Saturday afternoons and Sundays." '[14] In its final report in 1930, the Royal Commission remained 'deeply impressed by the great Science Museum at Munich' and was emphatic in callings:

> ... particular attention to the German Science Museum, not only because it is in itself a remarkable example of how a modern Museum can be made a great instrument of technical as well as of popular instruction, but because it is a symbol of national efficiency. It reveals the intense concentration of Germany to-day on the scientific means of industrial progress, a concentration which we believe has its sharp significance for this country.[15]

Implicit in the Commission's praise for the Deutsches Museum is a critique of the Science Museum, and increased ties to industry along German lines were one of the main recommendations of their final report. One response was a reconstituted Advisory Council, enlarged to include additional members of industry. A more visible response was the special exhibitions programme, which made the Science Museum's galleries, and even staff, available to particular industries for promotional displays. This programme began in 1933 with exhibitions by the plastics industry, following which the Advisory Council was happy to report that the Science Museum had wrested some leadership from the Germans in this area. In addition to benefits to the British plastics industry, the annual report noted that the exhibition also helped 'widen the sphere of influence of the Museum' and pointed to the subsequent arrangement of copycat exhibitions in the United States and Australia.[16]

In addition to the important role of industry, Science Museum visitors to Munich were especially impressed by the Deutsches Museum's innovative hands-on and moveable demonstrations of scientific and technological principles. Glazebrook, for example, concludes his 1930 report by singling out the Deutsches Museum's extraordinary efforts 'to make the Museum of educational value,' in particular the prominence of movable displays of apparatus by which the visitor can 'actually repeat the experiments for himself.'[17] The Science Museum's Children's Gallery, which opened in 1931, would be especially influenced by these innovations. The lengthy early reports of the gallery's progress by the Advisory Council repeatedly mention the important role of working models and hand-on exhibits: 'All of these [lifting appliance] exhibits can be worked by the child himself, and they enable him to make direct comparative tests between different methods of lifting, and to experiment as he wishes....'[18]

Yet, while it is true that impressions made in Munich were key to the development of the Children's Gallery, it is also true that Science Museum offices were not uncritical in their adoption of the Deutsches Museum's more interactive features. Reporting just before the initiation of the Children's Gallery project, Plummer registered his opinion that:

> ... there are far too many [working exhibits] at Munich. ... To my mind, it would be a far better plan to set out any apparatus requiring more than one or two buttons for its operation in rooms or places where it can be demonstrated by intelligent Attendants and ... to show a somewhat smaller number of working exhibits which should illustrate phenomena of fundamental importance.[19]

Plummer's feelings are echoed in another travel report from Munich by an anonymous contemporary, who says: 'The visitor is rather overwhelmed by the tremendous number of working experiments shown without, in my opinion, sufficient statement of the phenomena they illustrate and their relation to each other and with little indication of their consequences and industrial applications....'[20]

Ultimately, the Children's Gallery struck just this balance between working theoretical models and historical interpretation. While the Children's Gallery systematically introduced working displays of scientific principle, hands-on exhibits, and educational films to the Science Museum for the first time, the balance of the gallery remained firmly historical in nature. Artefacts, texts and especially dioramas – often displayed in 'past' and 'present' pairings – made up the bulk of the displays. As the Children's Gallery's official guide explained, 'although there are a number of exhibits illustrating elementary principles, the Gallery's chief purpose is to show what things *mean* ... rather than how they *work*.' The very arrangement of the gallery was meant to reinforce this. Soon after the opening of the Children's Gallery, its working exhibits were rearranged to be grouped near the entrance, allowing visitors to pass relatively quickly through to the more practical, artefactual and historical displays in the back of the gallery and in the main museum. In this way the Science Museum was able to incorporate German innovations without sacrificing what it viewed as its stronger commitment to historical exposition.[21]

The compromises of the Children's Gallery are more broadly reflective of the Science Museum's relationship with the Deutsches Museum. During the interwar years, Munich was much on the minds of Science Museum staff, advisers and governors. Yet the relationship between the Science Museum and the Deutsches Museum was always one of dialogue rather than simple influence. As we will see, this dialogue continues until the present day.

The Smithsonian Institution

The Science Museum has also enjoyed a long history of friendly competition and collaboration with its American cousin, the Smithsonian Institution in Washington, DC. Indeed, the Smithsonian itself has strong British roots, owing its very existence to the generosity of a little-known Englishman. In 1829, James Smithson, the illegitimate son of the Duke of Northumberland and a minor naturalist and Fellow of the Royal Society, died and bequeathed the better part of his considerable estate of more than $500,000 to the United States of America.[22]

Smithson himself had never been to America, and the circumstances of his decision to endow the United States with such a great fortune are a subject of great debate. The most plausible explanation is that Smithson made the gift out of spite. Though English inheritance laws allowed him to inherit his mother's fortune, they did not allow him his father's title. Smithson is known to have been outraged by this, and his gift 'to the United States of America, to found at Washington, under the name of the Smithsonian Institution, an establishment for the increase and diffusion of knowledge among men,' may have been made to spite the homeland that had denied him his perceived birthright. A more generous explanation is that Smithson was motivated by admiration for the Enlightenment ideals embodied in the new American Democracy and the inventive potential of its people. Whatever the truth, after several years of waiting for Smithson's nephew to die heirless (a stipulation laid down in Smithson's original will) in 1835, a debate in the Congress over whether such a gift could be received under the US Constitution, and some years in probate at the Court of Chancery in London, the funds, along with a few rarities and books collected by Smithson himself, finally arrived in Washington in 1838.

The debate was not finished. The Congressmen now in possession of Smithson's bounty quickly realised that his vague instruction for 'an establishment for the increase and diffusion of knowledge among men' could mean many things, and each began to argue an interpretation that would best benefit his own and his constituents' interests. Asher Robbins, a powerful senator from Rhode Island, advocated a national university. Senator Rufus Choate of Massachusetts argued for a national library. John Quincy Adams, former President of the United States, who had chosen to spend his twilight years representing his home district of Braintree, Massachusetts in the House of Representatives, vociferously advocated the establishment of an astronomical observatory. Other proposals ranged from a scientific society, to a school of natural history, to a teacher training institute. In the end, after eight years of impassioned debate and ultimate compromise, the 1846 'Act to Establish The Smithsonian Institution' was a document that completely satisfied no-one. The Smithsonian Institution that emerged was at once all, and none, of the things the various contestants had hoped it would be.[23]

When people today think of the Smithsonian, they imagine the vast complex of nearly twenty museums stretching westward from the US Capitol towards the Washington Monument and then beyond into neighbourhoods across Washington, DC. For its first thirty years, however, the Smithsonian was patently not a museum. Rather, in those early years, the Smithsonian operated variously as a funding agency for scientific expeditions, a printing press, a lecture facility, a chemical laboratory, a weather station, and much else. Moreover, unknown to Smithson, who had never seen the young nation first-hand, Washington in the first half of the nineteenth century was little more than a swampy way station for politicians, albeit with a magnificent Capitol building rising on a small knob above the Potomac River. The Mall, where the Smithsonian would ultimately be built, was not the two-mile-long grassy thoroughfare it is today, but a muddy creek running into the Potomac River. It was on this muddy strip that the purpose-built

Illustration 13.2   The Smithsonian Institution Castle, ca. 1860, on an otherwise empty National Mall. Courtesy of the Smithsonian Institution Archives

headquarters of the hybrid organisation opened in 1855. The 'Smithsonian Castle' still houses the main administrative offices of the Institution. Its asymmetrical towers and Victorian gothic facades sit somewhat uneasily amidst the later, pre-dominately neoclassical, buildings of governmental Washington. Even today the building suggests an institution not completely confident in its identity.

Though the original Castle incorporated a few small demonstration galleries, the Smithsonian's emergence as a museum really dates to 1876 and Philadelphia's Centennial Exposition. Philadelphia was the United States' first major international exhibition. Timed to celebrate the 100th anniversary of the Declaration of Independence, the Centennial Exposition was a conscious effort to position the United States on an equal footing – industrially, culturally and geopolitically – with the great powers of Europe. The exhibition was a tremendous success, attracting exhibits from nearly forty nations and nearly ten million visitors. By the time the fair closed, however, exhibiting nations found themselves in the midst of a global economic recession, and many decided to abandon ('donate') their exhibits to the United States government, which in turn gave them over – along with exhibits staged by the US Department of War, Department of the Navy, Treasury Department, Agriculture Department, and Post Office – to the Smithsonian Institution. Indeed, the Smithsonian had staged its own exhibits alongside these other government departments, comprising several boxcars full of objects and displays. The Institution returned from Philadelphia with more than forty boxcars worth of material.

The donations came at an opportune time for the Smithsonian. While the Smithsonian's first Secretary (chief executive), Joseph Henry, was more interested in fostering original scientific research than conserving and displaying museum collections, the Institution's second Secretary, Spencer Baird, was a natural collector. Baird took over as Secretary upon Henry's death in 1878, just in time to take full advantage of the collections from Philadelphia.

There was, however, nowhere to put them. Never intended as a museum, the Castle was ill suited for storing and displaying such diverse and extensive collections. After early lobbying by Henry (who didn't want the collections in his Castle in the first place) and later lobbying by Baird (who aimed to refashion the Smithsonian as a collecting institution), Congress finally appropriated funding for a dedicated museum building in 1879. The opening in 1881 of the United States National Museum, known today as the Arts and Industries Building, beside the Castle on the Mall marks the real beginning of the Smithsonian's existence as a museum.[24]

Like the South Kensington Museum, however, the United States National Museum was a catch-all facility. The materials acquired in Philadelphia were not only scientific and technological, but also ethnographic, historical, artistic and natural historical. In addition, the United States National Museum was charged with the various objects and apparatus accumulated in the course of the Smithsonian's earlier research activities and even the small mineralogical collection that came with the original Smithson bequest. With Baird at the helm, moreover, new collections were added from subsequent international exhibitions, and soon even the new Arts and Industries Building was bursting at the seams. In 1911, a new National Museum of Natural History opened across the Mall to house the natural history collections. Soon afterwards, a chorus for a dedicated science museum emerged. The early leader of this chorus was a young curator named Carl W. Mitman.

Trained as an engineer at Lehigh and Princeton, Mitman had arrived at the Smithsonian in 1911, and in 1918 became chief curator of the Institution's technology collections. In this position, Mitman worked to consolidate these collections and organise them under an independent Department of Mechanical and Mineral Technology, of which he became the first head. Then in 1920 Mitman went further. Amidst competing proposals for a National Gallery of Art, Mitman added a call for the Smithsonian Regents (its board of trustees) to build what would soon be styled the National Museum of Engineering and Industry.

Mitman's interwar efforts to found a National Museum of Engineering and Industry ultimately fell flat in Congress. Yet calls for a dedicated science museum for the Smithsonian re-emerged after the Second World War. In 1946, one of Mitman's protégés, Frank Taylor, resumed the campaign. The Cold War's climate of indirect international confrontation proved a more favourable one for arguing the merits of a national science museum, and in 1964 the National Museum of History and Technology opened beside the Washington Monument. Though owing much to Mitman's original vision, the new Museum of History and Technology was very much a Cold War foundation, intended to demonstrate the basis of American superiority in a context of historic scientific and technological achievement. In 1976, the atmosphere of Cold War competition on the Mall was heightened with the opening of the National Air and Space Museum, and during the 1980s the nationalist intentions of the Smithsonian's science museums were made even more explicit when the Museum of History and Technology was renamed the National Museum of American History.[25]

Two key incidents from the early twentieth century demonstrate both the close ties between the Science Museum and the Smithsonian and the keen competition that has often marked the relationship between these two claimants to leadership among English-speaking science museums.

Today, visitors to both the Smithsonian in Washington and the Science Museum in London can view the famous Flyer with which Orville and Wilbur Wright made their first flight in Kitty Hawk, North Carolina in 1903. The story of how one object can be displayed in galleries an ocean apart begins in the very year the Wright brothers ushered in the era of heavier-than-air manned flight. Orville and Wilbur were not alone in their efforts to build a flying machine. In the first years of the twentieth century there were several competing airplane projects, including one based in Washington and led by the Secretary of the Smithsonian Institution at the time, Samuel P. Langley. Langley was a well-regarded physicist and an early expert in aerodynamics. Beginning in the 1890s, Langley built and successfully tested several models of unmanned 'aerodromes', some of which managed to stay aloft on their own power for more than a minute and travel nearly a mile. Langley's success with unmanned flying machines attracted the attention of the United States War Department, which provided encouragement and funding for Langley to attempt additional tests with larger machines in hopes of achieving heavier-than-air manned flight. Despite more than $70,000 in government funding ($50,000 from the War Department and $20,000 from the Smithsonian itself), Langley never achieved his aim. After dropping two manned aerodromes into the Potomac River in 1903, Langley abandoned his aeronautical research for good. Just nine days after Langley's final test, the Wright brothers made their historic first flight.

The significance of these events was not immediately clear, however. It took several years for the engineering community to take full measure of the Wright brothers' achievement. For example, it was 1914 before the Franklin Institution in Philadelphia became the first scientific organisation to formally recognise the Wright Flyer as the first heavier-than-air manned flying machine. In the meantime, Langley's successor as Secretary, Charles Walcott, had decided to resurrect Langley's aerodrome, reconstructing his predecessor's invention according to its original specifications. This time around, the aerodrome flew, and in 1914 the Smithsonian announced that what had previously been labelled 'Langley's Folly' was in fact 'the first man-carrying aeroplane in the history of the world capable of sustained free flight'. When the reconstructed aerodrome was placed on display at the Smithsonian with this claim, it so outraged Orville Wright that in 1928 he sent his own plane to South Kensington to be displayed at the Science Museum. It was only after a formal appeal by Walcott's own successor, Secretary Charles Abbot, and Orville Wright's death in 1948 that the Wright Flyer was returned to the United States and placed on display in Washington. Today the original Wright Flyer hangs in the Smithsonian's National Air & Space Museum, while a replica – made from drawings made of the machine while it lived in South Kensington – sits in the Science Museum. By this convoluted path, both the Smithsonian and the Science Museum can claim to have held the Wright brothers' airplane in their collections. Remarkably, a similar story can be told of Thomas Edison's original 1877 phonograph, which came into the Science Museum's collections via the Patent Office Museum's collections in 1883. Ultimately, this too returned to the United States – to the Henry Ford Museum in Dearborn, Michigan – but only after a bitter struggle was resolved with the involvement of the Foreign Office and the US Department of State.[26]

This competition for objects disguises a much more collaborative relationship, one which has often brought the Smithsonian and the Science Museum together on more friendly terms. Even while all this was happening, the relationship between

Illustration 13.3   Herman Shaw, Director of the Science Museum, London, speaking during the ceremony held at the Museum to mark the return of the Wright Flyer to the Smithsonian Institution in November 1948

the Science Museum and the Smithsonian and its surrogates remained exceedingly collegial and productive. For instance, throughout the period of the Wright and Edison controversies, key proponents of a national science museum at Washington travelled to London to meet and receive advice from Lyons and his staff.[27]

More recently, co-operation between the Science Museum and the Smithsonian has only increased. The 1960s and 1970s began a period of increased scholarly activity among curators at both institutions. Under the leadership of Director Robert Multhauf, the Smithsonian's Museum of History and Technology became a primary locus of scholarly history of science and technology in the United States, playing host to the editorial board of *Isis*, the international journal of the History of Science Society, of which Multhauf was editor for more than a decade. Himself a PhD in History of Science, Multhauf populated his curatorial staff with other trained historians of science, such as Bernard Finn and Deborah Warner. These scholars maintained Multhauf's participation in international scholarly circles. Just as Multhauf had relocated *Isis* to the Smithsonian in the 1970s, the Society for the History of Technology's journal, *Technology and Culture*, was moved to Washington in the 1980s.[28]

At the same time, a very similar story was playing out at the Science Museum, which likewise began recruiting doctoral candidates in the history of science in the 1970s.[29] Not surprisingly, shared scholarly interests among staff members at the Smithsonian and the Science Museum – reinforced by increasing seriousness about material cultural studies among the broader community of academic historians of science and technology – translated into closer individual ties between curators. Early on in the period relationships were mainly personal, such as that between Finn and Science Museum Director Margaret Weston. In more recent years these relationships have become more institutionalised. The best example of this is the *Artefacts* collaboration. Co-organised by curators at the National Museum of American History, the Science Museum and the Deutsches Museum (which, as mentioned above, experienced a similar shift towards serious scholarship in the period), *Artefacts* provides a showcase for object-centred historiography of science and technology and aims to strengthen the Museum as a location for serious scholarly research. Since 1996 the organisers of *Artefacts* have hosted nearly a dozen workshops, not only in Washington, London and Munich, but also in Paris, Vienna and Utrecht, and have published seven edited volumes of original scholarship on topics ranging from medical devices and scientific imagery to military and space technology.[30]

Conclusion: Science Museums around the world

In a 1933 memorandum entitled 'Technical Museums: Their Scope and Aim', Henry Lyons wrote that, while the recent emergence of like-minded institutions in '12 or 15 countries' had made the category of science museums sufficiently coherent for profitable examination, they were still essentially local institutions, best understood as products of 'local conditions, local requirements and the resources available.'[31] This is still very true today, and a volume much like this one could surely be (and in most cases, has been) written about each of Lyons's twelve or fifteen museums. But, as we have seen, for most of their histories, these institutions have also enjoyed vigorous international engagement, both competitive and collaborative. This international community of museums includes the Science Museum, the Deutsches Museum and the Smithsonian's National Museum of American History. Yet, as both Lyons's statement and the locations of recent *Artefacts* workshops show, the community includes at least a dozen other science museums in capitals around the world. Even in Britain, Germany and America, there are important members of this community that have not received attention in this brief account. In Britain, world-class collections of scientific instruments reside at the Museum of the History of Science in Oxford and the Whipple Museum in Cambridge. In Berlin the Deutsches Technikmuseum, which opened a major new extension in 2003, presents a facade and collections to rival its cousin in Munich. And in Chicago the Museum of Science and Industry and the Adler Planetarium taken together provide citizens of that city science museum facilities equal to any in Washington, America or the world.[32]

Outside these three countries, first among institutions getting the attention they deserve in this volume is the Musée des Arts et Métiers at the Conservatoire National des Arts et Métiers (CNAM) in Paris. In fact, the CNAM is the oldest of the national science museums. Founded in 1794, more than a century before the anniversary this book celebrates, the CNAM has alternately served as a point of inspiration and contrast. As Robert Bud points out earlier in this volume, many contemporaries of 1851's Great Exhibition, especially those in the scientific and educational fields, had hoped that the Crystal Palace and its exhibits would be preserved as a grand British competitor to the French Conservatoire.[33] By the 1920s, however, the CNAM had stagnated, leading Plummer to describe the CNAM as 'moribund' and 'of interest only to those visitors with a sufficient historical knowledge of the subject.' 'There are some things wrong with the Deutsches Museum,' he wrote, 'but at the Conservatoire, everything is wrong.'[34] Having undergone a major renovation and expansion in the 1990s, however, the CNAM has now largely dispelled this image, regaining its rightful place among the first rank of world science museums.[35] A strikingly similar situation has presided over much of the twentieth-century history of Vienna's Technisches Museum. For a brief time after its opening in 1918, the Technisches Museum was the largest science museum in the world. Yet it quickly fell into a long period of decline as Austria struggled with the aftermath of two devastating world wars. However, like the CNAM, the renovated Technisches Museum Wien has re-emerged in the 1990s and 2000s with a new educational focus.[36] Leiden (the Boerhaave Museum), Florence (the Istituto e Museo di Storia della Scienza), Milan (the Museo Nationale della Scienza e della Tecnologia, the 'Leonardo da Vinci Museum') and Coimbra (Science Museum of the University of Coimbra) number among other European cities with major science museums. In addition, many national museums, though not science museums per se, boast significant science collections. These include the Rijksmuseum in Amsterdam and the British Museum in London.[37]

Neither should more recent foundations in Commonwealth countries be neglected. The Science Museum provided a model for founders of the Canadian Museum of Science and Technology in Ottawa during their early planning discussions in the 1950s and 1960s, and two Science Museum curators – Margaret Weston and Frank Greenaway – received strong consideration for the new Canadian national museum's first directorship. Meanwhile, the Powerhouse Museum, which like the Science Museum is a descendant of nineteenth-century international exhibitions, recently provided the Science Museum with a Director in Lindsay Sharp. Museums in some other Commonwealth countries do not share such close ties. For example, the museums of India's robust National Council of Science Museums network, which includes the Birla Industrial and Technological Museum in Kolkata, the Visvesvaraya Industrial and Technological Museum in Bangalore, the Nehru Science Centre in Mumbai, and the National Science Centre in New Delhi, were all founded after decolonisation.[38]

For most of its first 100 years, the Science Museum has stood at the forefront of a closely knit international science museum community. As its long relationships with its closest cousins – the Deutsches Museum and the National Museum of American History – show, this community has been at once highly competitive, consistently collaborative and tremendously productive. Let us share in Queen Victoria's prayer that this 'friendly and honourable rivalry' continues throughout the Science Museum's second century, 'encouraging the arts of peace and industry' and 'strengthening the bonds of union among the nations of the earth.'

Notes

1. George Sarton, 'War and Civilization', *Isis* 2 (1919), pp. 318–19.
2. 'Exhibition of the Industry of All Nations', *The Times*, 18 October 1849, p. 6.
3. 'The Opening of the Great Exhibition', *The Times*, 2 May 1851, p. 5.
4. 'The Great Exhibition', *The Times*, 16 October 1851, p. 2.
5. 'The Great Exhibition', *The New York Times*, 16 October 1851.
6. 'Close of the Great Exhibition', *The New York Times*, 31 October 1851.
7. Sources in English for the history of the Deutsches Museum are limited. The two best are probably Otto Mayr et al., *The Deutsches Museum: German Museum of Masterworks of Science* (London: Scala Publications, 1990) and Connie Moon Sehat, 'Education and Utopia: Technology Museums in Cold War Germany', unpublished PhD thesis, Rice University, 2006. Other useful sources include Kenneth Hudson, *Museums of Influence* (Cambridge: Cambridge University Press, 1987) and Bernard Finn, 'The Museum of Science and Technology', in Michael Steven Shapiro, ed. *The Museum: A Reference Guide* (New York: Greenwood Press, 1990), pp. 59–83. The German sources include Wilhelm Füssel and Helmuth Trischler, eds, *Geschichte des Deutsches Museum* (Munich: Deutsches Museum, 2003).
8. Science Museum Annual Report for 1930. Data for the size of the Deutsches Museum galleries is included in a report of a visit to Munich by Advisory Council member Sir Richard Glazebrook, which is appended to the 1930 Report. Glazebrook's numbers conform to other contemporary estimates.
9. Sehat, 'Education and Utopia: Technology Museums in Cold War Germany', Introduction, Chapter 1. Also see Otto Mayr, *Wiederaufbau: Das Deutsches Museum, 1945–1970* (Munich: Deutsches Museum, 2003).

10. H.W. Dickinson, 'Presidential Address: Museums and Their Relation to the History of Engineering and Technology', *Newcomen Society Transactions* 14 (1933–4), pp. 1–12, on p. 9.

11. David H. Follett, *The Rise of the Science Museum under Henry Lyons* (London: Science Museum, 1978), p. 93; David Rooney, *The Events Which Led to the Building of the Science Museum Centre Block*, Science Museum Papers in the History of Technology, No. 7 (London: Science Museum), pp. 5, 9–12.

12. *Oral Evidence, Memoranda and Appendices to the Interim Report* (London: Royal Commission on National Museums and Galleries, 1928); *Oral Evidence, Memoranda and Appendices to the Final Report* (London: Royal Commission on National Museums and Galleries, 1929); *Final Report, Part II* (London: Royal Commission on National Museums and Galleries, 1930).

13. W.G. Plummer, 'Report on Visit to the Deutsches Museum, Munich on 6th December, 1929', 23 January 1930, SMD, Z 193/1.

14. Science Museum Annual Report for 1930.

15. *Final Report, Part II* (London: Royal Commission on National Museums and Galleries, 1930), pp. 48–9.

16. Science Museum Annual Report for 1933. See also Follett, *The Rise of the Science Museum under Henry Lyons*, pp. 74, 119–21.

17. Science Museum Annual Report for 1930.

18. Science Museum Annual Report for 1933. See also Science Museum Annual Report for 1931.

19. Plummer, 'Report on Visit to the Deutsches Museum, Munich on 6th December, 1929'.

20. 'Report on Visits to European Museums', undated, SMD, Z 210.

21. F. St A. Hartley, *The Children's Gallery* (London: HMSO, 1935), p. 3; Lewis W. Phillips, *Outline Guide to the Exhibits* (London: Science Museum, 1937).

22. 'Last Will and Testament of James Smithson', October 23, 1826, Smithsonian Institution Archives, http://siarchives.si.edu/history/exhibits/documents/smithson-will.htm (accessed 11 February 2010).

23. United States Congress, 'An Act to Establish the "Smithsonian Institution" for the Increase and Diffusion of Knowledge Among Men', August 10, 1846, http://www.sil.si.edu/exhibitions/smithson-to-smithsonian/1846act.htm (accessed 10 February 2010). Other sources for the foundation and early history of the Smithsonian include William Jones Rhees, *The Smithsonian Institution: Documents Relative to Its Origin and History* (Washington: Smithsonian institution, 1879) and Paul Henry Oehser, *The Smithsonian Institution* (New York: Praeger, 1970).

24. George Brown Goode, *The Genesis of the National Museum* (Washington, 1892).

25. Frank A. Taylor, 'The Background of the Smithsonian's Museum of Engineering and Industries', *Science* 104(2693) New Series (9 August 1946), pp. 130–2; Marilyn Cohen, 'American Civilization in Three Dimensions: The Evolution of the Museum of History and Technology of the Smithsonian Institution' (PhD, The George Washington University, 1980); Arthur P. Molella, 'The Museum That Might Have Been: The Smithsonian's National Museum of Engineering and Industry', *Technology and Culture* 32(2) (1991), pp. 237–63.

26. The story of the Langley dispute is told in greater detail in 'Langley Dispute Ends', *The Science News-Letter* 42(19) (7 November 1942), p. 292. See also Paul Henry Oehser, *Sons of Science: the Story of the Smithsonian Institution and Its Leaders* (New York: Greenwood Press, 1968) and Follett, *The Rise of the Science Museum Under Henry Lyons*, pp. 154–5. The Edison dispute is covered briefly by Follett on p. 154. See also Frank W. Hoffmann and Howard Ferstler, 'Edison, Thomas Alva', in *Encyclopedia of Recorded Sound* (2005), pp. 349–51, and 'American Government Honors Edison', *The Science News-Letter* 14(394) (27 October 1928), pp. 253–4, in which Edison's Congressional

Medal of Honor ceremony at which the phonograph was returned to him from London is described.

27. See, for example, Charles R. Richards, *The Industrial Museum* (New York: The Macmillan Company, 1925).
28. Robert P. Multhauf, 'A Museum Case History: The Department of Science and Technology of the United States Museum of History and Technology', *Technology and Culture* 6(1) (1965), pp. 7–58; Arthur Molella, 'The Research Agenda', in *Clio in Museum Garb: The National Museum of American History, The Science Museum, and the History of Technology* (London: Science Museum, 1996), pp. 37–46; Pamela M. Henson, ' "Objects of Curious Research": The History of Science and Technology at the Smithsonian', *Isis* 90 (1999), pp. S249–S269; Robert C. Post, ' "A Very Special Relationship": SHOT and the Smithsonian's Museum of History and Technology', *Technology and Culture* 42(3) (2001), pp. 401–35.
29. Robert Bud, 'History of Science and the Science Museum', *British Journal for the History of Science* 30(1) (1997), pp. 47–50.
30. See the 'Series Preface' in *Manifesting Medicine: Bodies and Machines* (Amsterdam: Harwood Academic Publishers, 1999), the first volume in the series, and the 'Introduction' to *Exposing Electronics* (Amsterdam: Harwood Academic, 2000), the second volume in the series.
31. Henry Lyons, 'Technical Museums, Their Scope and Aim', 1933, SMD, Z 183/1.
32. J.A. Bennett, 'Museums and the Establishment of the History of Science at Oxford and Cambridge', *British Journal for the History of Science* 30(1) (1997), pp. 29–46; Sehat, 'Education and Utopia: Technology Museums in Cold War Germany'; Jay Pridmore, *Museum of Science and Industry, Chicago* (New York: Harry Abrams in association with the Museum of Science and Industry, Chicago, 1997).
33. Hudson, *Museums of Influence*; Stella V.F. Butler, *Science and Technology Museums*, Leicester museum studies series (Leicester: Leicester University Press, 1992).
34. Plummer, 'Report on Visit to the Deutsches Museum, Munich on 6th December, 1929'.
35. A multimedia account of the renovations of CNAM's historic accommodation in the Priory of Saint-Martin-des-Champs can be viewed on the museum's website, 'Musée des arts et métiers: Chronicle of the restoration', http://www.arts-et-metiers.net/musee.php?P=142&lang=ang&flash=f (accessed 10 February 2010).
36. Helmut Lackner, Katharina Jesswein and Gabriele Zuna-Kratky, *100 Jahre Technisches Museum Wien* (Vienna: Ueberreuter, 2009).
37. Hudson, *Museums of Influence*; Finn, 'The Museum'; Robert P. Multhauf, 'European Science Museums', *Science* 128(3323) New Series (5 September 1958), pp. 512–19.
38. Sharon Babian, 'CSTM Origins: A History of the Canada Science and Technology Museum', http://www.sciencetech.technomuses.ca/english/about/CSTM_Origins.cfm (accessed 10 February 2010); Wade Chambers and Rachel Faggetter, 'Australia's Museum Powerhouse', *Technology and Culture* 33(3) (1992), pp. 548–59.

Afterword: A Speech made at the Dinner to Celebrate the Centenary of the Science Museum on 11 June 2009 by Professor Chris Rapley CBE, Director of the Science Museum

Good evening distinguished guests, ladies and gentlemen, colleagues. It is my privilege to welcome you to this historic celebration of the one-hundredth anniversary of the establishment of the Science Museum.

I would like to thank Michael Wilson, one of our trustees, and his Dana and Albert R. Broccoli Charitable Foundation for supporting tonight's dinner, and for his ongoing commitment to both the Science Museum and the National Media Museum. Without his support we would not have been able to hold this event tonight, and so we are especially grateful.

I would also like to thank L'Oréal and Bollinger for their support tonight. As you will see from our programme, L'Oréal are celebrating their centenary this June, and as part of their ongoing education work they have a particular interest in encouraging women to develop a career in the sciences. So thank you both for your generous contributions.

Finally in this vein, I should let you know that this evening in New York another dinner will take place. Life Technologies, a US-based company, has recently agreed to contribute funds to our project to update and relaunch our biomedical gallery 'Who am I', and they will make the announcement at their dinner tonight. This is great news for the Science Museum as it means that through the support of the Wellcome Trust, GlaxoSmithKline, and now Life Technologies, we have achieved our £4.5 million funding target to enable us to complete this major project.

As you all know, the Science Museum traces its roots back to the 1851 Great Exhibition. The exhibition generated a substantial financial surplus which was used to establish a number of institutions in this area, including our forerunner, the South Kensington Museum, which later became the V&A. But it was the efforts of the astronomer Norman Lockyer, the chemist Henry Roscoe, and the great educationalist Civil Servant Sir Robert Morant around the turn of the last century that led to the Science Museum being established as an independent institution on 26 June 1909. These influential individuals championed science as key to the educational and commercial strength of the nation.

As an aside it is interesting that Norman Lockyer was the founding editor of the journal *Nature*, which remains the most prestigious science journal today. He was also the discoverer of the element and noble gas helium, which he achieved by recognising that a mysterious yellow spectral line that he and the French astronomer Jules Janssen independently observed in the 1868 solar eclipse was the signature of a previously unknown element, which he named helium because of its discovery in the light from the Sun (from the Greek 'Helios'). Thirty-seven years later huge reserves of helium were found in natural gas fields in the USA and his prediction was vindicated. So an element, a journal and the Science Museum – an impressive legacy I think you will agree – and I would hope that in a show of hands to vote on which is the most important that there would be no doubt that it is the Science Museum!

But what about our role? A report to the government in 1911 on the purpose of the newly established Science Museum recommended:

> So far as is possible by means of exhibited scientific instruments and apparatus, machines and other objects, the Collections of the Museum ought to afford illustration and exposition of the various branches of Science... and of their applications in the Arts and Industries.

One hundred years later, that echo from the past still rings true. Science and engineering remain key to the commercial strength of our nation as well as the future well-being of humanity. Indeed they are seen by many as the key to 'UK plc' powering its way out of the recession. Britain remains the sixth largest manufacturing economy in the world, providing a firm foundation on which to build. But to do so requires a stream of young people enthused to become the scientists, engineers, designers and entrepreneurs of the future, and a public informed of the nature of science and supportive of its goals. The Science Museum may be one hundred years old, but it has never been more relevant.

Given these opportunities and challenges, we have defined the role of the Museum as being 'To make sense of the science that shapes our lives', and our vision: 'To be the best place in the world for people to enjoy science'. We are committed to providing life-enhancing experiences which raise curiosity, release creativity and shape the future. To achieve this, learning is at the core of all we do, to underpin and enable the nation's knowledge economy and to do so by offering insight and inspiration to the innovators of tomorrow.

In our treatment of the great contemporary issues, such as decarbonising the world's energy system to address the threat of climate change, we see the Museum as an agent of social change, with the objective of not just engaging our visitors in an enjoyable day out, but inspiring them to change their private, professional and political lives in a way which shapes a better future.

I am pleased to say that these views of our role and its importance received strong endorsement at our Centenary press launch yesterday from Lord Mandelson, the Secretary of State for the newly formed Department of Business, Innovation and

Skills (D-BIS). I should say that I permitted Lord Mandelson to stand on the foot-plate of Stephenson's Rocket for a photo-call, something even our energy hall curator is not allowed to do! But a picture speaks a thousand words and I judge it will be greatly to our benefit. My only concern is the nature of the caption that *Private Eye* may apply!

What was the Science Museum like in 1909? One hundred years ago its collections were displayed and stored in overcrowded temporary buildings and it was almost twenty years before the objects and displays were housed to a satisfactory standard. The building of the East Block began in 1913 but was not finally completed until its official opening by George V in 1928. But there was much that we would find familiar. There were amazing moving machines, and exhibits that visitors could activate with a single touch. These working exhibits and demonstrations attracted a far wider audience than those directly concerned with science. Indeed, the South Kensington Museum's first official report in 1858 commented that 'In the evening, the working man comes to the South Kensington Museum accompanied by his wife and children.' So we have been a destination for 'doing as well as seeing,' and for families right from the start.

Even so, the approach to public engagement was often rather technical, and it was Colonel Sir Henry Lyons, the Director in the 1920s, who insisted on putting the needs of the 'ordinary visitor' ahead of the specialists. As a result, a few years

Illustration A.1 The First Secretary of State, the Rt. Hon. Lord Mandelson taking the controls of the 'Rocket' locomotive at the opening of the Science Museum's centenary celebrations on 9th June 2009. The Museum's centenary banner can be seen on the upper right of the photograph. © Justin Sutcliffe / Science Museum.

later *The Times* was able to report that more children visited the Science Museum than any other museum in London. This led in 1931 to the Museum opening the 'Children's Gallery', which pioneered the use of interactive exhibits to engage young people with the underlying principles of science and technology. This rapidly became one of the Museum's most popular attractions and the inspiration for later interactive galleries worldwide, and was the forerunner of our own LaunchPad, the most visited gallery of its kind in the world.

We see the start of our second century as a major opportunity to raise our game. We have an ambitious plan to transform the Museum, its contents and our Large Object Storage Facility at Wroughton and thereby fulfil our vision of 'The Museum of the Future'. We have thought carefully about the changes to this building and its contents, and we have aimed to maximise the impact and benefit at the minimum cost. Even so, the price tag is of the order of £100,000,000. To spur the corporate world, the charities, the government and private sponsors to assemble this investment, we are launching a public appeal for £1,000,000 to demonstrate popular support for ourselves as a cherished national institution and icon.

Now I will draw to a close by telling you briefly about some of the highlights of our year of Centenary celebrations. Firstly, as you see, we have already dressed the Museum and launched our Centenary Journey. This new interpretive route through the ground floor of the Museum features ten iconic objects that have shaped the modern world. Examples are Stephenson's Rocket, which transformed personal mobility and the transport of goods, and in so doing accelerated the Industrial Revolution, and at the other extreme the Apollo 10 command module, which provided a view of our planet credited with a major shift in human thinking about sustainability and the stewardship of nature.

The birthday party itself will be a three-day extravaganza on 26 to 28 June [2009], starting with a day of inspirational activities for schools and followed by a weekend of science shows, demonstrations, performances and drama for families and general visitors. We will celebrate a Space Season, which will also commemorate the fortieth anniversary of the Apollo moon landings. This will begin with an evening premiere of a new live arrangement of Brian Eno's seminal 1983 album *Apollo*, and the opening on 23 July of the first of our new exhibitions, 'Cosmos and Culture'. 'Cosmos and Culture' – which is our contribution to the International Year of Astronomy – will trace the impact over 400 years of the instruments, observations and resulting insights that have determined our knowledge of the universe and our place within it.

Finally, the end of the centenary year will be celebrated on 26 June 2010. On this day two major refurbished and reinvented galleries devoted to contemporary and future science will open in the Wellcome Wing. 'Who Am I?' and 'Antenna' will constitute major strides on our journey to be the best place in the world to enjoy science and will continue the Museum's aim of the past hundred years – to help us make sense of the science that shapes our lives.

I will conclude by showing you the 'Museum of the Future' video, which illustrates visually what we are committed to turn into a reality by 2015. Before we run the video, I will explain some key points, asking you to keep in mind that what you will see is a concept, not a design. Our proposed transformation of the building takes advantage of the partial pedestrianisation of Exhibition Road, that will be completed next year, and is designed to achieve maximum benefit for minimum cost. To achieve this we will alter the front of the building to increase its visibility, attractiveness and accessibility. We will create an orientation area, where our visitors can plan their visit; we will renew the galleries at the heart of the Museum – the so called 'Treasury' Galleries – and we will completely update our contemporary wing, the Wellcome Wing. We will provide a new 'SkySpace' to house our space, astronomy and cosmology exhibits and a new public venue. We will also improve our lift facilities to make the whole Museum more accessible, and we will take the opportunity to further improve our environmental credentials.

So, without further ado, I am delighted to take you on a journey to and through the Museum of the Future. [At this point, Professor Rapley played the 'Museum of the Future' video and commented on the main elements of the plan as shown in the video. Images of 'The Beacon' on the Exhibition Road side of the East Block, the interior of 'SkySpace' Gallery and the golden canopy of SkySpace' can be seen in the plates section of this volume.]

Ladies and gentlemen, I hope you will agree that we are indeed the Museum of The Future.

Appendix 1: Temporary (Special) Exhibitions, 1912–1983

Peter J.T. Morris and Eduard von Fischer

The year given is the year the exhibition opened; it may have continued into the following calendar year. The main source before 1939 is Appendix I of E.E.B. Mackintosh, 'Special Exhibitions at the Science Museum' (SMD, Z 108/4), which has been followed even when the exhibitions do not appear in the Sceince Museum Annual Reports, supplemented by the list in Follett, *The Rise of the Science Museum*, pp. 122–3. Otherwise the exhibitions have been taken from the Annual Reports.

1912	History of Aeronautics
1914	Gyrostatics
1914	Science in Warfare

First World War

1919	Aeronautics
	James Watt Centenary
1923	Typewriters
1924	Geophysical and Surveying Instruments
	Kelvin Centenary
	Centenary of the Introduction of Portland Cement
1925	Stockton and Darlington Railway Centenary
	Centenary of Faraday's Discovery of Benzine [sic]
	Wheatstone Apparatus
	Seismology and Seismographs
1926	Adhesives Board, DSIR
	Centenary of Matthew Murray
	Fiftieth Anniversary of the Invention of the Telephone
1927	British Woollen and Worsted Research Association
	British Non-Ferrous Metals Research Association
	Solar Eclipse Phenomena
	Newton Bi-centenary
1928	George III Collection of Scientific Apparatus
	Cartography of the Empire
	Modern Surveying and Cartographical Instruments
	Weighing
	Photography

1929   British Cast Iron Research Association
            Newcomen Bicentenary
            Historical Apparatus of the Royal Institution
            Centenary of the Locomotive Trials at Rainhill
            Development of Weighing Machines
            Metallurgy
1930   North of Ireland Linen Research Association
            Modern Astronomy
            Rayleigh Scientific Apparatus
            Liverpool and Manchester Railway Centenary
1931   Geophysical Instruments and Methods
            Modern Glass Technology
            British Optical Instruments
            Rafts, Canoes and Boats
            Cadastral Maps of the Empire
1932   Electrical Measuring Instruments
            Meteorological Instruments and Cloud Photographs
            Seventy-five Years of Progress in Technical Industry, 1857–1932
            Optical Phenomena and Optical Instruments
            Manufacture of Artificial Dyes
            British Fishing Boats and Coastal Craft
            Weighing and Linear Measuring Equipment
1933   Photoelectric Cells and their Applications
            Centenary of Richard Trevithick
            Native Boats
            Plastic Materials and their Uses
1934   Refrigeration
            Rubber
            Parsons Turbine Jubilee
            Launch of RMS 'Queen Mary'
            Air Race to Australia
1935   King George V Jubilee
            James Watt Bicentenary
            Modern Welding
            Noise Abatement
            Electrodeposition
            Empire's Airway
            Surveying Exhibition
1936   London and Greenwich Railway Centenary
            Very Low Temperatures
            Maiden Voyage of RMS 'Queen Mary'
            Royal Aeronautical Society's 70th Anniversary
            British Fishing Boats
            Smoke Abatement
            Electric Illumination

1937   Practical Marine Screw Propulsion
         Television
         London and Birmingham and Grand Junction Railways Centenary
         Lubrication and Lubricants
         Atom Tracks
         Timber Research and Utilization
1938   The Centenary of Transatlantic Steam Navigation
         Microphotography of Documents
         Science in the Army
         Chinese Junks
1940   Aircraft in Peace and War

Second World War

1946   German Aeronautical Developments
         Science Exhibition
         Naval Mining and Degaussing
         91st Annual Exhibition of the Royal Photographic Society
1947   Pasteur
         Centenary of the Chemical Society [also called Chemical Progress]
         Centenary of the Institution of Mechanical Engineers
         Fiftieth Anniversary of the Electron
         Home and Factory Power
         History of Surgery
1948   Darkness into Light
         Science in Building
         Mechanical Music
1949   King George III Collection of Scientific Instruments
         Special Exhibition for the Blind
1950   French Scientific Instruments
         Metals in the Service of Mankind
         One Hundred Years of Submarine Cables
         Science of Weather
         Special Exhibition for the Blind
1951   Science Museum – Past and Future
         British Standards – The Measure of Industrial Progress
         Razors Past and Present
1952   The British Clockmaker's Heritage
         Ramsay Centenary
         One Hundred Alchemical Books
         Penn-Gaskell Collections
1953   Navigation Today
         Royal Photographic Society Centenary
         Historic Books on Machines
         Wright Brothers Jubilee

1954	The Story of Oil
	Historic Astronomical Books
	Centenary of the Ministry of Education Library
	Netherlands Scientific Books
1955	Working Model of a Complete Steelworks
	Electrical Standards
	Centenary of Sir Charles Vernon Boys
	Frequency-modulated System of Sound Broadcasting
	Centenary of the Bunsen Burner
	Display of Dividing Engines
1956	250th Anniversary of the Birth of Benjamin Franklin
	Perkin Centenary Celebrations
	Tercentenary of the Pendulum Clock
1957	International Geophysical Year
1958	Sir William Siemens FRS
	Controlled Fusion of Atoms
1959	British Railways Modernisation
1960	150 Years of Canned Food
	Tercentenary of the Royal Society
	Historic Books on Mining and Kindred Subjects
1961	Research Rockets
	Corrosion Prevention
	The Dondi Astronomical Clock of 1364
	Modern Methods of Transport Control
	Diamond Jubilee of Transatlantic Radio
	Cross-Channel Link and Thames Crossing
1962	Colonel Glenn's Space Capsule 'Friendship 7'
	'Aerial' Satellite
	Atoms at Work
	Aluminium in Electrical Engineering
	The New Photography
1963	TRANSIT Navigational Satellite
	Centenary of the Birth of Frederick R. Sims
	The Gas Industry Today
1964	Motorways in Progress
	S. Z. de Ferranti
	Collectors' Pieces – Clocks and Watches
	Ministry of Public Building and Works
	The Beginnings of Cinematography
1965	Alan B. Shepard's Mercury Space Capsule 'Freedom 7'
	Heat Engines
	Hovercraft
	Freedom from Hunger
	Oil Painting of a Watch
1966	French Scientific and Technical Books

	The Engineer's Day
1967	Welding Technology
	Wasa [The Vasa warship]
1968	Golden Jubilee of the Royal Air Force
	The Tradition of Czech and Slovak Technical Development
	Computing Past and Present
	The Evolution of Security [locks and fastenings]
	R E B Crompton
1969	Bicentenary of the Birth of Sir Marc Isambard Brunel
	James Watt and Richard Arkwright
	Alcock and Brown
1970	Moon Fragments
	First World War Sopwith Snipe Aircraft
	Mercedes-Benz CIII Experimental Motor Car
1971	1955 Rolls-Royce Chassis
	Short SC1 VTO Aircraft
	Charles Babbage
	Centenary of the Institution of Electric Engineers
	Bicentenary of the Birth of Henry Maudsley
	Natural Gas
	Scientific Toys and What They Teach
1972	Search [the National Research Councils]
	Advancing Railway Technology [the APT]
	Special Effects
	Edwardian Trains – the Pre-Grouping Railways at their Zenith
	Bicentenary of the Birth of Luke Howard
1973	A Word to the Mermaids [submarine telegraphy]
	Quincentenary of the Birth of Nicolaus Copernicus
1974	Phonographs to Holographs
	The Breath of Life [discovery of oxygen]
	You and Your Analytical Chemist
	'I will put a girdle about the earth...' [centenary of the birth of Marconi]
	Tercentenary of George Ravenscroft's Patent for Lead Glass
	Tower Bridge Observed
	The Modern Lifeboat
1975	Dondi Astrarium
	Centenary of the Death of Sir Charles Wheatstone
	William Boyer, Photographer of the Chain
	A Trick of Light
	Shield Tunnelling
	New Materials in Design
	Mary Rose [Tudor warship]
1976	Science and Technology of Islam
	Apollo 10

Centenary of the Telephone
Caxton Quincentenary
Electrifying Time
1977   The Jet Age [70th birthday of Frank Whittle]
Sun Pictures [centenary of the death of W.H. Fox Talbot]
Lasers
Star Wars
The Trumpet Shall Sound [Edison's invention of the phonograph]
1978   The Story of Everest [25th anniversary of first ascent]
More than Meets the Eye
William Harvey and the Circulation of the Blood
Centenary of the Birth of Lord Nuffield
Ships and Seamen of Gwynedd
Josiah Wedgwood – The Arts and Sciences United
Medicine through the Artist's Eye
Right Way Up [the self-righting lifeboat]
You Press the Button ... [snapshot photograph]
The Medical Picture Show
1979   The Art of the Engineer
Stanley Spencer in the Shipyard
Coal Today and Tomorrow
Bennet Woodcroft and the Heritage of British Patents
The Photography of Walter Nurnberg – An Image of Industry
Musical Boxes
The History of Wine
Gilbert Daykin (1886–1939)
150th Anniversary of the Rainhill Trials
Centenary of David Hughes' Invention of the Audiometer
Going Places
New Acquisitions by the Wellcome Museum
1980   Challenge of the Chip
The Great Optical Illusion [50th anniversary of first Baird TV sold]
A Visit to the Dentist [centenary of the British Dental Association]
Medicines for Man
The Mini Comes of Age
Henry Wellcome: Pioneer of Portable Medicines [centenary of Burroughs   Wellcome]
Words about Wires
Trade Tokens
Gandolfi – A Family of Camera Makers
Murchison Oilfield Production Platform [scale model]
Printer at Work
Just Add Water [public water supply]
Photographic Items
1981   All Stations

Seeing the Invisible [50th anniversary of the electron microscope]
Menai Bridges
Centenary of Service [public electricity supply]
The Natural World of Great Britain
Twelve Points of View
Chasing Rainbows
The Photography of Space

1982   The Great Cover-Up Show
This is IT
Science in India

1983   Coal – British Mining in Art, 1680–1980
Beads of Glass
Science and Conscience [Max Born and James Franck]
Light Dimensions [holography]

Appendix 2: Table of Senior Staff at the Science Museum, 1893–2000

Surname	Year ent.	Year left	Highest position
Anderson, Robert Geoffrey William	1975	1984	Deputy Keeper
Annas, Percy James	1934	1948	Chief Warder
Austin, Jillian Frances	1980	1984	Asst Keeper
Bagley, John Arthur	1976	1989	Asst Keeper
Ball, Ian Michael	1975	1994	Head of Operations and House Management
Barclay, Alexander	1921	1959	Keeper
Barrett, James	1865	1900	Keeper
Bathe, Basil Wroughton	1949	1976	Asst Keeper
Baxandall, David	1898	1934	Keeper
Becklake, (Ernest) John Stephen	1972	1994	Keeper
Bevin, John Frederick	1985	in post	Acting director of corporate services
Bonnyman, George Sim	1918	1934	Chief Warder
Boon, Timothy Martyn	1982	in post	Chief Curator
Bowers, Brian Peter	1967	1998	Deputy Keeper
Bracegirdle, Brian	1977	1989	Acting Head of Collections Management
Bradford, Samuel Clement	1899	1938	Keeper
Brown, (Christopher) Neil	1976	2006	Principal Curator
Bryden, David John	1979	1987	Asst Keeper
Bud, Robert Franklin	1978	in post	Principal Curator
Butcher, Alan Edward	1974	1991	Asst Keeper
Cain, John Clifford	1959	1961	Guide Lecturer
Caine, (Elizabeth) Anne	1993	in post	Head of Finance
Calvert, Henry Reginald	1934	1967	Keeper
Caunter, Cyril Francis	1950	1959	Asst Keeper
Chaldecott, John Anthony	1949	1976	Keeper
Chew, (Victor) Kenneth	1958	1978	Keeper
Chilton, Donovan	1938	1974	Keeper
Clark, V. Charles	1973	1980	Principal
Clowes, Geoffrey Swinford Laird	1924	**1937**	Asst Keeper
Coiley, John Arthur	1973	1992	Head, National Railway Museum
Collop, Allan Digby	1948	1953	Asst Keeper
Comper, John Herbert*	1953	1956	Museum Superintendent

Continued

Appendix 2   Continued

Surname	Year ent.	Year left	Highest position
Corry, A. Kenneth	1972	1984	Deputy Keeper
Cossons, Sir Neil	1986	2000	Director
Crawhall, Thomas Currah	1926	1938	Asst Keeper
Creasey, George Wright	1938	1946	Asst Keeper
Crosley, Alton Swift	1924	1955	Chief Draughtsman
Darius, Jon	1981	1989	Senior Curator
Davison, Charles Saint Clair Buick	1938	1972	Asst Keeper
Davy, Maurice John Bernard	1920	**1950**	Keeper
Day, Lance Reginald	1951	1987	Keeper
Defries, Jeffrey Jack	1988	1998	AD and Head of Resource Management Division
Denman, Roderick Peter George	1922	1930	Asst Keeper
Dickinson, Henry Winram	1895	1930	Keeper
Dingley, Pauline Olive	1986	2003	Head of Library
Duff, George Alexander	1930	1954	Chief Warder
Durant, John Robert	1989	2000	AD and Head of Science Communication Division
Fahy, Martin C.	1958	1987	Chief Warder
Farmelo, Graham Paul	1990	2003	Head of Exhibitions
Festing, Major-General Edward Robert	1864	1904	Director
Follett, Sir David Henry	1937	1973	Director
Forward, Ernest Alfred	1901	1937	Keeper
Fox, Robert	1988	1988	AD and Head of Research and Information Services Division
Fryer, Robert Paul	1978	**2000**	Chief Warder and Head of Security
Fulcher, Lionel William	1885	1926	Keeper
Garratt, Gerald Reginald Mansel	1934	1971	Keeper
Geddes, (William) Keith Elliott	1968	1986	Deputy Keeper
Gilbert, Keith Reginald	1948	**1973**	Keeper
Gosling, Christopher	1989	1998	Head of Personnel and Legal Services
Gosling, Robbie	1955	1974	Asst Keeper
Greenaway, Frank	1949	1980	Keeper
Griffiths, John	1980	2003	Senior Curator
Groom, Sydney Herbert	1930	1958	Guide Lecturer
Hall-Patch, Anthony	1979	1989	Senior Curator
Hartley, Frederick St Aubyn	1920	1957	Keeper

Continued

Appendix 2   Continued

Surname	Year ent.	Year left	Highest position
Hill, Kenneth Gordon*	1956	1963	Museum Superintendent
Hutchison, Cdr John Kenneth Douglas	1937	**1944**	Asst Keeper
Ireson James Wenman Bassell*	1948	1953	Museum Superintendent
Jackson, (William) Roland Cedric	1993	2002	Head of Science Museum
Janson, Stanley Eric	1935	1969	Keeper
Keene, Suzanne Victoria	1992	2000	Head of Collections Management
Kelly, Sydney Thomas	1907	1948	Museum Superintendent
Kenedy, Robert Christopher	1960	1963	Asst Keeper
Kent, Valler Amos	1924	1966	Senior Draughtsman
King, Alfred Charles	1864	1895	Keeper
Lacey, George William Brian	1954	1986	Keeper
Lancaster-Jones, Ernest	1920	**1945**	Keeper
Last, William Isaac	1890	**1911**	Director
Law, Rodney James	1963	1985	Asst Keeper
Lawrence, Ghislaine Mary (formerly Skinner)	1979	2003	Principal Curator
Lebeter, Fred	1937	1967	Keeper
Lyons, Col. Sir Henry George	1912	1933	Director
Mackintosh, Col. Ernst Elliott Buckland	1933	1945	Director
Mann, Peter Robert	1973	1993	Senior Curator
Mather, Ann	1999	2004	Head of Personnel and Development
Mayfield, Heather Margaret	1979	in post	Head of Content and Deputy Director, Science Museum
McCleave, William John Ogden	1954	1973	Chief Warder
McConnell, Anita	1985	1988	Senior Curator
McWilliam, Robert Coutts	1991	2003	Senior Curator
Molloy, Timothy	1993	in post	Head of Design
Morris, Peter John Turnbull	1991	in post	Principal Curator
Morrison-Scott, Sir Terence Charles Stuart	1956	1960	Director
Morton, Alan Queen	1979	2003	Acting Head of Physical Sciences and Engineering
Mummery, G. Roger	1985	1987	Senior Curator
Nahum, Andrew	1980	in post	Principal Curator
Newmark, Mrs Ann Katharine	1975	2003	Head of Documentation and Registrar

Continued

Appendix 2   Continued

Surname	Year ent.	Year left	Highest position
O'Dea, William Thomas	1930	1966	Keeper
Ogilvie, Sir Francis Grant	1911	1920	Director
Overton, George Leonard	1898	1935	Keeper
Parker, Miss Hannah Joyce	1948	1978	Deputy Keeper
Parkinson, Thomas Frank	1893	1925	Keeper
Pemberton, C. Mark	1987	2000	AD and Head of the Public Affairs Division
Phippen, Miss Helen Dorothy	1955	1976	Asst Keeper
Pledge, Humphrey Thomas	1928	**1960**	Keeper
Plummer, William Graham	1928	**1936**	Asst Keeper
Preston, Michael Richard	1964	1987	Keeper
Pretty, William Edward	1931	1947	Asst Keeper
Price, Roger	1981	1991	Senior Curator
Raimes, Mrs Jane Mary (née Pugh)	1969	1979	Asst Keeper
Reeve, J.A.*	1963	1970	Museum Superintendent
Rhodes, Kenneth J	1987	1988	Museum Administrator
Richards, George Tilghman	1928	1953	Guide Lecturer
Robinson, Derek Anthony	1974	1999	Acting Head of Collections
Robinson, John Corin	1973	1994	Asst Keeper
Rolfe, Mrs Eileen Maitland	1975	1978	Asst Keeper
Rowles, Arthur L.	1973	1983	Asst Keeper
Sahiar, Aderji Burjorji	1961	1979	Deputy Keeper
Semmens, Peter William Brett	1974	1987	Asst Keeper
Sharp, Lindsay Gerard	1976	1978	Asst Keeper (Director, 2000–2005)
Shaw, Herman	1920	**1950**	Director
Simmons, Lt Col Thomas Mortimer	1958	1973	Deputy Keeper
Skinner, Frederick George	1922	**1955**	Deputy Keeper
Smith, Engineer Captain Edgar Charles	1924	1929	Guide Lecturer
Sneed, Geoffrey Colin	1972	1984	Asst Keeper
Spencer, Arthur John	1900	**1934**	Deputy Keeper
Spratt, Hereward Philip	1930	1967	Deputy Keeper
Stephens, Peter Donald	1974	1991	Asst Keeper.
Stowers, Arthur	1930	1962	Keeper
Strimpel, Oliver Bernard Raphael	1979	1983	Asst Keeper
Sumner, Philip Lawton	1948	1974	Deputy Keeper
Suthers, Terry	1988	1992	AD and Head of Public Services Division

Continued

Appendix 2   Continued

Surname	Year ent.	Year left	Highest position
Suthers, Terry	1988	1992	AD and Head of Public Services Division
Swade, Doron David	1982	2002	Head of Collections
Taylor, Frank Sherwood	1950	1956	Director
Thoday, Cdr Alfred George	1961	1974	Deputy Keeper
Thomas, David Bowen	1961	1979	Keeper
Thomas, Gillian M.	1993	1997	AD and Head of Project Development Division
Tuck, Lt Cdr Walter James	1961	1976	Deputy Keeper
Tucker, Jonathan Leslie	1999	2009	Acting Director, NMSI
Urquhart, Donald John	1938	1948	Asst Keeper
van Riemsdijk, John Theodore	1954	1984	Keeper
Vaughan, Denys	1970	1992	Deputy Keeper
Wall, Major Victor Clement	1955	1976	Guide Lecturer
Ward, Francis Alan Burnett	1931	1970	Keeper
Wartnaby, John	1951	1982	Keeper
West, Mrs Lesley A (née Burdett)	1962	1978	Asst Keeper
Westcott, George Foss	1921	1957	Keeper
Weston , Dame Margaret Kate	1955	1986	Director
White, Ernest William	1920	1959	Deputy Keeper
Will, Leonard Duncan	1978	1994	Head of Library and Information Services
Wilson, Anthony Walford	1977	1994	Head of Publications
Wilson, George Buckley Laird	1954	1973	Deputy Keeper
Winton, Walter	1950	1980	Keeper
Wood, Sidney	1894	1914	Senior Keeper
Woolfe, Hyman	1964	1983	Asst Keeper
Wright, Thomas	1977	1998	Head of Collections

Note: *seconded from the Department of Education (dates are for the period of secondment only)
†Between 1978 and 1985 Stephens was Director of the London Transport Museum.
Bold means that they died in service
AD = Assistant Director

Appendix 3: Visitor Figures for the Science Museum, 1909–2008

Year	Visitor Figures	Source of Data
1909	514,954	Science Museum Register of Admissions 1909–1952 – SMD, Z 100/7
1910	461,930	ditto
1911	397,707	ditto
1912	416,412	ditto
1913	345,289	ditto
1914	331,212	ditto
1915	346,740	ditto
1916	88,995	ditto
1917	46,741	ditto
1918	279,463	ditto
1919	366,915	ditto
1920	477,522	Science Museum Report – 1920
1921	445,658	Science Museum Report – 1921/1922
1922	494,055	Science Museum Report – 1921/1922
1923	474,277	Science Museum Report – 1923
1924	437,816	Science Museum Report – 1924
1925	429,558	Science Museum Report – 1925
1926	576,734	Science Museum Report – 1926
1927	709,166	Science Museum Report – 1927/1928
1928	900,053	Science Museum Report – 1927/1928
1929	1,061,754	Science Museum Report – 1929
1930	1,132,761	Science Museum Report – 1930
1931	1,170,981	Science Museum Report – 1931
1932	1,241,528	Science Museum Report – 1932
1933	1,255,818	Science Museum Report – 1933
1934	1,142,472	Science Museum Report – 1934
1935	1,327,190	Science Museum Report – 1935
1936	1,281,338	Science Museum Report – 1936
1937	1,271,599	Science Museum Report – 1937
1938	1,137,635	Science Museum Report – 1938
1939	712,287	Advisory Council Report – 1953 (Museum open from January to August)
1940	76,000	Advisory Council Report – 1953 (Museum open from February to June)
1941	Museum closed	
1942	Museum closed	
1943	Museum closed	
1944	Museum closed	
1945	Museum closed	
1946	1,288,464	Advisory Council Report – 1952; Science Museum Report 1946 (Museum reopened mid-February)
1947	748,183	Advisory Council Report – 1952; Science Museum Report 1947
1948	980,094	Advisory Council Report – 1952; Science Museum Report 1948
1949	900,032	Advisory Council Report – 1952; Science Museum Report 1949
1950	1,039,522	Advisory Council Report – 1952; Science Museum Report 1950
1951	1,024,809	Advisory Council Report – 1952; Science Museum Report 1951
1952	1,062,473	Advisory Council Report – 1952; Science Museum Report 1952
1953	1,062,402	Science Museum – 1953

Continued

Appendix 3   Continued

Year	Visitor Figures	Source of Data
1954	1,200,394	Science Museum Report – 1954
1955	1,143,815	Science Museum Report – 1955
1956	1,191,303	Science Museum Report – 1956
1957	1,317,000	Science Museum Report – 1957
1958	1,292,000	Science Museum Report – 1958
1959	1,274,000	Science Museum Report – 1959
1960	1,147,609	Science Museum Report – 1960; summary sheet in SCM/2006/0329/005
1961	1,057,733	Science Museum Report – 1961; summary sheet in SCM/2006/0329/005
1962	1,130,686	Science Museum Report – 1962; summary sheet in SCM/2006/0329/005
1963	1,210,344	Science Museum Report – 1963; summary sheet in SCM/2006/0329/005
1964	1,365,703	Science Museum Report – 1964; summary sheet in SCM/2006/0329/005
1965	1,507,722	Science Museum Report – 1965; summary sheet in SCM/2006/0329/005
1966	1,700,000	Science Museum Report – 1966; summary sheet in SCM/2006/0329/005 shows a total of 1,184,079 but is more likely to be 1,684,079
1967	1,910,976	Science Museum Report – 1967; summary sheet in SCM/2006/0329/005
1968	2,172,187	Science Museum Report – 1968; summary sheet in SCM/2006/0329/005
1969	2,126,077	Science Museum Report – 1969; summary sheet in SCM/2006/0329/005
1970	2,120,599	Science Museum Report – 1970; summary sheet in SCM/2006/0329/005
1971	1,941,727	Science Museum Report – 1971; summary sheet in SCM/2006/0329/005
1972	1,936,242	Science Museum Report – 1972; summary sheet in SCM/2006/0329/005
1973	2,311,711	Science Museum Report – 1973; summary sheet in SCM/2006/0329/005
1974	2,051,908	Science Museum Report – 1974; summary sheet in SCM/2006/0329/005
1975	2,404,232	Science Museum Report – 1975; summary sheet in SCM/2006/0329/005
1976	2,507,880	Science Museum Report – 1976/77; summary sheet in SCM/2006/0329/005
1977	3,360,624	Science Museum Report – 1976/77; summary sheet in SCM/2006/0329/005
1978	3,486,228	Science Museum Report – 1978; summary sheet in SCM/2006/0329/005
1979	3,710,108	Science Museum Report – 1979; summary sheet in SCM/2006/0329/005
1980	4,224,027	Science Museum Report – 1980; summary sheet in SCM/2006/0329/005
1981	3,847,718	Science Museum Report – 1981; summary sheet in SCM/2006/0329/005

Continued

Appendix 3   Continued

Year	Visitor Figures	Source of Data
1982	3,306,338	Science Museum Report – 1982; summary sheet in SCM/2006/0329/005
1983	3,345,822	Science Museum Report – 1983; summary sheet in SCM/2006/0329/005
1984	3,019,892	Science Museum Review – 1987; summary sheet in SCM/2006/0329/005
1985	2,723,947	Science Museum Review – 1987; summary sheet in SCM/2006/0329/005
1986	2,994,451	Science Museum Review – 1987; summary sheet in SCM/2006/0329/005
1987	3,166,294	Summary sheet in SCM/2006/0329/005
1988	2,261,048	Summary sheet in SCM/2006/0329/005
1989	1,121,103	Summary sheet in SCM/2006/0329/005
1990	1,303,345	Summary sheet in SCM/2006/0329/005
1991	1,327,503	Summary sheet in SCM/2006/0329/005
1992	1,212,504	Summary sheet in SCM/2006/0329/005
1993	1,277,117	Summary sheet in SCM/2006/0329/005
1994	1,274,253	Science Museum Topline Visitor Numbers from 1994 to 2008 provided by Tim Neal, Visitor Insight Executive (NMSI), dated 30 June 2009
1995	1,556,368	ditto
1996	1,548,366	ditto
1997	1,537,289	ditto
1998	1,550,211	ditto
1999	1,481,235	ditto
2000	1,335,432	ditto
2001	1,352,649	ditto
2002	2,722,154	ditto
2003	2,886,850	ditto
2004	2,169,138	ditto
2005	2,149,856	ditto
2006	2,421,473	ditto
2007	2,684,945	ditto
2008	2,705,677	ditto

Index

Objects associated with specific scientists or inventors, for example Sir William Herschel's prism, are indexed under the person concerned with the object given in brackets after the name, viz. Sir William Herschel (prism). Pictures are given in bold type, so a picture of an object is indexed under the specific object rather than the person. Plates are spread between pages 193–194.